Geology and Mineralogy of Gemstones

Geology and Mineralogy of Gemstones

Editor: Elisha Price

Callisto Reference
www.callistoreference.com

Callisto Reference,
118-35 Queens Blvd., Suite 400,
Forest Hills, NY 11375, USA

Visit us on the World Wide Web at:
www.callistoreference.com

ISBN: 978-1-64116-805-2 (Hardback)

Cataloging-in-Publication Data

Geology and mineralogy of gemstones / edited by Elisha Price.
p. cm.
Includes bibliographical references and index.
ISBN 978-1-64116-805-2
1. Precious stones. 2. Mineralogy. 3. Geology. I. Price, Elisha.
QE392 .G46 2023
553.8--dc23

Table of Contents

Preface

A gemstone refers to a cut and polished piece of mineral crystal that can be utilized to create jewelry or other adornments. Gemstones are created underneath the surface of the Earth and may contain traces of other minerals known as inclusions. The majority of gemstones are found in alluvial gravels and igneous rocks but metamorphic and sedimentary rocks can also contain gem materials. Gemstones are classified into different species and groups on the basis of chemical composition and crystal structure. The most significant gemstone species include corundum, quartz, beryl, tourmaline and garnet. The majority of gemstones are hard but some soft minerals are also utilized in jewelry due to their luster or other physical qualities which hold aesthetic value. This book unravels the recent studies on the geology and mineralogy of gemstones. With its detailed analyses and data, it will prove immensely beneficial to professionals and students involved in the study of gemstones at various levels.

This book is the end result of constructive efforts and intensive research done by experts in this field. The aim of this book is to enlighten the readers with recent information in this area of research. The information provided in this profound book would serve as a valuable reference to students and researchers in this field.

At the end, I would like to thank all the authors for devoting their precious time and providing their valuable contribution to this book. I would also like to express my gratitude to my fellow colleagues who encouraged me throughout the process.

Editor

1

Mineralogy, Geochemistry and Genesis of Agate

Jens Götze [1,*], Robert Möckel [2] and Yuanming Pan [3]

1 Institute of Mineralogy, TU Bergakademie Freiberg, Brennhausgasse 14, 09599 Freiberg, Germany
2 Helmholtz-Zentrum Dresden-Rossendorf, Hemholtz Institute Freiberg for Resource Technology, Chemnitzer Str. 40, 09599 Freiberg, Germany; r.moeckel@hzdr.de
3 Department of Geological Sciences, University of Saskatchewan, Saskatoon, SK S7N 5E2, Canada; yuanming.pan@usask.ca
* Correspondence: jens.goetze@mineral.tu-freiberg.de

Abstract: Agate—a spectacular form of SiO_2 and a famous gemstone—is commonly characterized as banded chalcedony. In detail, chalcedony layers in agates can be intergrown or intercalated with macrocrystalline quartz, quartzine, opal-A, opal-CT, cristobalite and/or moganite. In addition, agates often contain considerable amounts of mineral inclusions and water as both interstitial molecular H_2O and silanol groups. Most agate occurrences worldwide are related to SiO_2-rich (rhyolites, rhyodacites) and SiO_2-poor (andesites, basalts) volcanic rocks, but can also be formed as hydrothermal vein varieties or as silica accumulation during diagenesis in sedimentary rocks. It is assumed that the supply of silica for agate formation is often associated with late- or post-volcanic alteration of the volcanic host rocks. Evidence can be found in association with typical secondary minerals such as clay minerals, zeolites or iron oxides/hydroxides, frequent pseudomorphs (e.g., after carbonates or sulfates) as well as the chemical composition of the agates. For instance, elements of the volcanic rock matrix (Al, Ca, Fe, Na, K) are enriched, but extraordinary high contents of Ge (>90 ppm), B (>40 ppm) and U (>20 ppm) have also been detected. Calculations based on fluid inclusion and oxygen isotope studies point to a range between 20 and 230 °C for agate formation temperatures. The accumulation and condensation of silicic acid result in the formation of silica sols and proposed amorphous silica as precursors for the development of the typical agate micro-structure. The process of crystallisation often starts with spherulitic growth of chalcedony continuing into chalcedony fibers. High concentrations of lattice defects (oxygen and silicon vacancies, silanol groups) detected by cathodoluminescence (CL) and electron paramagnetic resonance (EPR) spectroscopy indicate a rapid crystallisation via an amorphous silica precursor under non-equilibrium conditions. It is assumed that the formation of the typical agate microstructure is governed by processes of self-organization. The resulting differences in crystallite size, porosity, kind of silica phase and incorporated color pigments finally cause the characteristic agate banding and colors.

Keywords: agate; quartz; chalcedony; silica minerals; micro-structure; trace elements; O-isotopes; paragenetic minerals

---

1. Introduction

Agates belong to the most fascinating mineral objects in nature because of their wide spectrum of colors and spectacular morphologies. Therefore, they play a dominant role as gemstones and cut stone since antiquity. The name "*Agate*" can be dated back to ca. 350 B.C. (Theophrast) and was probably related to the discovery of agates in the river *Achates* (recently *Drillo*) in Sicily. Today, agate deposits and agate treatments are known from historical and recent sites all over the world [1–4].

Both in historic and recent times, many speculations have been made about the processes leading to the formation of agates. First descriptive reports and discussions about agates are known from the 18th century [5,6]. Serious scientific investigations concerning chemical composition, properties and genesis of agates started in the middle of the 19th century [7]. As a result of these numerous studies a wide spectrum of theories, both serious and speculative, about the formation of agates exists today (see compilations in e.g., [1,8]).

Early theories hypothesizing a formation of agates from circulating SiO_2-rich solutions in volcanic rocks, assume that the characteristic agate banding is the result of rhythmic silica supply to natural voids and cavities ("outer rhythm" [7]). Several years later, Daubrée [9] published interesting experiments about the alteration of silicate glasses by alkaline solutions (>400 °C, 1000 atm). These reactions resulted in concentric chalcedony layers, similar to observed structures in natural agates. Comparable results were later published by Nacken [10]. Liesegang [11,12] presented his fundamental results about the formation of rhythmic color bands due to diffusion processes of metal ions in silica gels He concluded that a zonation and banding can be created by "inner processes" such as simple chemical reactions. However, later studies revealed that banding and color distribution in agates are not always identical [8]. The characteristic color banding in agates seems to be more likely related to the micro-structure and the distribution of finely dispersed color pigments within the SiO_2 matrix.

Recent mineralogical and geochemical investigations during the last decades significantly increased our knowledge about the mineralogy and genesis of agates. Numerous publications about certain agate occurrences or specific aspects of agate formation have been published (e.g., [1,8,13–60]).

According to these recent studies, agates can be regarded as banded chalcedony, which is intergrown or intercalated with other silica phases. Moreover, agates often contain considerable amounts of mineral inclusions and water (1–2 wt%) as both interstitial molecular H_2O and silanol groups [13,21,40]. These impurities may form spectacular internal structures or may be responsible for the different coloration of agates [13,29,40,61].

The chemical and mineralogical complexity of agate is a result of its complex formation history. In principle, agates occur in practically all rock types however, their formation in volcanic rocks is the most abundant. Therefore, the majority of agate deposits worldwide is related to SiO_2-poor (andesites, basalts) and SiO_2-rich (rhyolites, rhyodacites) volcanic rocks. Other types of agate may occur as hydrothermal vein agate or due to silica accumulation in sediments and sedimentary rocks [1,8,15,49]. The chemical composition of agates as well as the presence of specific paragenetic minerals (carbonates, clay minerals, zeolites, iron compounds) indicate that the formation of volcanic agates seems to be associated with late- or post-volcanic alteration processes within the host rocks (e.g., [24,37,62,63]).

Although general ideas concerning the formation of agates exist, the origin of agate remains incompletely understood. The discussions are controversial especially due to the fact that no one has unambiguously documented agate formation in real time and agates have never been successfully synthesized in the laboratory. Only certain silica mineralization in active geothermal regions or at the sea-floor (white smokers) is comparable to agates [64].

The present paper provides a global review of what is known on agate properties, based on the literature and completed by numerous additional data. On this basis, an attempt is made to establish a preliminary model of agate formation.

2. Materials and Methods

In the present paper results of extensive studies on agate samples of different origin and type are summarized and compared with data from literature. The analyzed material includes more than 300 agate samples from all over the world and from different parent rocks (basic and acidic volcanic rocks, hydrothermal vein agates, agates from sedimentary host rocks—see Table S1). The agate samples as well as the surrounding host rocks were visually described and documented, and selected aliquots prepared for further investigations.

The mineral and chemical composition of the agate host rocks was generally analyzed by a combination of *X*-ray diffraction (XRD) and *X*-ray fluorescence (XRF) measurements on selected and prepared sample material. The qualitative and quantitative phase compositions were analyzed using an URD 6 (Seifert/Freiberger Präzisionsmechanik, Freiberg, Germany) with Co Kα-radiation in the range 5–80° (2θ). Data evaluation was realized using Analyse RayfleX v.2.352 software (GE Sensing & Inspection Technologies GmbH, Ahrensburg, Germany) and subsequent Rietveld refinement with Autoquan v.2.7.00 (GE Sensing & Inspection Technologies GmbH, Ahrensburg, Germany) [65]. XRF measurements on powdered samples mixed with Li-tetraborate were carried out with a PANalytical Axios Minerals spectrometer and WROXI package (PANalytical, Almelo, The Netherlands) [37].

The identification of different SiO_2 phases in the agate samples was done by a combination of X-ray diffraction and Raman spectroscopy [29]. Additional results were obtained from microscopic investigations by polarizing, cathodoluminescence (CL) and scanning electron microscopy (SEM) on polished thin sections (30 µm). Conventional polarizing microscopy was made with a Zeiss Axio Imager A1m microscope (ZEISS, Thornwood, NY, USA). These investigations were completed by SEM studies using a JEOL 6400 SEM with EDX detector (JEOL Ltd., Akishima, Japan). CL measurements were made with an optical CL microscope HC1-LM (LUMIC, Bochum, Germany) on carbon-coated, polished thin sections [66]. CL microscopy and spectroscopy was performed with a Peltier cooled digital video-camera (OLYMPUS DP72, OLYMPUS Deutschland GmbH, Hamburg, Germany) and an Acton Research SP-2356 digital triple-grating spectrograph with a Princeton Spec-10 CCD detector (OLYMPUS Deutschland GmbH, Hamburg, Germany), respectively [67].

The paramagnetic centers of powder agate samples were analyzed by electron paramagnetic resonance (EPR) spectroscopy using a Bruker EMX spectrometer (Saskatchewan Structural Science Centre, Saskatoon, Canada). The equipment operated with microwave frequencies of ~9.63 GHz and 9.39 GHz at 295 K and 85 K, respectively. Experimental conditions included modulation frequency of 100 kHz, modulation amplitude of 0.1 mT, and microwave powers from 0.2 to 20 mW to obtain optimal conditions for different center types. The spectral resolutions were ~0.146 mT for wide scans from 50 to 650 mT and 0.024 mT for narrow scans from 300 to 350 mT.

Trace-elements in agates were analyzed by instrumental neutron activation analysis (INAA—XRAL Laboratories, Mississauga, ON, Canada), inductively coupled plasma mass spectrometry (ICP-MS) and laser ablation ICP-MS (LA-ICP-MS), respectively [30,68]. Additional sections perpendicular to the agate banding were prepared for selected agate samples to analyze the spatial distribution of trace elements. The thick sections were analyzed using a Thermo Finnigan Element 2 mass spectrometer with a DUV 193 laser ablation system [69] and alternatively with a double-focusing sector field mass spectrometer ELEMENT XR with a NewWave 193 nm excimer laser probe (Thermo Scientific, Waltham, MA, USA) [70]. Additional measurements were made with conventional ICP-MS using a Perkin Elmer Sciex Elan 5000 quadrupole instrument with a cross-flow nebulizer and a rhyton spray chamber (Perkin Elmer Inc., Baesweiler, Germany) [71] to facilitate the quantification of the very low contents of REE and some other elements in the agate samples.

Isotope and fluid inclusion studies on quartz and chalcedony samples were carried out to get information about the formation temperatures of the agates. The oxygen isotope composition of SiO_2 phases was examined by a laser-based micro-analytical method in the Stable Isotope Laboratory of the University Lausanne, Lausanne, Switzerland. Separate chalcedony and macrocrystalline quartz pieces were prepared, chemically treated and analyzed with a Thermo Fisher Scientific MAT 253 mass spectrometer [72]. Fluid inclusions in macrocrystalline quartz of the agates (polished 100 µm thick sections) were investigated using a Linkam THMS 600 heating-freezing stage. For the calibration of homogenization and freezing measurements, two synthetic fluid inclusion standards (SYN FLINC; pure H_2O, mixed H_2O-CO_2) were used [37]. In addition, the composition of gaseous and liquid fluids in the SiO_2 matrix of the agates was analyzed using a thermogravimetry-mass-spectrometry system (NETZSCH STA 409; Netsch, Selb, Germany), which was coupled to a QMS 403/5 quadrupole mass spectrometer (Pfeiffer Vacuum, Aßlar, Germany) [73].

Last but not least, the investigations included mineralogical and geochemical studies on parageneti minerals and organic materials in agates to get more detailed information about alteration, accumulatior and crystallisation processes. In general, associated minerals were identified by a combination o polarizing and scanning electron microscopy (SEM), *X*-ray diffraction and Raman spectroscopy. Ramar measurements were carried out using a confocal Jobin Yvon (Horiba) LabRam-HR spectrometer (Horiba Bensheim, Germany) on an Olympus BX41 optical microscope, with an 1800-lines-per-millimete diffraction grating and a Peltier-cooled, Si-based CCD detector. The analyses were complemented by Infrared (IR) absorption spectra acquired using a Bruker Hyperion microscope interfaced to a Bruke Tensor 27 Fourier transform infrared spectrometer [35].

The trace-element and isotopic composition of paragenetic calcite in certain agates was analyzec because of its ubiquitous occurrence with agates in the form of intergrowths and pseudomorphs Trace elements were measured by INAA (XRAL Laboratories, Toronto, ON Canada). The C and C isotope composition of paragenetic calcite was analyzed with the method of acid digestion using a Finnigan MAT 252 mass spectrometer in the Stable Isotope Laboratory of the University of Lausanne (Lausanne, Switzerland) [74].

3. Geological Occurrences and Types of Agates

In general, agate occurrences are distributed around the world on all continents, and agates have already been formed very early in the Earth's history. The oldest known occurrence, the Warrawoona agate in Western Australia, was found in 3.48 Ga old metamorphosed rhyolitic tuffs [75]. More thar one billion years old agates are also known from the basalts of the Lake Superior region in the USA and Canada.

A closer view reveals that agate occurrences are in particular connected with geological periods o strong volcanic activities such as huge basaltic lava flows or eruptions of acidic lava from Permiar to Tertiary. Chemical and mineralogical analyses of the host rocks show that most global agate occurrences are related to both SiO_2-poor (andesites, basalts) and SiO_2-rich (rhyolites, rhyodacites volcanic rocks (Figure 1).

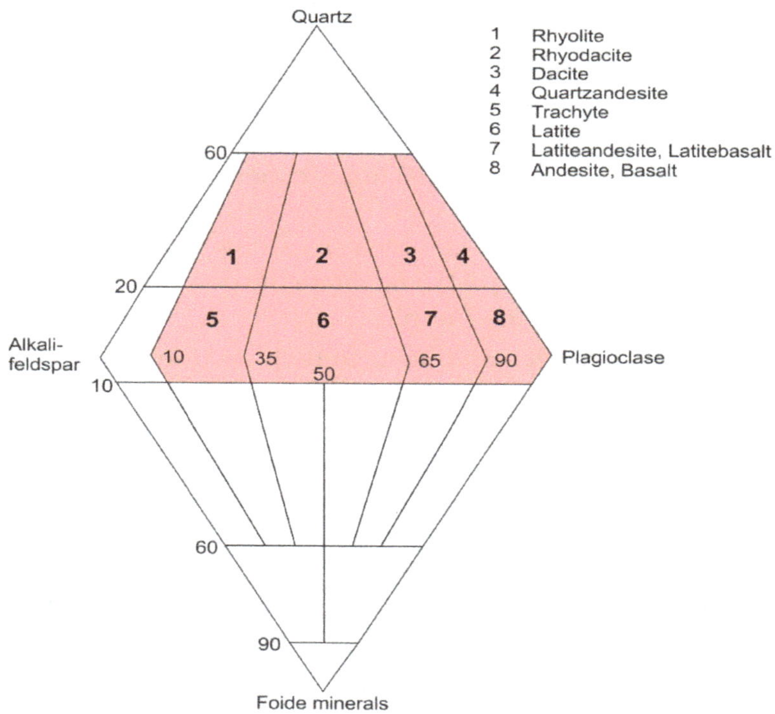

Figure 1. Streckeisen diagram showing the field (**red**) of agate-bearing volcanic rocks.

In acidic volcanics, agates originate from the infill of silica into cavities of spheroidal aggregates, so called lithophysae (thundereggs), whereas agate formation in basic volcanic rocks happens in former vesicular cavities (Figure 2a–d). The occurrences of *volcanic agates* are in particular a result of alteration processes of the volcanic host rocks. Therefore, they are preferentially formed in the marginal parts of volcanic bodies and ignimbrite layers, which are enriched in water and other volatiles [22,37].

Figure 2. Compilation showing the different main agate types; (**a**) lithophysa (thunderegg) within the flow texture of the rhyolitic host rock from Nesselhof, Thuringia (Germany); (**b**) large thunderegg with agate from rhyolitic ignimbrite of St. Egidien (Saxony, Germany); (**c**) andesitic host-rock with both empty and agate-filled vesicles at the Rancho Coyamito (Chihuahua, Mexico); (**d**) finely banded agate from a trachyandesitic host-rock of Xuanhua (Hebei province, China); (**e**) hydrothermal vein agate in metamorphic host rocks (gneisses) from Röthenbach (Saxony, Germany); (**f**) sedimentary banded microcrystalline granular quartz from a clay seam in carbonate host-rocks from Dulcote (Mendip, UK); (**g**) clastic agate pebble from river sediments of the Yangtze river near Nanjing (Jiangsu province, China); Scale bar is 5 cm.

It is very likely that the process of agate formation in many cases already started during the volcanic activities. Several investigations indicated the participation of residual magmatic fluids in the agate formation and the heating of meteoric fluids, which are necessary for the alteration of the rock and the release of silica and other chemical elements forming secondary minerals (e.g., [22,24,37]). However, in certain occurrences such as the Paraná Continental Flood Basalt Province, Brazil or the Deccan Trap Province, India strong indications were found that the agate formation took place remarkably later than the emplacement of the surrounding volcanic host rocks [76–78].

Besides the preferred formation in volcanic rocks, agates can also be formed by other processes and in other rock types. In different host rocks, vein agates and agate-like structures occur in breccia, cracks and fissures, which are often mineralized by hydrothermal solutions. *Hydrothermal vein agates* can be found in mm- and cm-sized veins (Figure 2e) but can also reach thicknesses of several dm and lengths of several hundreds of meters. The vein agates are often related to hydrothermal ore deposits. The formation of vein agates is not restricted to volcanic host rocks but can also be found in igneous and metamorphic host rocks.

The formation of agates in sedimentary environments is relatively rare compared to that in volcanic host rocks. *Sedimentary agates* have been found especially as irregular forms in stratigraphic sequences of carbonate rocks and clastic sediments (Figure 2f). In addition, silicification of residues of animals and plants is common in the surface region of certain sediments and volcano-sedimentary units. Agate structures are well known from silicified wood remains worldwide and silicified dinosaur bones from Utah, USA or silicified corals from Florida, USA [79–81]. Another common feature is the secondary deposition and redistribution of agates from primary deposits. Agates in clastic sediments (river gravel and marine sediments) are known from the surroundings of many agate occurrences worldwide (Figure 2g).

4. Agate Properties

4.1. Mineralogy and Micro-Structure of Agates

4.1.1. SiO_2 Phases in Agate

Numerous mineralogical investigations have shown that most agates consist of more than one silica phase [23,29,40,82]. The most common mineral in agate is trigonal α-quartz. It is present both as macrocrystalline and granular micro-crystalline variety, as well as in form of the disordered fibrous quartz varieties chalcedony (length-fast) and quartzine (length-slow).

Chalcedony is the common and dominant form of silica in agates. One of the most conspicuous features under the polarizing microscope is the presence of characteristic chalcedony "fibers" (Figure 3a), which was first reported by Brewster [83]. Investigations by X-ray diffraction confirmed the assumption that these structures consist of α-quartz, and the dimensions of the "fibers" were measured with diameters of 0.1–1 μm and lengths of up to several mm [84].

However, electron microscopic studies could not confirm such a fiber structure of chalcedony at high magnification [8,18,85]. Instead, the chalcedony "fibers" represent an intergrowth of quartz microcrystals of ca. 0.1 up to 3 μm in size (Figure 3b), which are oriented with the c-axis perpendicular to the "fiber" direction (length-fast), whereas their orientations are parallel to the c-axis in quartzine (length-slow). The fibrous "chalcedony crystals" are polysynthetically twinned according to the Brazil-twin law, with the c-axis slightly twisted around the fiber axis [23,86].

Heaney [23] developed a model of "dislocation growth ", which can explain the formation of the fibrous structure of chalcedony and the high frequency of defects, Brazil twinning and twisting in the chalcedony "fibers". Accordingly, the formation of dislocations in the crystal structure during crystal growth promotes the arrangement of atoms/molecules into spiral layers, which is energetically favoured compared to the incorporation into plane crystal faces. Such dislocations can be initiated by point defects or incorporated impurity ions in the crystal structure. The varying torsion of the c-axis (the optical axis) of the quartz crystallites causes different optical orientation and results in the formation of so called "Runzelbänderung" in polarized light (Figure 3a).

Quartzine is rare in agates and occurs only on the outer edge of the geodes or as interlayers between chalcedony and macrocrystalline quartz (Figure 4). Folk and Pittmann [87] concluded that the presence of quartzine may point to a deposition in sulphate-rich solutions or evaporitic conditions. However, up to now the specific conditions leading to the formation of quartzine are not completely understood.

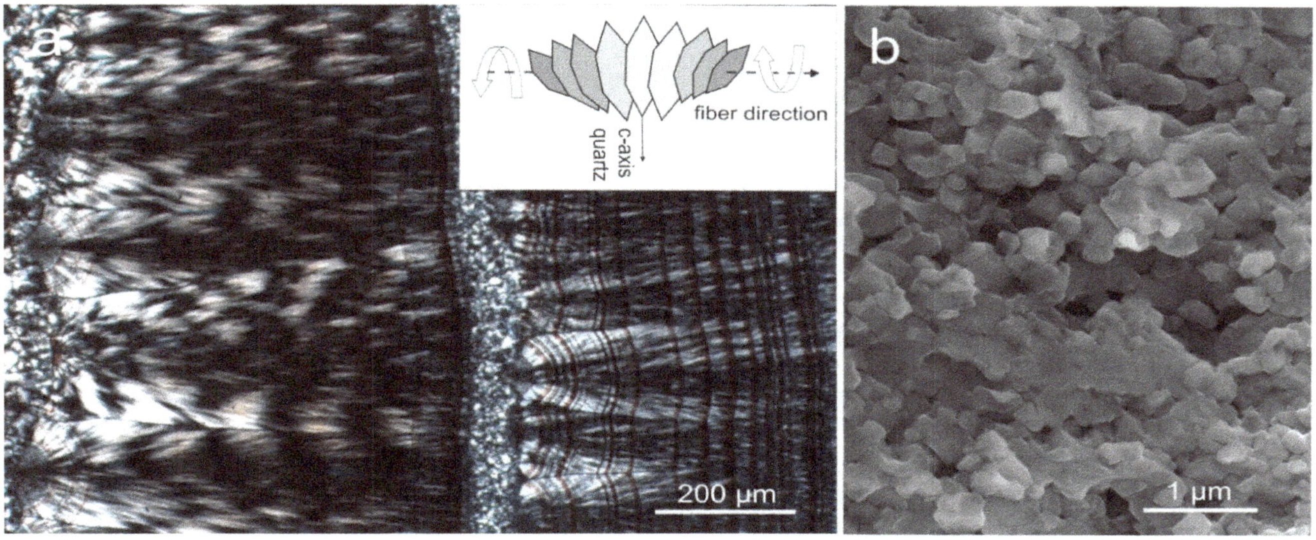

Figure 3. (**a**) Micrograph of an agate from Schlottwitz, Germany in transmitted light (crossed polars) showing chalcedony "fibers" with the characteristic "Runzelbänderung" caused by twisting of the optical axis of the quartz crystallites oriented with their c-axes perpendicular to the "fiber" axis; the inset schematically shows the formation of chalcedony "fibers"; (**b**) SEM micrograph of the same area revealing numerous quartz microcrystals <1 µm in size.

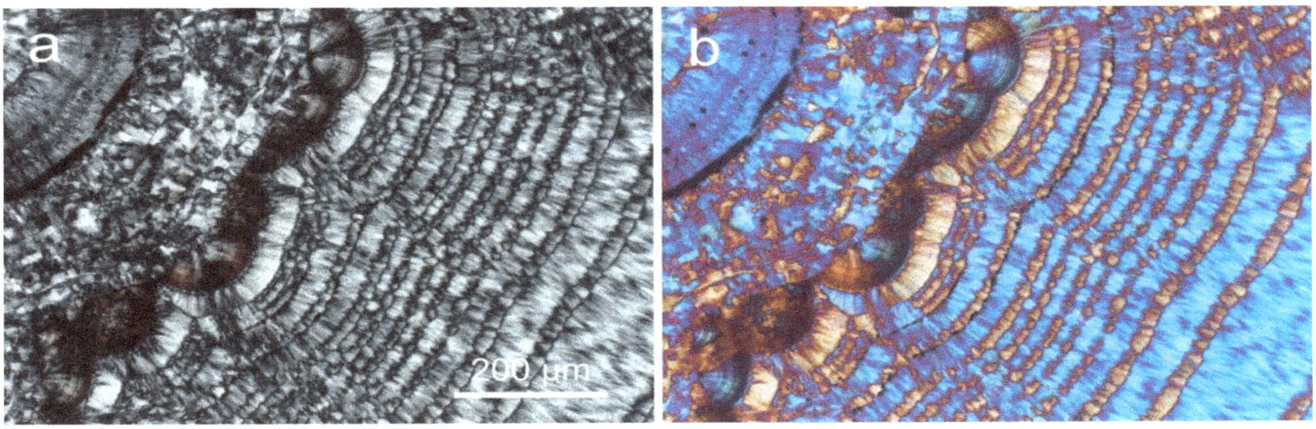

Figure 4. Agate from Xuanhua (Hebei province, China) in transmitted light (crossed polars) (**a**) and with λ compensator (**b**) showing alternating bands of fibrous chalcedony (blue) and quartzine (yellow).

Recent studies have shown that opal-CT, opal-C, cristobalite and moganite are also common in many agates worldwide [29,40,45,54]. Opal-CT represents an intermediate stage between pure amorphous and crystalline silica. It often occurs as tiny spherical aggregates (lepispheres) formed by thin crystal blades [82]. The crystal structure consists of fine interlayers of crystalline cristobalite and tridymite sometimes intergrown with amorphous opal-A (Figure 5). Opal-CT was detected e.g., in agates from the Richardson Ranch, Oregon (USA), Los Indios (Cuba), Sidi Rahal (Morocco) and Jalgaon/Savda, India [45,49].

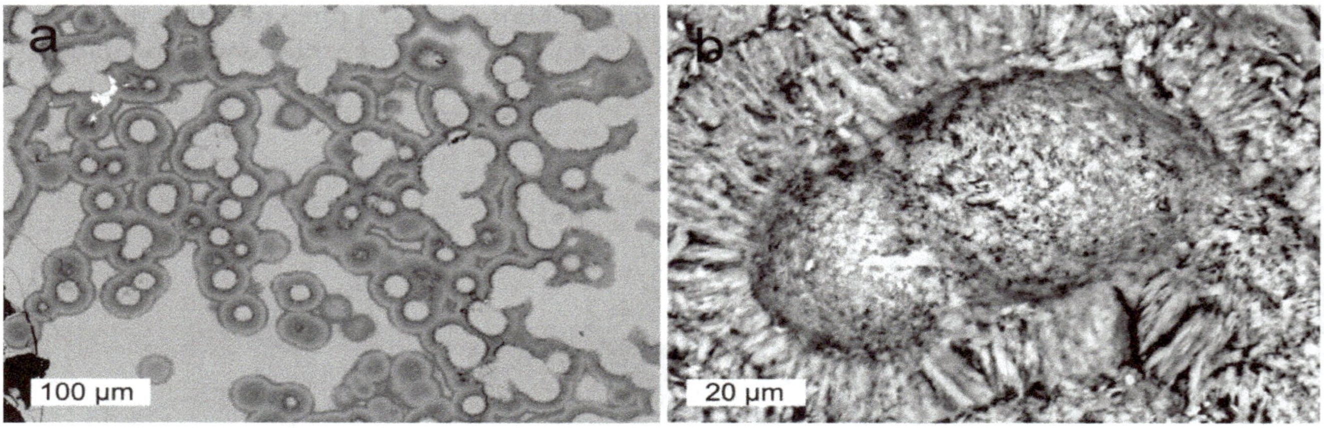

Figure 5. (**a**) SEM migrographs of spherical opal-CT aggregates within a chalcedony matrix in an agate from El Picado, Cuba; (**b**) close-up of (**a**) showing lepispheres of opal-CT consisting of radially arranged crystal blades.

Moganite was first described by Flörke et al. [88] from ignimbrites of the Mogan formation Gran Canaria. The crystal structure of this monoclinic SiO_2 modification is characterized by alternate stacking of layers of right- and left-handed quartz, with a periodic twinning according to the Brazil-twin law at the unit-cell scale [86]. Although the occurrence of pure moganite is rare, the intergrowth of chalcedony with moganite is frequent in agates, chert and flint [29,54,59,82,89,90].

Because of the similar optical properties and the narrow intergrowth with chalcedony it is difficult to reveal the presence of moganite in agate. However, Raman spectroscopic studies showed that moganite and quartz can be distinguished based on their different spectral characteristics [91]. Using these results, Götze et al. [29] first documented the variations in moganite content in different parts of agates and even within the chalcedony banding. Accordingly, Raman profiles or Raman mapping [92] can provide detailed information about the quantitative spatial distribution of moganite in agates (Figure 6).

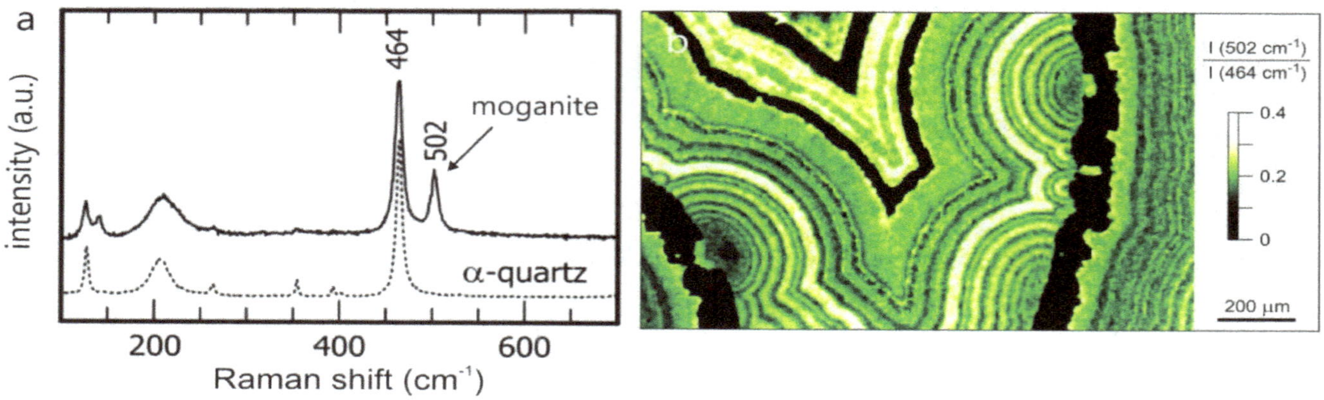

Figure 6. (**a**) Raman spectra of α-quartz (dotted line) and moganite-bearing chalcedony showing the characteristic peaks at 464 cm^{-1} and 502 cm^{-1}, respectively; (**b**) Raman mapping revealing the moganite distribution in an agate from St. Egidien (Saxony, Germany) based on the intensity ratio of the two characteristic Raman bands (modified after Nasdala et al. [92]—courtesy Lutz Nasdala).

In contrast to the microcrystalline SiO_2 species, amorphous opal-A is rare in agate. Previously it was concluded that agates contain about 10% of opal-A [93,94]. However, Flörke [95] first suggested that most agates do not contain any opal-A. It can be assumed that most of the identified amorphous silica reported in the older literature is likely caused by the powder preparation process for XRD measurements. Amorphous opal-A is mainly restricted to recent siliceous sinters, where it relatively rapidly converts to opal-CT [96].

The presence and spatial distribution of different silica phases is not only a result of the primary crystallisation processes but can also be influenced by secondary processes such as ageing, metamorphosis and/or temperature. These secondary processes can significantly change the primary phase composition, especially due to the conversion of unstable or metastable silica phases (opal-A, opal-CT, moganite) into quartz. The comparison of natural agate samples with synthetically treated agates showed that the conversion of metastable silica phases into quartz is accompanied by coarsening of crystallite size and loss of structural water [19,21,75,97,98]. Moxon and Carpenter [61] reported that in 300 to 1100 Ma old agate samples from 9 regions six had moganite contents below 6 wt% and three contained 1 wt% or less. The systematic change in phase composition and structural parameters may have potential to estimate the age of agates.

4.1.2. Microstructure of Agates and Agate Banding

One of the most conspicuous features of almost all agates is the typical agate banding. For the characterization of the general microstructure of agates, at least two types of agate banding have to be distinguished: (I) wall-lining agates ("fortification agates") with agate banding that is more or less conforming the outline of the host cavity (Figure 7a), and (II) horizontally layered agates ("Uruguay agate", water line agate) with silica layers parallel to the previous horizontal orientation (Figure 7b). There are certain intermediate states and combinations of these two main agate types (Figure 7c).

Figure 7. Examples for different types of banding in agates; (**a**) wall-lining ("fortification") agate from Ojo Laguna (Chihuahua, Mexico); (**b**) horizontal banding ("Uruguay agate") in an agate from Bightbridge, Simbabwe; (**c**) mixed type with both wall-lining and horizontal banding as well as macrocrystalline amethyst in the center from Tres Pinheiros-Fontoura Xavier (Rio Grande do Sul, Brazil); Scale bar is 2 cm.

The general structure of wall-lining agates often consists of three main parts/zones: 1. outermost layer of either microcrystalline, granular quartz or spherulithic chalcedony (or sometimes quartzine) in many cases including other minerals like e.g., calcite, hematite and/or pseudomorphs, 2. main zone of fibrous chalcedony, which develops from the spherulitic chalcedony, and 3. central part, where chalcedony passed over to macrocrystalline quartz, sometimes containing a hollow centre (Figure 8). Repeated layers with varying crystallite size and/or intercalations of fibrous chalcedony and granular quartz (or quartzine) are possible.

Figure 8. Cross section of a wall-lining agate from Wiederau (Saxony, Germany) from margin (left) to the center (right) in transmitted light (crossed polars) representing the general textural succession of spherulitic and micro-granular chalcedony—fibrous chalcedony—macrocrystalline quartz (width of the micrograph is 10 mm).

The formation of horizontally layered "Uruguay agates" can be explained by the gravitationa sedimentation of sols with flocculated, large SiO_2- particles. Under conditions of fast coagulatior (flocculation) of the monomeric silicic acid (H_4SiO_4) horizontal layers are formed [99]. The formatior of sol particles requires a high silica concentration and therefore, horizontally layered agates are ofter characteristic features for specific agate occurrences (Figure 9a). Walger [62] reported that horizonta layers in agates from Permian volcanic rocks in Germany are strictly parallel to the surface anc therefore, can be used as geological level. The sporadic presence of discordant silica layers in agate (Figure 9b,c) indicates that the orientation of the nodule has been changed during the formation of th agate bands.

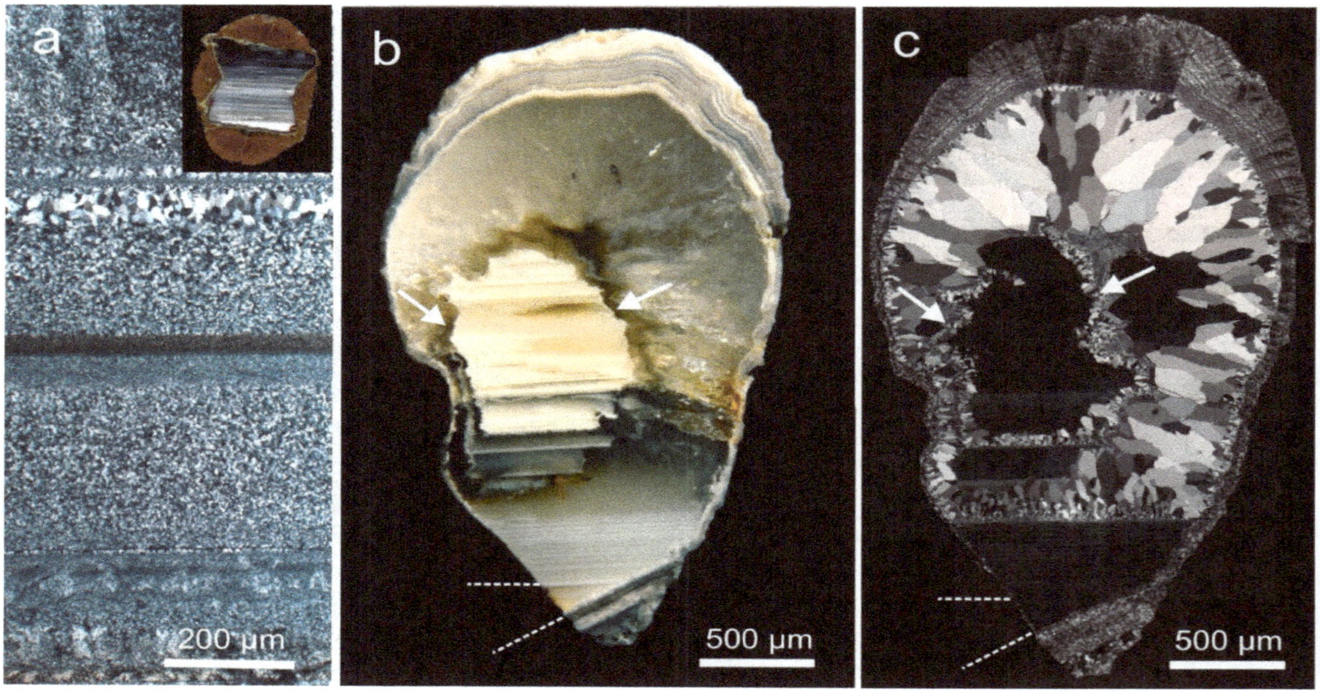

Figure 9. (**a**) Transmitted light (crossed polars) micrograph of an "Uruguay agate" from the Richardson Ranch (Oregon, USA, see inset); the different horizontal layers show strong variations in crystallite size (**b**,**c**) agate from Gröppendorf (Saxony, Germany) as original sample (**b**) and in transmitted light (crossed polars, (**c**)) with tilted horizontal layers and ongoing re-crystallisation of quartz replacing former horizontal bands (see arrows).

Microscopic investigations with high magnification revealed that visible macroscopic bands may consist of up to 1000 micro-bands per mm. A closer look reveals that agate banding appears due to periodic change in the characteristics of the microscopic sub-particles. Accordingly, the visible banding is caused by variations in the type of silica phases, their crystal size and habit, as well as porosity and the distribution of micro-inclusions of minerals (e.g., Fe-oxides/-hydroxides) within the SiO_2 matrix (Figure 10, compare Figures 4 and 6). Holzhey [100] published diameters of the submicroscopic SiO_2 particles between 0.07 to 0.7 μm for wall-lining agates and 0.2 to 2.5 μm in horizontally banded agates, which is similar to the diameters of microscopic globules in opal (0.25 to 0.4 μm; [101]).

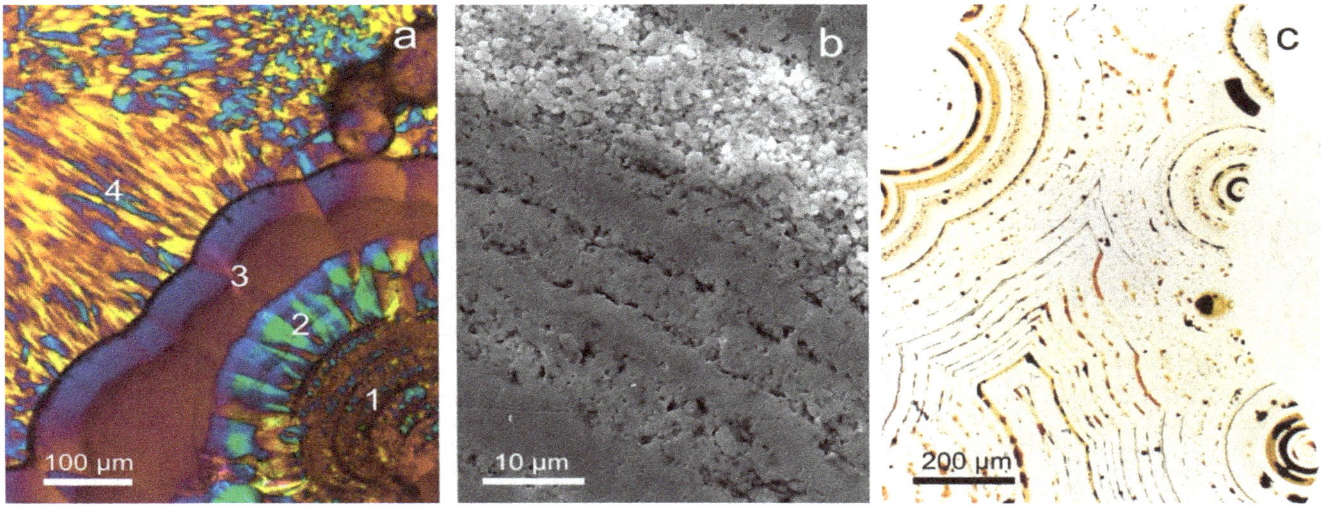

Figure 10. (**a**) Micrograph in transmitted light (crossed polars, λ-compensator) of an agate from the Whale Bay, Iceland showing a sequence of microcrystalline, granular quartz (1) quartzine (2) opal-C (3) fibrous chalcedony (4); (**b**) SEM micrograph of an agate from St. Egidien (Saxony, Germany); the high magnification reveals significant variations in crystal size and porosity in the different chalcedony layers; (**c**) transmitted light micrograph of an agate from Wiederau (Saxony, Germany); fine-dispersed Fe-oxides/-hydroxides are aligned along the micro-banding and cause the visible color banding.

Beside the characteristic agate banding, additional characteristic microtextures can provide information concerning the physico-chemical conditions of formation. For instance, primary growth textures such as colloform or botryoidal textures indicate precipitation of silica gel in free space, whereas comb-like, crustiform or zonal textures in quartz crystals point to a direct crystallisation from hydrothermal fluids [102]. Growth lines in euhedral quartz crystals, so called "Bambauer lamellae", are preferentially formed under strongly varying physico-chemical conditions during agate formation [46,52,55,60]. Most of these features can be related to epithermal conditions of formation and frequently occur in hydrothermal vein agates.

Other microtextures (feathery/flamboyant, mosaic/jigsaw-puzzle) are significant indications for their generation during recrystallisation of metastable silica phases such as amorphous silica precursors or chalcedony [46,49,55,60,102]. Replacement textures (lattice-bladed, pseudo-acicular) result from the secondary replacement of soluble phases (e.g., calcite, barite) by quartz. In general, such quartz textures due to the recrystallisation of amorphous silica or replacement textures are good indications of boiling conditions in epithermal environments. An excellent compilation about these microtextures in quartz is given by Dong [102].

4.1.3. Point Defects in Agate

Chalcedony, quartzine and macrocrystalline quartz are the most frequent SiO_2-forms in agates. Despite their different structural specifics, all three species show the same basic crystal structure of trigonal, low-temperature alpha-quartz that is composed exclusively of $[SiO_4]^{4-}$ tetrahedra with all oxygens joined together in a three-dimensional network. Different factors (e.g., concentration of SiO_2 and impurities in the mineralizing solution, temperature, growth velocity, etc.) can result in the formation of defects in the quartz lattice during crystallisation. Besides extended defects such as micro-inclusions of minerals and fluids or twinning and dislocations, defects at the atomic scale of the crystal lattice (zero-dimensional point defects) are present [103–106].

First investigations of the defect structure of agates from different parent rocks of certain world-wide localities by a combination of electron paramagnetic resonance (EPR) spectroscopy, cathodoluminescence (CL) and trace element analysis showed that defects related to silicon and/or oxygen vacancies are the most frequent point defects in agate [107]. In addition, the substitution of Si atoms by Al, Ge and/or Fe is common. In agate samples from different localities the following paramagnetic centers were detected: O_2^-, O_2^{3-}, O_3^-, E'_1, $[AlO_4]^0$, $[FeO_4/M^+]^0$, and $[GeO_4/M^+]^0$. In contrast, defect centers of the type $[TiO_4/Li^+]^0$ or $[TiO_4/H^+]^0$, which were common in quartz of the parent volcanic rocks, were not detected in agate [107]. In general, the abundance of the O_2^-, O_2^{3-}, O_3^-, centers (related to silicon vacancy) and E'_1 centers (oxygen vacancy) in chalcedony is remarkably higher than in macrocrystalline quartz. This high defect density indicates rapid growth of silica under non-equilibrium conditions probably with a non-crystalline precursor.

Figure 11a shows the spectra of three selected agate samples from Germany in a wide scan measured at a microwave power of 2 mW. The EPR spectra contain a characteristic oxygen-vacancy electron center E'_1 [108] and an orthorhombic Fe^{3+} center at the effective g value of 4.28 [103,109]. A closer look at the central magnetic region (Figure 11b) shows further multiple species of silicon-vacancy hole centers (e.g., superoxide and ozonide radicals such as B, B′, C, and C′; [110–115]). The agate from St. Egidien, Saxony (Germany) is particularly interesting due to abundant B/B′ and C/C′ centers even more than quartz samples from high-grade uranium deposits [109,115]. Weak signals at $g = 1.997$ and 1.994 belong to the Ge (B) and Ge E'_1 centers [108]. The simulated g values of the $[AlO_4]^0$ center are shown in the experimental spectrum in Figure 11c compared with the analyzed spectrum of an agate from Hausdorf (Saxony) containing 709 ppm Al.

The comparison of the EPR spectra of worldwide agate samples shows, that the oxygen-vacancy electron center E'_1 is the dominant defect center but in varying concentrations (Figure 12). In particular, chalcedony in agates from acidic volcanic rocks is characterized by elevated concentrations of the E'_1 center, much more than chalcedony in agates from basic volcanic rocks and macrocrystalline quartz within the agates. In general, quartz in agates has much more abundant oxygen-vacancy electron centers (by several orders of magnitude) than hydrothermal and pegmatite quartz that were measured for comparison (e.g., from Minas Gerais, Brazil, and Sichuan, China).

In addition, EPR spectra of the chalcedony parts of the agates often show high intensities of silicon-vacancy hole centers (Figures 11 and 12), whereas the number of these centers is lower in the macrocrystalline quartz part of the agates (Figure 11d). Nevertheless, the EPR analysis of clear macrocrystalline quartz crystals in agates provided powder-like spectra indicating that the quartz crystals are not really homogeneous but consist of numerous "micro-crystals".

Luminescence methods are suitable analytical tools for the visualization of point defects and therefore, provide useful information about the defects in the agate structure and their spatial distribution [117]. The common use of UV light (photoluminescence—PL) produces luminescence images, which reveal the heterogeneous microtexture of agates (Figure 13). The frequently observed bright green luminescence is caused by activation due to uranyl ($(UO_2)^{2+}$) ions [36] and therefore clearly indicates trace amounts of uranium in the SiO_2 network (Figure 13b).

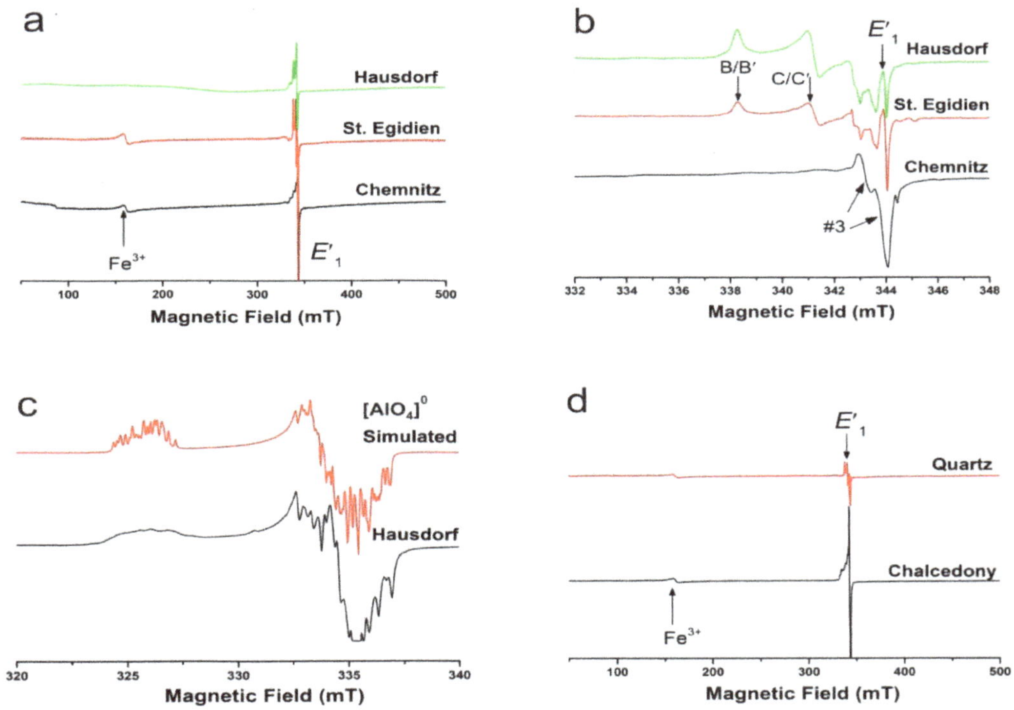

Figure 11. Powder EPR spectra of agate samples from Hausdorf, St. Egidien and Chemnitz (Saxony, Germany); (**a**) wide scans (microwave power of 2 mW and 295 K) showing a rhombic Fe^{3+} center as well as a pronounced E'_1 center in all samples; (**b**) narrow scan for the central magnetic field region (microwave power of 20 mW and 295 K) showing multiple species of silicon-vacancy hole centers (B/B′, C/C′, and #3); (**c**) narrow scan of the central magnetic field region (microwave power of 20 mW and 85 K) showing the $[AlO_4]^0$ center and silicon-vacancy hole centers in the agate from Hausdorf; also shown for comparison is a simulated spectrum of the $[AlO_4]^0$ center (data from Walsby et al. [116]); (**d**) wide scans (microwave power of 2 mW and 295 K) showing that the E'_1 center in chalcedony part of the agate sample from St. Edigien has a markedly higher intensity than its counterpart in macrocrystalline quartz from the center of the agate geode.

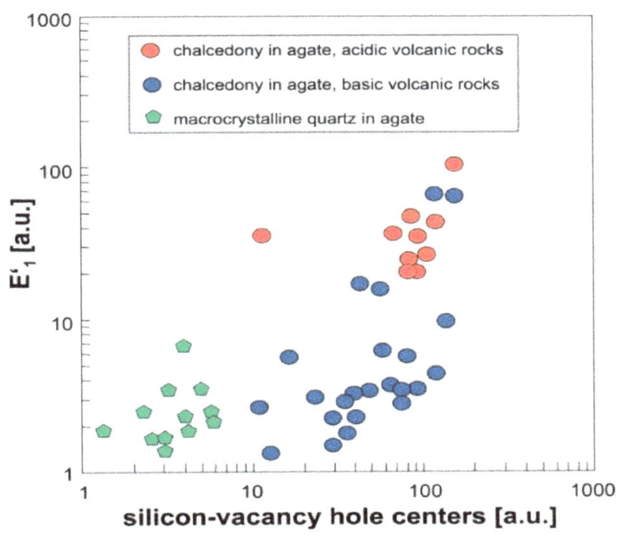

Figure 12. Relative abundances of silicon-vacancy hole centers (O_2^-, O_2^{3-}, O_3^-) and the oxygen-vacancy electron center E'_1 in chalcedony of agates from acidic and basic volcanic rocks, respectively, as well as macrocrystalline quartz in agates (data from Götze et al. [107]).

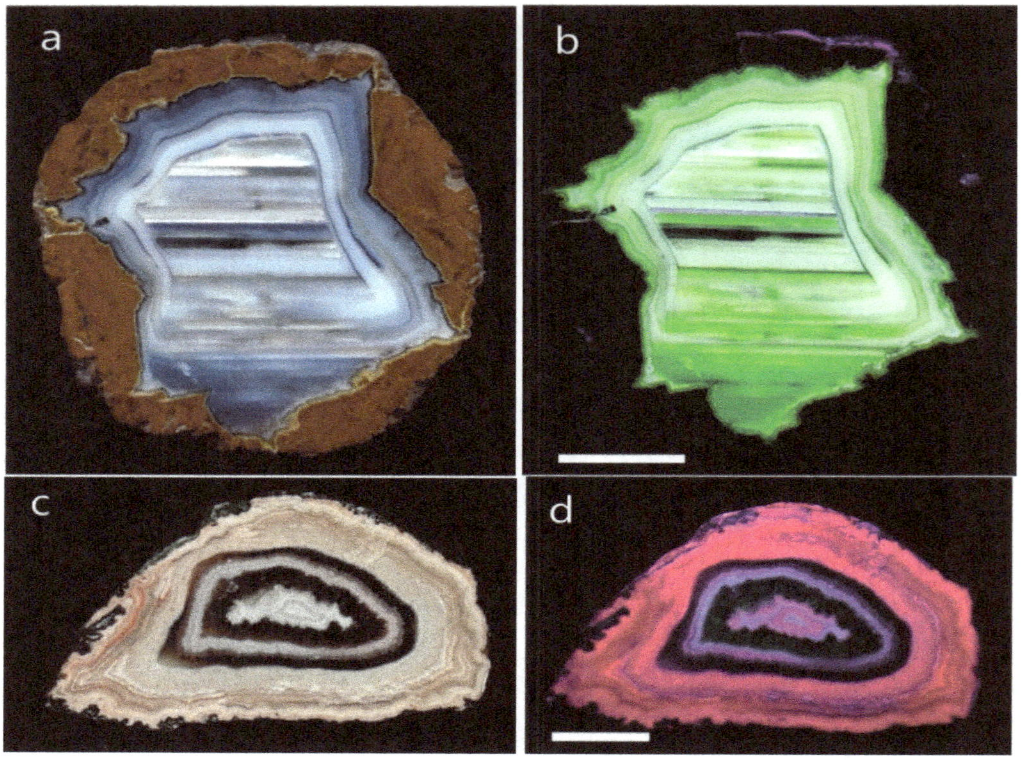

Figure 13. Agates from the Richardson Ranch, Oregon (USA, (**a**,**b**)) and from Idar-Oberstein (Germany, (**c**,**d**)) in day light (left) and under short-wave UV excitation (Photoluminescence, right); scale bar is 2 cm; the green luminescence is caused by uranyl ions (UO_2^{2+}), whereas the dominant reddish PL can be related to structural defects in the SiO_4-network.

Cathodoluminescence microscopy and spectroscopy can also be used to reveal differences in the microstructure of macrocrystalline quartz and chalcedony in the agates. Cathodoluminescence microscopy enables the visualization of zoning and other internal structures in agates and quartz incrustations at the micro-scale, which often markedly different to observations with conventional polarizing microscopy (Figure 14). For instance, most of the apparent homogeneous quartz crystals show a wide variety of internal textures under CL including oscillatory zoning, sector zoning or skeletal growth (see Figure 14b,f). These irregular internal textures and sector zoning point to silica crystallisation under non-equilibrium conditions.

The visible CL of agates can vary drastically (Figure 14). On one hand, agate samples can exhibit more or less one dominant CL colour, whereas in other agates SiO_2 phases exhibit multiple CL colors with different shades of blue, violet, green, yellow, and red. The visible luminescence colours are caused by different CL emission bands from the blue to the red spectral region with variable intensities which are associated to various structural defects (e.g., oxygen/silicon vacancies or broken bonds) and/or trace elements (e.g., Al, Li, Fe) in the lattice [117]. Moreover, CL can help to visualize different SiO_2 generations (Figure 14g) or secondary alteration effects such as radiation damage by the detection of radiation halos (Figure 14h). These characteristic CL properties can often be used to reconstruct specific geological processes.

Information concerning the lattice defects causing the different luminescence colors can be obtained by spectral measurements. Based on the fact that the defects causing the different CL emissions reflect the variable physico-chemical conditions of formation, CL spectra of agates often differ from those of quartz from crystalline rocks [107,118,119]. In the CL spectra of agates at least three broad emission bands can be detected: a dominating red band at 650 nm, a yellow band at about 570 nm and a blue band of mostly low intensity (Figure 15).

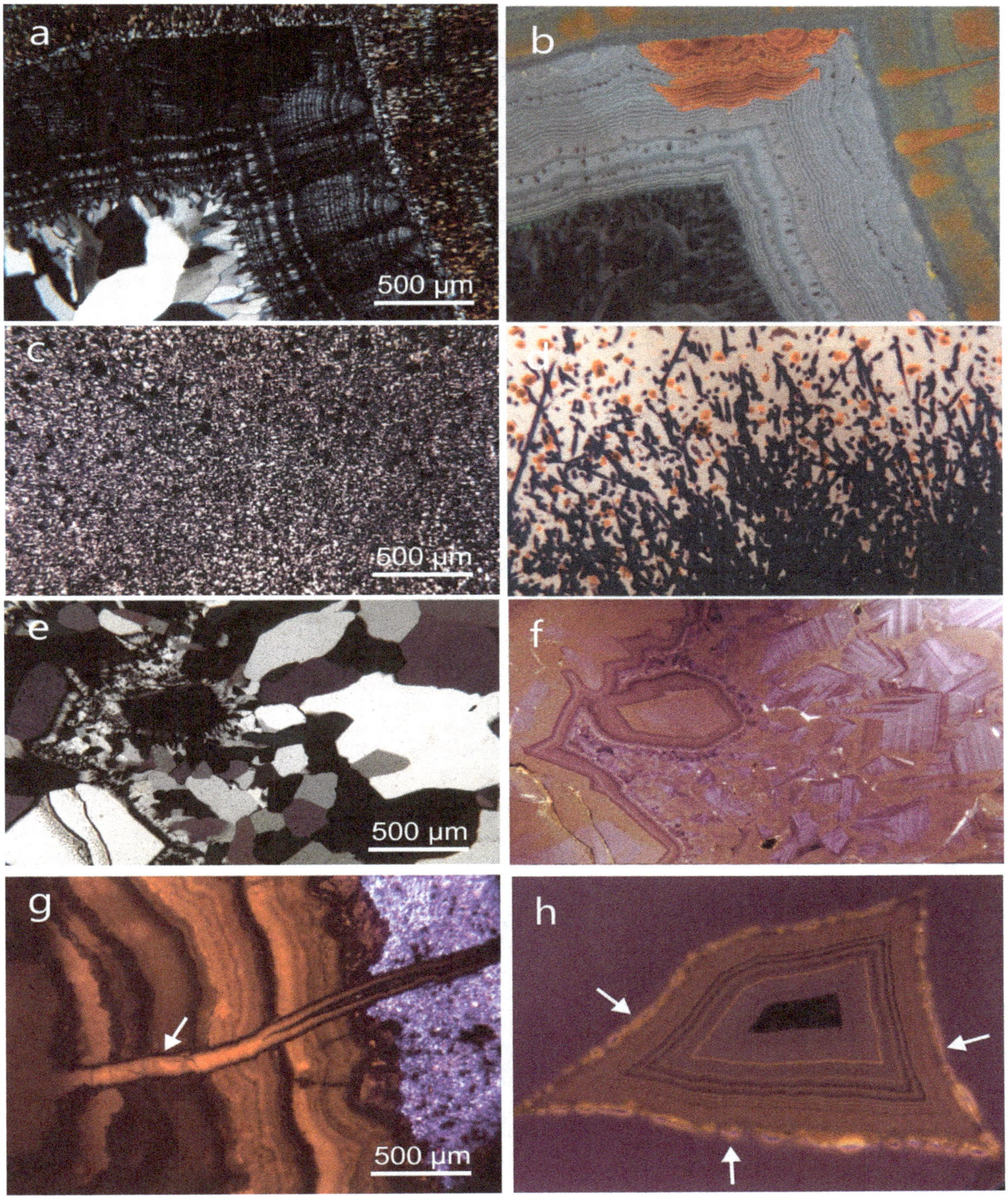

Figure 14. Micrographs of various agates in transmitted light (crossed polars) and cathodoluminescence (CL); (**a**,**b**) agate in liparite from Beipiao (Liaoning province, China) with multiple luminescence colors; note the heterogeneous internal CL textures of quartz crystals (**b**) appearing homogeneous in polarized light (**a**); (**c**,**d**) hydrothermal vein agate from Halsbach (Saxony, Germany) exhibiting spectacular microtextures in CL (**d**) not visible in polarized light (**c**); (**e**,**f**) agate from Gröppendorf (Saxony, Germany); the macrocrystalline quartz crystals appear homogeneous in polarized light (**e**) and show remarkable oscillatory as well as sector zoning in CL (**f**); (**g**) agate in rhyolite from Gehlberg (Thuringia, Germany); CL reveals a flow channel crosscutting the agate banding (arrow); (**h**) CL micrograph of an agate from Chemnitz (Saxony, Germany); the arrows point to a growth zone with numerous radiation halos due to radioactive micro-inclusions.

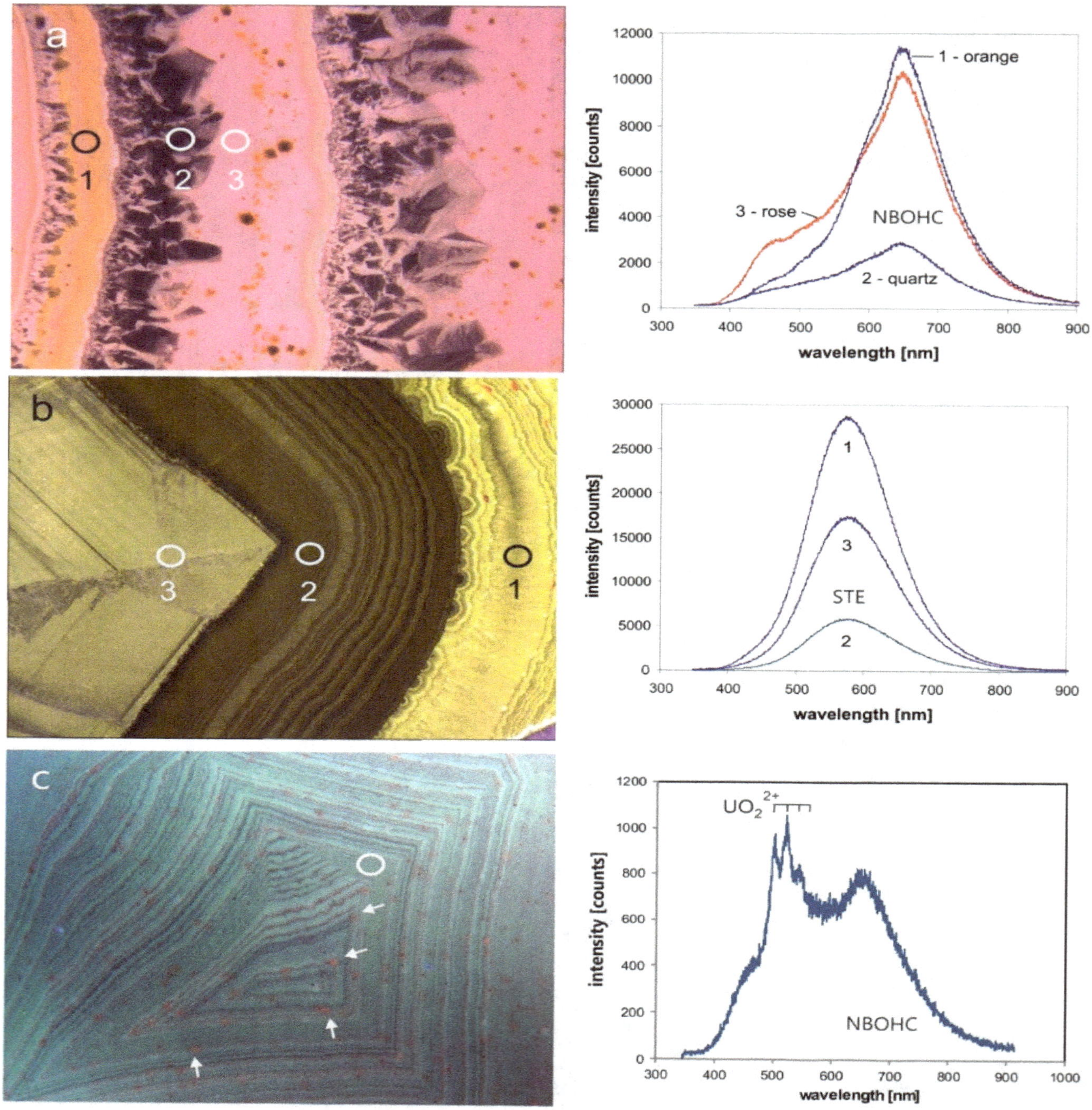

Figure 15. CL micrographs and related CL spectra of agates from different locations and host rocks; the circles mark the analytical spots of spectral measurements; (**a**) sedimentary "Dryhead agate" from the Pryor mountains, Montana (USA) showing growth zones with variable CL dominated by the 650 nm emission (non-bridging oxygen hole center, NBOHC); (**b**) agate in ignimbrite from St. Egidien (Saxony, Germany) with yellow CL; the spectra exhibit a broad emission band around 570 nm (self-trapped exciton, STE) with varying intensities; (**c**) agate in basalt from Soledade, Rio Grande do Sul (Brazil) showing greenish CL; the visible CL color results from the overlapping of a broad band at 650 nm (NBOHC) and multiple peaks between 500 and 600 nm of the uranyl ion (UO_2^{2+}); note the radiation halos around U-bearing inclusions (arrows).

The emission bands at 450 nm (2.69 eV) and 650 nm (1.91 eV) are caused by lattice defects [105,118,120]. The 650 nm band (Figure 15a) is the most common CL emission in chalcedony and can be related to the non-bridging oxygen hole center (NBOHC; [121]). The 650 nm emission is very sensitive to electron irradiation resulting in an increase of the band intensity during electron bombardment due to the conversion of different precursors (e.g., silanol groups: Si-O-H, Na impurities: Si-O-Na) into hole centers [105]. The specific CL behavior in chalcedony can be explained by elevated values of silanol groups [13,118].

Another defect induced luminescence is yellow CL (emission band at 570 nm—2.17 eV), which preferentially appears in macrocrystalline quartz and chalcedony of agates in acidic volcanic rocks (Figure 15b). The 570 nm emission band is related to high oxygen deficiency and local structural disorder in quartz. Götze et al. [68] related this specific CL emission to fast crystallisation from a non-crystalline precursor at low-temperature (mostly <250 °C).

The other type of luminescence active defects in agates is attributed to trace elements. Aluminium is the most frequent trace element in quartz and chalcedony (up to a few 1000 ppm). Several studies showed that the alkali (or hydrogen) compensated [AlO_4/M^+] center is responsible for a transient blue emission band at ~390 nm (3.26 eV) [122–124]. The 390 nm emission is very sensitive to electron irradiation and the strong decrease under an electron beam is caused by the dissociation and electromigration of the charge compensating cations out of the interaction volume under the influence of the irradiation induced electrical field.

A striking greenish luminescence in agate is related the uranyl ion (UO_2^{2+}) and can be excited both by short-wave UV light (<300 nm) or an electron beam [36]. The visible greenish luminescence in agates is characterized by a typical emission line at ~500 nm accompanied by several equidistant lines due to the harmonic vibrations of oxygen atoms in the uranyl complex (Figure 15c). The mechanism of the uranyl luminescence is very effective and can be detected in agates with U contents as low as 1 ppm.

4.2. *Geochemistry of Agates*

4.2.1. Trace Elements

Trace elements represent impurities in the SiO_2 matrix of agates and may be structurally incorporated into the regular lattice, in inter-lattice positions or can be related to micro-inclusions of fluids and minerals [106,125]. The chemical elements were mostly incorporated during agate formation and therefore, are important geochemical indicators of geological processes reflecting the origin of mineral-forming fluids and/or the specific conditions of crystallisation.

Table 1 presents a compilation of trace-element data from the literature representing different kinds of agates from worldwide occurrences. The results show a broad range of trace-element concentrations reflecting the influence of the geological background of the specific locations and the agate type. In principle, trace element concentrations are low and the content in macrocrystalline quartz is mostly lower than that in the associated chalcedony. Only certain elements are relatively enriched in agates. These are in particular Al, Ca, Fe, K and Na, which are probably mobilized and transported together with SiO_2 from the host rocks during agate formation. Moreover, elevated concentrations of halogens (up to 380 ppm F and up to 154 ppm Cl; [63]) indicate that these two elements play an important role in fluids of the alteration and transport processes. In contrast, Sc, Nb, Ta, Th, Ti and other immobile elements are present in very low concentrations, often below the detection limit of trace-element analysis; only Powolny et al. [55] reported significant amounts of Zr in agates from Borówno, Poland.

Table 1. Variations of selected trace-element concentrations (in ppm) of chalcedony and macrocrystalline quartz in agates of different origin (data from [1,26,27,30,31,36,37,58,68,69,126]).

Element	Agates in Acidic Volcanic Rocks		Agates in Basic Volcanic Rocks		Vein Agates		Agates in Sedimentary Rock	
	Chalcedony	Quartz	Chalcedony	Quartz	Chalcedony	Quartz	Chalcedony	Quartz
Al	109 ... 7300	34 ... 1380	15 ... 1170	30 ... 700	640 ... 7000	75 ... 546	103 ... 809	1 ... 13
Ca	<10 ... 540	<10 ... 27	<10 ... 710	<10 ... 120	20 ... 400	<16 ... 45	350 ... 1400	630 ... 1200
Fe	30 ... 4500	<30 ... 978	7 ... 3900	<30 ... 670	150 ... 3000	150 ... 1700	26 ... 1600	<1 ... 30
Mn	<1 ... 120	<1 ... 10	<1 ... 90	<1 ... 46	1 ... 27	3 ... 6	<1	<1
K	40 ... 3640	30 ... 600	20 ... 580	66 ... 420	150 ... 1360	25 ... 480	na	na
Na	90 ... 850	22 ... 600	85 ... 750	20 ... 140	50 ... 350	100 ... 240	48 ... 323	45 ... 158
Mg	5 ... 556	3 ... 70	1 ... 364	4 ... 287	16 ... 220	46 ... 105	na	na
F	<25 ... 380	<25 ... 52	<25 ... 228	<25 ... 270	79 ... 156	31 ... 200	na	na
Ge	3.2 ... 22	0.9 ... 30	0.5 ... 8.5	0.5 ... 15	1.3 ... 4.8	0.6 ... 4.5	<1 ... 2	<1
B	<10 ... 135	<2 ... 155	3 ... 146	<2 ... 120	<2 ... 18	6 ... 46	na	na
U	<1 ... 21	<1 ... 14	<1 ... 21	<1 ... 4	<1 ... 15	<1 ... 4	<1 ... 72	5 ... 18

na = not analyzed.

Chemical analyses of separate agate zones or element profiles perpendicular to the bandin provided more detailed information concerning the trace-element distribution within the microstructur of the agates [1,26,27,30,31,36,37,58,68,69,126]. The concentration profiles show variations for mos elements and only some elements behave relatively constant (compare Figure 16, Table 2).

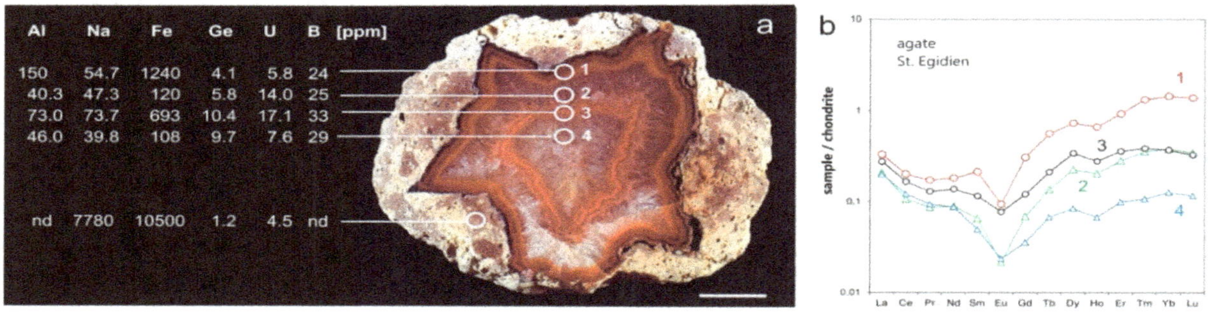

Figure 16. (**a**) Trace-element data from ICP-MS analyses of different zones (1, 3 = chalcedony; 2, 4 = macrocrystalline quartz) in an agate from St. Egidien, Saxony (Germany) compared to the composition of the rhyolitic host rock (the circles mark the areas of chemical analyses; nd—not detected); (**b**) chondrite-normalized REE distribution patterns (normalization according to data of Mason [127]) of the same agate zones shown in (**a**).

Table 2. Concentrations of trace elements (in ppm) from LA-ICP-MS local chemical analyses of a profile across an agate from St. Egidien, Saxony (Germany) (data from Götze et al. [68]); the numbers relate to the analytical points indicated in Figure 17.

Spot	Ge	Al	Fe	B	Ga	Ca	Li	Na	K	Rb
1	77.4	-	11.5	36	0.07	-	0.98	77.4	41.1	0.45
2	16.4	366	46.9	34.2	0.71	62.4	1.66	101	162	1.86
3	22.2	329	48.2	33.3	0.81	58.3	1.89	67.4	168	1.5
4	94.9	-	11.3	33.4	0.16	-	0.89	-	61.8	0.5
5	53.5	-	12.4	23.5	0.04	-	0.86	-	49.2	0.32
6	93.6	-	14.5	31.3	0.12	-	0.51	-	37.4	0.44
7	26.8	-	18.7	8	0.14	-	0.59	-	32	0.15
8	68.9	-	12.8	19.2	0.12	-	0.47	-	33	0.24
9	31.6	103	154	28.7	0.38	9.7	0.49	37.9	89.2	0.58
10	36.3	<6.0	47.7	29.2	0.17	9.4	0.39	-	51.9	0.33
11	28.9	-	47.8	41	0.11	-	0.49	-	55.1	0.43
12	20.4	-	61.9	35.7	0.12	-	0.59	-	60.7	0.33
13	13.2	<6.0	55.4	45.6	0.22	-	0.35	-	60.2	0.31
14	21.2	162	66.9	25.8	0.98	-	0.30	-	149	0.97
15	0.5	-	1.9	<1.0	0.18	53.3	0.30	-	1.2	-

(-) = not detectable; in general Ti < 1.2 ppm, Ba < 0.1 ppm, Sr < 0.1 ppm, Mn < 0.6 ppm.

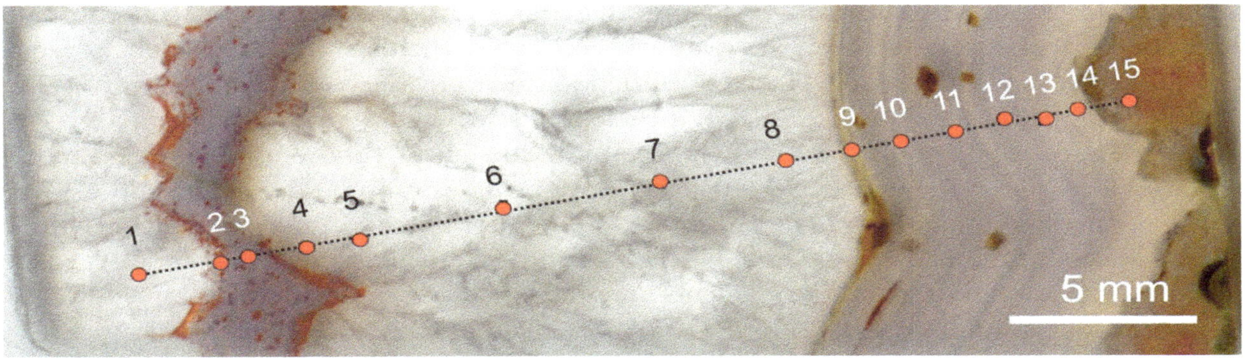

Figure 17. Micrograph of an agate from St. Egidien, Saxony (Germany) with a profile showing the analytical spots (spot numbers relate to analytical data in Table 2) of trace element analysis by LA-ICP-MS.

Results of these studies revealed a remarkable behavior for specific chemical elements. Remarkable high concentrations of Ge (>90 ppm), B (46 ppm) and U (>20 ppm) were detected in agates, which often exceed the average concentration of the Earth's crust (for Ge = 1.4 ppm, B = 12 ppm) and sometimes the element concentration in the surrounding host rocks (Figure 16). Interestingly, the concentrations of Ge, B and U in macrocrystalline quartz are sometimes higher than in associated chalcedony (Figure 16). High Ge contents can be explained by the similar geochemical character of Si and Ge, which causes the common transport and incorporation of Ge into the quartz structure [125,128]. The occurrence of Ge and B defect centers in quartz of agates was confirmed by recent EPR measurements [68,107].

The observed high concentrations of U within some of the agates are noteworthy. In the literature, U contents of up to 1200 ppm are published for chalcedony in agates from acidic volcanic rocks [129]. The U contents of sedimentary agates from the Dryhead area (Montana, USA) are amongst the highest values (compare Table 1). The location is close to a Uranium deposit, and therefore there is high availability of uranium in the vicinity [31]. The mobility of uranium during the alteration of volcanic rocks was investigated by Zielinski [130], who observed a parallel accumulation of Si and U. Recent investigation showed that a substitution of Si ions in the quartz lattice by U is unlikely due to the different crystal-chemical properties, and instead the incorporation as uranyl silicate complex is favoured [36,131]. First, uranium can be adsorbed from solution by natural silica compounds (silica colloids) and then trapped as uranyl silicate complex in a stable silica matrix [131]. Luminescence studies confirmed the existence of uranyl compounds in quartz and chalcedony of agates [36] (compare Section 4.1.3).

First results were also published for rare earth elements (REE) in agates [30,37,55], which are preferentially bound to fluid inclusions and therefore, reflect the composition of the agate mineralizing fluids. The chondrite normalized REE patterns of different agate zones in Figure 16 are characterized by pronounced negative Eu anomalies and a slight slope of the light REE (La—Sm), whereas the shape of the heavy REE patterns (Gd—Yb) is increasing. Moreover, agates may show slight "tetrad effects", i.e., a subdivisions of the REE series into four concave-upward groups [132,133]. Such REE distribution patterns can be explained by a primary crustal signature that is overprinted by an enrichment of heavy REE due to the preferred complexation of released HREE by carbonate- and F-complexes during the alteration and transport processes [134]. However, various scenarios for mineralizing SiO_2-bearing fluids for agate formation are possible, sometimes also resulting in positive Eu-anomalies [55]. Indications for the mixing of extremely different compounds (e.g., volatiles, meteoric water) are also provided by the presence of Ce anomalies in some agates [34,37,135].

The measured trace-element profiles illustrate, that there can be a wide scatter of concentrations for certain trace elements, whereas others are relatively constant in their spatial distribution. For instance, balanced concentrations at a low level were recorded for the elements Li, Rb, Ba, Sr, Ga or Ti (Table 2). Contrast variations can be detected for Fe concentrations, which are mostly related to the color of the agates (Figure 16). Strongly colored (especially reddish and yellow) chalcedony bands have the highest Fe contents due to micro-inclusions of certain Fe-compounds (e.g., hematite, goethite). These zones can also contain elevated concentrations of other transition elements such as Mn, Ni or Cr [1,30]. Concentrations of up to 400 ppm Ni and 500 ppm Cr are reported by Barsanov et al. [136] for agates from Timan and Caucasus. The elements are probably incorporated in fine-grained iron oxides, where concentrations of up to 14 wt% Mn, >1000 ppm Ni and >550 ppm Cr were measured [1].

The general trend of the lateral trace element distribution from the agate host rock to the center shows a slight decrease in element concentrations within chalcedony without a direct relation between trace-element contents and banding, and with lowest contents of most elements in macrocrystalline quartz. These variations can probably be explained by a "self-purification" process during agate formation. Certain chemical elements are firstly precipitated together with the primary amorphous silica and accordingly are finely dispersed in the silica matrix. The following chalcedony growth may initially incorporate high contents of trace elements. During further crystallisation, the concentration of impurities decreases. Heaney [23] and Merino et al. [27] stated that the formation of frequent defects, Brazil twinning and the typical twisting in chalcedony are caused by the high trace-element contents in silica. Chalcedony fibers in agates may therefore be the result of dislocation growth. The continuous lowering of trace elements in the crystallisation medium may then finally cause the formation of coarse-grained quartz.

4.2.2. Isotope Composition

Isotope data of the SiO_2 phases in agates may provide information about the isotopic composition of fluids from which the agates were formed and allow a first estimation of formation temperatures. The composition of the oxygen isotopes ($\delta^{18}O$) in agates from various locations and of different types reveals a rather wide range from 13 to 33‰ (Figure 18). In most cases macrocrystalline quartz from the respective specimens has slightly lower $\delta^{18}O$ values than chalcedony of the same sample and more closely matches the values of hydrothermal quartz precipitated by late magmatic fluids [137]. The difference between the $\delta^{18}O$ values of chalcedony and quartz can possibly be explained by the lower equilibrium constants in the system quartz–water compared to the systems tridymite–water and cristobalite–water [138,139]. Accordingly, amorphous SiO_2, chalcedony, cristobalite and tridymite may have higher $\delta^{18}O$ values than quartz formed under the same temperature conditions.

During crystallisation of quartz from a mineral-forming fluid the initial $^{18}O/^{16}O$ ratio changes due to the preferred incorporation of the heavy ^{18}O-isotope into the quartz lattice [137]. Because of the temperature dependence of this fractionation the measured isotope ratios can be used to calculate formation temperatures on the basis of the fluid oxygen isotopic composition [139]. However, no data exist regarding their oxygen isotope composition, since a direct analysis is mostly not possible because of the extreme small amount of fluid inclusions. Therefore, the temperatures of isotope fractionation were calculated assuming different fluid compositions of meteoric water (−10‰), oceanic water (0‰) and magmatic water (+8‰).

The calculation of agate formation temperatures based on these data provided rather low temperatures between 20 °C and 230 °C [30,31,37,140,141], depending on the isotopic composition of the involved waters. These temperatures cover a broad range between sedimentary origin and hydrothermal formation and have to be discussed in combination with other geochemical data. Fluid inclusion studies provided similar temperatures and confirm the results from the oxygen isotope investigations (see Section 4.2.3).

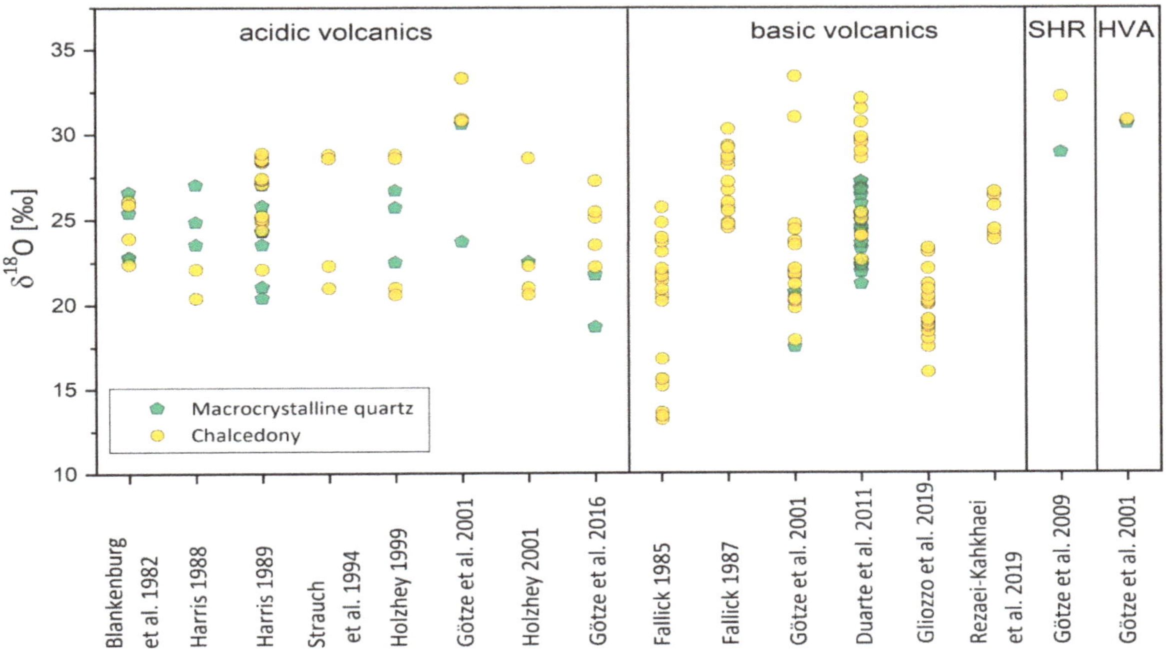

Figure 18. Compilation of oxygen isotope data ($\delta^{18}O$) of agates from different origins (SHR = sedimentary host rock; HVA = hydrothermal vein agate); data are compiled from [14,30,31,37,58,101,140–147].

Results of spatially resolved oxygen isotope analyses of different agate bands reveal strong variations of the $\delta^{18}O$ values (partially >10‰) over small scales in some agates, whereas others show a homogeneous isotopic composition (Figure 19). Assuming that the analyzed isotopic compositions are of a primary nature, the variations could be explained either by fluctuations of temperatures resulting in changes of the conditions of isotope fractionation, or variations in the primary isotope composition of the fluids, i.e., participation of fluids of different geochemical composition. Multiple temperature variations are not really likely, whereas the supply of SiO_2-bearing solutions by mixing of surface water, heated meteoric water and/or magmatic fluids was discussed by several authors [13,30,37,41]. Different ratios of the participating fluids could then cause variations in the isotope composition.

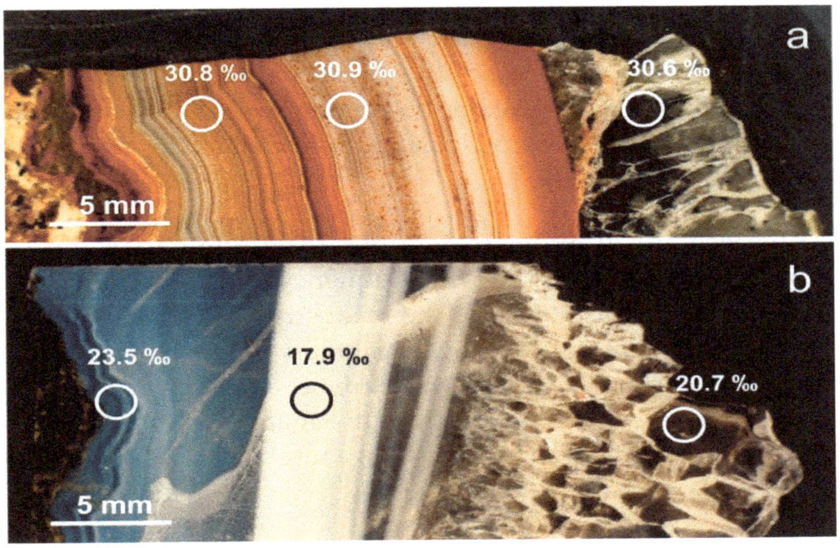

Figure 19. Hydrothermal vein agate from Schlottwitz, Germany (**a**) and agate from mafic lavas of the Ardownie quarry, Scotland (**b**) with results of spatially resolved oxygen-isotope analyses (circles).

Alternatively, isotope fractionation under non-equilibrium conditions could explain the observed isotopic variations due to various kinetic processes such as diffusion, adsorption, solution-precipitation phase transformations or boiling. Kita and Taguchi [148], for instance, found kinetic effects in the fractionation between SiO_2 and geothermal waters during precipitation of SiO_2 that resulted in variations of the $\delta^{18}O$ values of some ‰. Accordingly, the assumption of agate formation from silica-bearing geothermal waters via an amorphous precursor could also explain heterogeneities in the isotopic composition.

Strong variations were also reported for δD values in chalcedony (−130‰ to −40‰) and macrocrystalline quartz (−90‰ to −50‰) in agates from different localities including variations between different agate bands within single agate samples [30,145]. These data support the idea of fluid mixing and non-equilibrium conditions during agate formation. Therefore, the isotope data of agates have to be interpreted with care.

4.2.3. Fluid Inclusions and Water in Agate

Several studies have shown that agates may contain considerable amounts of water and other fluid components [1,13,15,21,34,37,40,73,149]. The H_2O content in the agate matrix varies between 0. and 3 wt% [13,21,40] and is preferentially present as OH (Si-OH silianol groups). This is in contrast to opal, which may contain up to 10 wt% H_2O, predominantly as molecular water. Flörke et al. [13 assumed that the incorporation of OH and H_2O in chalcedony is especially related to structurally damaged crystal boundaries.

Frondel [150] first stated that the content of structural water in different agate zones can differ due to variations in the amount of different SiO_2 phases and their microstructure within the agates. Graetsch et al. [151] investigated separate parts of chalcedony from Brazilian agates and determined water contents between 0.65 and 1.42 wt% in the wall-lining agate and 0.56 up to 0.92 wt% in horizontal agate layers. The ratio between silanol water and molecular water varied between 0.4 and 1.0. Based on these data they postulated temperatures of agate formation below 250 °C and pressures lower than 50 MPa.

Moxon and Rios [40] and Moxon [21] found a decreasing amount of silanol water with increasing geological age during their investigations of agates from worldwide occurrences in 23 to 2717 million years old parent rocks. The total water content varied in the agates between 0.23 and 1.78 wt% and the average amount of structural silanol water was found to be 76% of the total water content. They concluded that silanol groups were released during the recrystallisation of moganite to chalcedony and therefore, an estimation of the age of agates based on the amount of structural water should be possible. These conclusions supported previous assumptions that agates from Brazil and New Zealand were formed long after their host rocks [21].

Small amounts of gases and liquids from mineral-forming fluids can be incorporated into growing crystals during their formation. The type, frequency and chemical composition of such fluid inclusions can provide valuable information for the reconstruction of the mineral-forming processes. Several fluid inclusion studies in agates have shown that more or less no visible fluid inclusions exist in cryptocrystalline quartz (chalcedony) because of the extreme small crystal size. Therefore, analyses are only possible in macrocrystalline quartz crystals from the agate centers. The frequency of measurable primary inclusions in the quartz crystals of agates is in general very low (Figure 20) and the inclusions are extremely small (<20 μm).

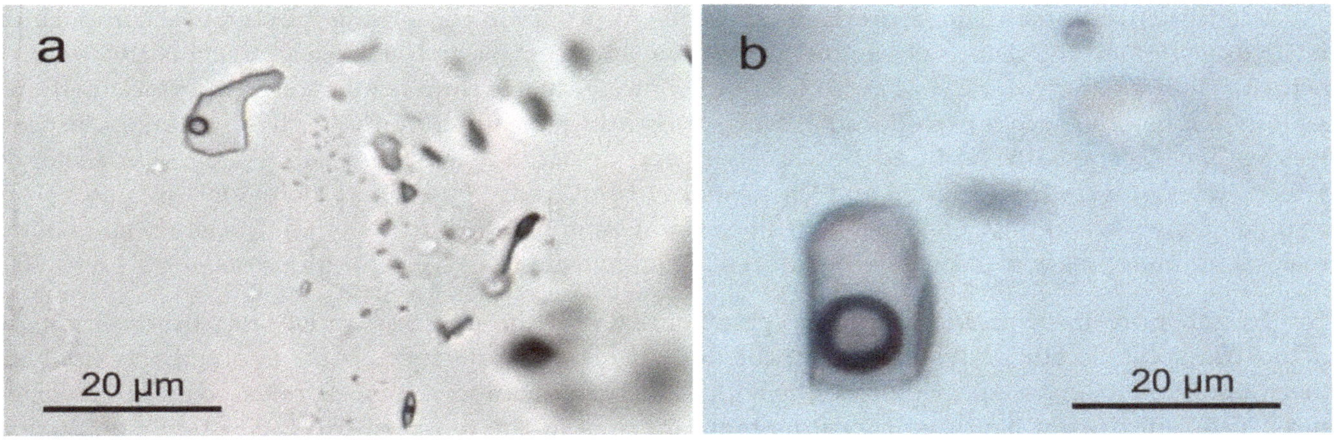

Figure 20. Primary two-phase (gas-liquid) inclusions in macrocrystalline quartz of agates from Saxony (Germany): (**a**) Burgstall, (**b**) St. Egidien.

Published homogenization temperatures of quartz from agates scatter in a very broad temperature range between <100 °C and >500 °C [15,37,149,152,153]. However, the interpretation of the measured homogenization temperatures has to be done very carefully. Different degrees of filling in the inclusions (i.e., ratio of gas to liquid phase) provide indications for fluid trapping under heterogeneous conditions. This conclusion is supported by a strong variation of the measured homogenization temperatures. Such heterogenization can be caused by splitting of fluids into an aqueous phase with high salinity and a gaseous phase with low salinity at high temperatures (>375 °C) due to drop of pressure [154,155].

Another possibility is phase separation in the subcritical temperature range (<375 °C) at low pressure due to boiling in an open system (aqueous phase with saturated water vapor). These conditions result in the occlusion of gas/liquid mixtures of varying ratios. Therefore, homogenization temperatures cannot directly be used as trapping temperatures of the fluids [154]. The minimum temperature of trapping is then represented by the lowest measured temperatures of fluid homogenization. Taking into account all these facts, a heterogeneous trapping of inclusions has to be assumed for most of the investigated agates from acidic and basic volcanic rocks. Table 3 summarizes minimum homogenization temperatures for agates of some German occurrences. These data confirm the results concerning agate formation temperatures obtained from oxygen isotope studies (compare Section 4.2.2).

Table 3. Minimum homogenization temperatures (T_{hom}) obtained from fluid inclusion studies of agates from different occurrences in Germany.

Agate Location	Host Rock/Agate Type	T_{hom}
Gehlberg (Thuringia)	rhyolite/lithophysa	95 °C
Chemnitz (Saxony)	ignimbrite/lithophysa	176 °C
St. Egidien (Saxony)	ignimbrite/lithophysa	186 °C
Mügeln (Saxony)	rhyolite/lithophysa	177 °C
Burgstall (Saxony)	rhyolite/lithophysa	172 °C
Gröppendorf (Saxony)	melaphyre/amygdale	134 °C
Schlottwitz (Saxony)	hydrothermal vein agate	80 °C

In addition, cryometric measurements provided information concerning the chemical composition of the investigated inclusions. In agates from basic and intermediate rocks, quartz mostly contains inclusions having low salinity (<4 eq% NaCl). In contrast, agates from acidic volcanic rocks sometimes contain inclusions with up to 30 eq% NaCl indicating the involvement of magmatic solutions. Besides NaCl, especially KCl and $CaCl_2$ have been detected as salt phases [37,149]. For instance, eutectic temperatures of −10.5 °C and the absence of salt hydrates in fluid inclusions of agates from Saxonian agates favored KCl as the main salt component in the fluid system [37]. Measurements of the freezing point depression according to Hall et al. [156] revealed salt concentrations of 0.65–2.4 wt% KCl.

To get more information about the geochemistry of the gaseous and liquid fluids involved in the transport of silica, systematic studies of agates were performed by evolved gas analysis (thermogravimetry-mass-spectrometry; [34,73]). The volatile components were released and transferred into a vacuum during heating at various temperatures in the temperature range between 25 and 1450 °C and identified according to their mass-charge-ratio and the corresponding intensity ratios of the detected volatile fragments.

The results provided information regarding the participation of different fluids during agate formation. The detected volatiles consisted of compounds of C, N, S, F and Cl with oxygen and/or hydrogen (e.g., H_2O, HF, NO, S, SO, CO, CO_3^{2-}). In particular in agates from volcanic rocks several hydrocarbon compounds as well as carbonic acids could be identified [34,44,73]. The incorporation of hydro-carbon compounds into the SiO_2 matrix points to co-precipitation of both compounds probably originating from the same source. Hydrothermal methane and/or other hydro-carbon compounds in the volatiles participating in the silica accumulation could have served as precursors for the detected organic fluids in the agates.

The data emphasize that a transport of elements and chemical compounds in aqueous solutions cannot be the only process governing agate formation. Therefore, the discussion of the agate genesis has to be focused not exclusively on the system SiO_2–H_2O but has to consider a complex multi-component system including both meteoric water and magmatic fluids (CO_2, HF, HCl, hydrocarbons, etc.).

4.3. Paragenetic Minerals in Agates

Many field observations and analytical results document that besides the different silica minerals numerous other mineral phases are formed pre-, syn- and post-genetically with the agates. Depending on the type of host rocks and the time of formation these minerals can occur as mineral inclusions in quartz, paragenetic minerals intergrown with SiO_2, or pseudomorphs, and their size can range from microscopic inclusions up to several cm in size. These differences can be related to the different geological background and differences in the conditions of mineralization of the agates.

The compilation of detected minerals in agates in Table 4 illustrates that all mineral classes occur. Furthermore, several organic phases were detected in agates such as carbonaceous material, solid bitumen or disordered graphite-like substances [35,44–46]. Carbonates (especially calcite), iron oxides/hydroxides and clay minerals are the most frequent and present in more or less all types of agates, whereas minerals of the zeolite group are typical phases in volcanic agates [1,15,49,78]. It is noteworthy, that certain chemical elements may be present in different chemical compounds in the same agate. For instance, native copper can occur together with copper oxide and copper sulfide, or pyrite/marcasite in association with iron oxides/hydroxides in agate geodes. This is a line of compelling evidence for highly variable physicochemical conditions (especially redox conditions) during mineral formation.

Table 4. Compilation of detected minerals in agates (unpublished data and data from [1,15,18,22,24,28,30,31,34,35,38,42,44–47,49,52–55,59,60,76,78,157–165]); minerals and mineral groups in bold are the most frequent phases in agates.

Mineral Group	Minerals
Elements	copper, lead, sulphur, graphite
Sulfides	pyrite, marcasite, sphalerite, galenite, chalkopyrite, covelline
Oxides/Hydroxides	**hematite, goethite**
	todorocite, ramsdellite, birnessite, pyrolusite, rancieite, hollandite, cryptomelane/psilomelane, manganite
	cuprite, rutile, anatase, spinel (magnesio-chromite)
Carbonates	**calcite**, aragonite, dolomite, siderite, ankerite, rodochrosite, strontianite, magnesite, smithsonite, malachite, azurite, bastnesite-(Ce)
Sulfates	barite, celestine, anhydrite, gypsum
Phosphates	apatite, monazite-(Ce), rabdophane-(Ce), xenotime-(Y)
Halides	fluorite
Silicates	**clay minerals** (kaolinite, illite, montmorillonite, beidellite, saponite, nontronite)
	zeolites (mordenite, heulandite/clinoptilolite, harmotome, chabasite, natrolite, analcime, scolezite, mesolite, stilbite, thomsonite, laumontite, brewsterite, philipsite)
	glauconite–celadonite series, chlorite, serpentine, talc, prehnite, datolithe, epidote, apophyllite, chrysocolle, orthoclase, albite, plagioclase

The observed mineral associations (and their stability fields) in volcanic agates indicate a preferred formation under hydrothermal and low temperature conditions. The most frequent mineral phases such as clay minerals, zeolites, iron oxides/hydroxides and carbonates can be related to the alteration processes of the volcanic host rocks and emphasize that besides enormous amounts of SiO_2 also Al, Fe, Ca, Na and K are released during these processes. The presence of water- and hydroxyl-bearing minerals, carbonates and fluorite in volcanic agates highlights the role of H_2O, CO_2 and F-complexes as main volatile compounds for the transport of substances [34].

In contrast to the agates from volcanic rocks, hydrothermal vein agates can be formed within veins and fissures of the crystalline host rocks directly from circulating hydrothermal solutions. Therefore, SiO_2 minerals can precipitate and crystallize together with a number of other hydrothermal minerals such as carbonates, sulphates, fluorite and/or certain ore minerals (oxides, sulphides).

In sedimentary rocks, where agates are formed by filling of cavities and empty pore space with SiO_2 or the silicification of concretions and pseudomorphism, these processes often take place during sedimentation or early diagenesis, when the sediments contain enough moisture. Characteristic mineral associations commonly include carbonates, sulphates, clay minerals, sulphides and oxides/hydroxides [15,31,42,157,158]. In addition, large crystals of quartz, calcite, celestine or barite can form in the center of the geodes.

Carbonates, especially calcite ($CaCO_3$), belong to the most frequent minerals in volcanic agates [30,38,76,78,159,160]. Coarse-grained calcite has often been formed as overgrowth on quartz or chalcedony in the central part of the agate geodes, but calcite layers intergrown with chalcedony are also possible (Figure 21). A special case is the so called "calcite agate" [38,161,162], an "agate" completely consisting of calcite (Figure 21d). The multiphase character of the calcite formation and several replacement processes can be proven by microscopic investigations (Figure 21b,c).

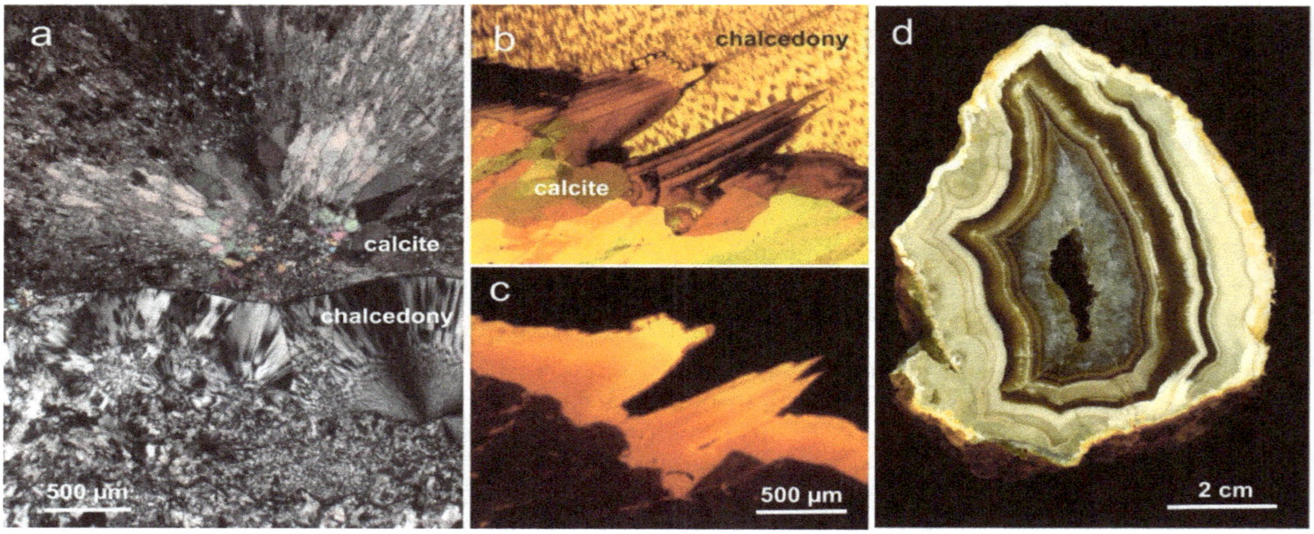

Figure 21. Appearance of calcite in agates: (**a**) micrograph in transmitted light (crossed polars) of the transition zone between chalcedony and calcite in an agate from Xuanhua, province Hebei (China); note the sharp boundary between the two phases and the spherulitic growth of calcite; (**b**,**c**) micrographs in transmitted light (crossed polars, (**b**)) and CL (**c**) of an agate from Agate Point, Canada showing two calcite generations (dark brown, orange CL) in chalcedony matrix (dark); the second calcite generation replaces the banded chalcedony and forms pseudomorphs; (**d**) "calcite agate" from Krásný Dvoreček (Czech Republic) with the typical banded appearance of agate but consisting completely of carbonate.

Geochemical investigations of calcite in agates sometimes showed several carbonate generation and proved a temperature of crystallisation in the range between ca. 20 and 230 °C [1,30,76,77,160] which confirms the estimated temperatures of agate formation. In addition, oxygen and carbon isotop studies of calcite in agates provided indications concerning participation of at least two different fluid during formation. Relatively high $\delta^{18}O$-values (up to +25‰) point to a sedimentary source, wherea $\delta^{13}C$-values between −5 and −15‰ can be related to a primary magmatic origin [1,30,78,159,160 The isotope signature of calcite can probably be explained by processes of fluid mixing durin formation, where carbon inherits its primary magmatic signature (volcanic H_2CO_3 as the dominan carbon species), whereas the signature of oxygen isotopes is overprinted by the secondary influence o meteoric water (H_2O).

Another remarkable feature in worldwide agate occurrences is the appearance o iron-bearing phases in agates, indicating high concentrations of iron in the mineralizin fluids [1,15,30,31,37,47,59,159,163]. Different types of iron oxides and hydroxides in the marginal part of the agates as well as within the chalcedony matrix cause the typical colorations in red, brown anc yellow (Figure 22). In addition, enrichment of iron compounds in the form of crusts or earthy masse in the central parts of agates can be observed.

Microscopic investigations revealed that at least three different types of iron oxides (hydroxides can be distinguished, which are closely connected with the agate formation (Figure 22). Iron oxide (e.g., hematite) of probably pre- or syngenetic formation often appear at the interface between hos rock and agate or in the outermost chalcedony layer. The crystals show spherulitic as well a acicular or dentaloid forms, the latter probably originating from the replacement of former existin carbonates (pseudomorphs).

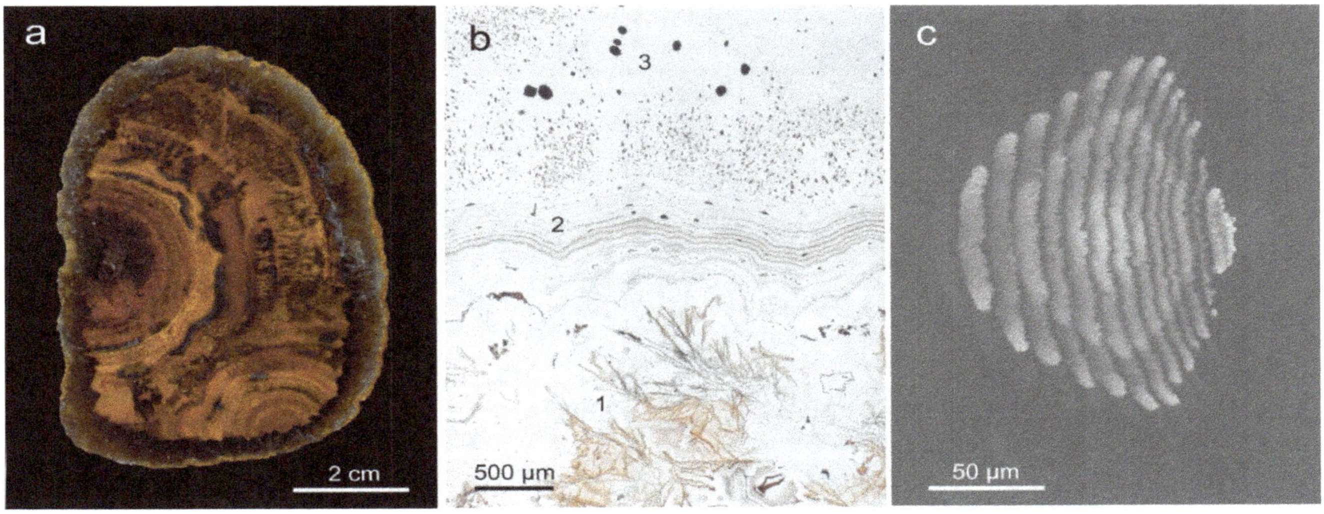

Figure 22. Fe-oxides/hydroxides in agates: (**a**) "agate" from Xuanhua, province Hebei (China) almost completely consisting of Fe-oxides/-hydroxides (hematite, goethite) showing a spherulitic growth; (**b**) micrograph in transmitted light of an agate from Xuanhua, province Hebei (China) showing the silica matrix with the distribution of Fe-oxides/-hydroxides: dentritic (1), aligned along the chalcedony banding (2) and large, irregular particles (3); (**c**) SEM-BSE image showing details of the microtexture of Fe-oxides/-hydroxides in an agate from St. Egidien, Saxony (Germany), which follows the microtexture of the calcedony banding indicating simultaneous crystallisation.

Other types of iron oxide inclusions have been observed within the agate matrix (Figure 22b). On one hand, laminated and lenticular agglomerates of very small hematite particles (<10 µm) occur. In contrast, the other type is represented by larger, sometimes globular or irregular aggregates of up to several 100 µm in size. The fine-grained particles were probably released from the silica matrix during crystallisation, moved together with the growth front and accumulated along the banding. In contrast, larger iron oxide inclusions show interesting structures consisting of several rounded disks. These structures indicate that the iron oxides neither existed before the crystallisation of the chalcedony, nor have been incorporated later. Most likely, iron oxides and silica precipitated simultaneously from a precursor sol.

The complex processes during agate formation and the accompanied variations of the physico-chemical conditions often result in a wide variety of inclusions and replacements of earlier existing phases in agates (e.g., pseudomorphism and perimorphism). Pseudomorphism of quartz is frequently detectable after carbonates (calcite, aragonite) and sulfates (barite, anhydrite/gypsum), but also after zeolites and pyrite [1,15,28]. The earlier existing minerals have been replaced by SiO_2-rich fluids (Figure 23). In most cases the original minerals are completely replaced and the former mineral is at most recognizable on basis of its preserved crystal shape. Therefore, relics of the pre-existing minerals are only detectable using highly advanced analytical methods. Figure 23 shows a sample of agate pseudomorphs from Liebgensmühle near Leisnig, (Saxony, Germany) with pseudomorphs of an elongated shape. Microscopic CL studies under high magnification revealed relics of primary dolomite and inclusions of kaolinite probably from alteration processes (Figure 23c). However, the agate almost completely consists of SiO_2. In other cases, interstices between pre-existing minerals can be filled by agate such as in the case of the well-known Paraiba agates in Brazil. In result, the agates develop polygonal outer shapes resembling crystal faces (so called "polygonal agate"; Figure 23d).

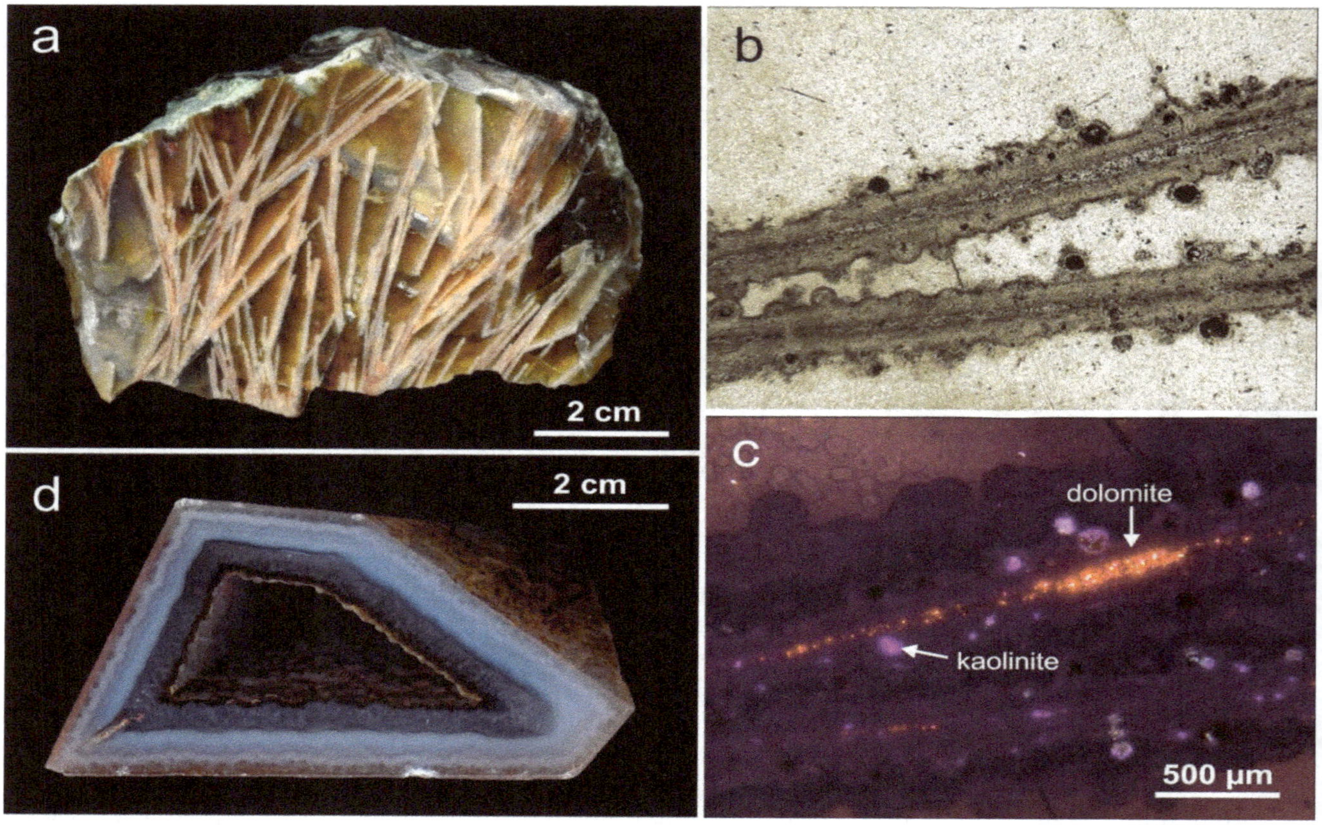

Figure 23. (**a**) Pseudomorphic mineral formation in an agate from Liebgensmühle (Saxony, Germany); (**b**) micrograph in transmitted light showing details of the lath-shaped crystals; (**c**) CL micrograph of the same area, which reveals relics of dolomite in the core of the lath-shaped crystals as well as tiny kaolinite from alteration processes; (**d**) agate from Cachoeira dos Índios, Paraiba (Brazil); the polygonal shape developed due to infill of silica into interstices between laminated calcite (or barite) crystals.

5. Discussion

5.1. Origin of Cavities for Agate Formation

A precondition for the formation of agates is the availability of cavities or open spaces, in which silica can be accumulated. Depending on the geological environment, different types of holes and fissures may exist, which significantly influence the appearance and characteristics of the associated agates. The outer shape of an agate is more or less a replica of the former existing cavity (compare Figure 2). Therefore, in different types of rocks the kind of cavities is variable.

In basic volcanic rocks (basalts, andesites), the accumulation of silica and the formation of agates take place in former gas/liquid bubbles of the solidified lava (amygdales). Because of the high temperatures and low viscosities of these basaltic melts, the lava is flowing over large areas and many gas bubbles are formed in the internal, coalescing during ascent and got stuck during cooling of the solidifying basalt (Figure 24a). These gas bubbles often exhibit rounded or drop-like shapes and form vesicular textures. The primary relations are often visible in the outcrop showing vesicles that are arranged in the flow structure of the former lava (Figure 24b).

Figure 24. (**a**) Vesicular trachyandesite in the agate quarry of Dishuya (Hebei province, China); the numerous cavities formed by fluid bubbles that got stuck in the cooling lava; the inset shows an agate-filled cavity (amygdale); (**b**) agate-filled former bubbles arranged in the flow texture of the solidified andesitic lava from the Steinkaulenberg, Idar-Oberstein (Germany).

The formation of cavities in acidic volcanic rocks occurs via several stages. The process starts with the formation of spherical forms of crystallisation, so called high-temperature crystallisation domains (HTCD) [166]. HTCD with one or more cavities are called lithophysae (thundereggs), which exclusively occur in SiO_2-rich lavas, ignimbrites or subvolcanic bodies and crystallize above the glass transition temperature T_g [167,168]. The position of the lithophysae within the flow texture of the volcanic rocks often provides evidence concerning their early crystallisation before the final solidification of the melt (Figure 2a). Cavity formation is assumed to be the result of transient tensional stress in the melt close to the HTCD crystallisation front (Figure 25a; [166]).

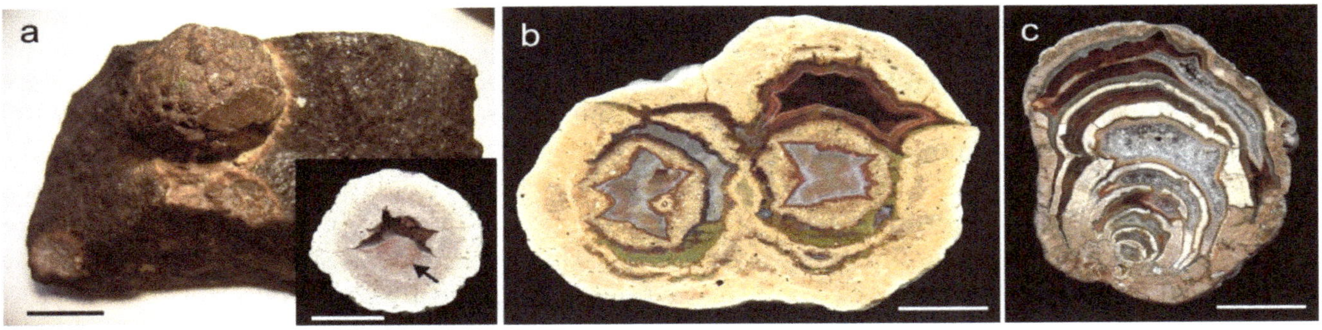

Figure 25. Different kinds of lithophysae with agates in silica-rich volcanic rocks: (**a**) lithophysa in an ignimbritic pitchstone from Zwickau, Saxony (Germany); the inset shows a lithophysa from Chemnitz, Saxony (Germany) with crystallisation center (arrow) and empty cavity; (**b**) double-lithophysa with multiple agate cavities from St. Egidien, Saxony, Germany; (**c**) thunderegg from the Baker Egg Mine, Deming (New Mexico, USA) with multiple crescent-shaped agate cavities; scale bar is 2 cm.

The complexity of the processes can result in the development of single, double and multiple lithophysae (Figure 25b) as well as in complex structures due to the twofold (or multiple) opening and filling with silica or alteration products such as clay minerals or iron compounds (Figure 25c). Lithophysae normally appear in sizes of several cm or dm, but examples with diameters of more than 4 m ("megaspherulites") have also been reported from several localities in the USA, Argentina and China [169].

Hydrothermal vein agates represent a completely different genetic type of agates. These agates form directly within veins and fissures in different crystalline host rocks, which have been formed by tectonic processes (Figure 2e). The vein-like structures are often connected with deep-seated faults with ascending hydrothermal solutions. Vein agates and veinlike silica mineralization often occur together with ore mineralization in hydrothermal systems, e.g., with carbonates, sulfates, sulphides or certain gold deposits worldwide (e.g., [102,170]). In the case of sufficient silica supply, the vein structures can be completely filled with agate and quartz. The sealing of the veins during continuous hydrothermal activity can cause increasing pressure in deeper levels and the fracturing of the primary vein structure [171]. These fractures are often filled with a secondary generation of agate or quartz which is visible by healed brecciation ("brecciated agate" or "Trümmerachat" Figure 26a).

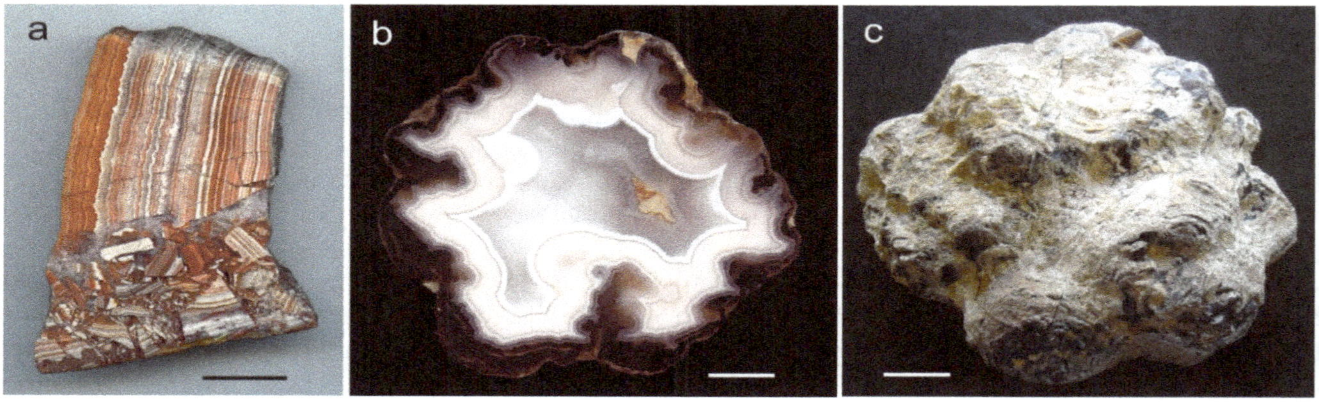

Figure 26. (**a**) Hydrothermal vein agate ("Trümmerachat") from Schlottwitz, Saxony (Germany) showing brecciation and healing by secondary silicification in the lower part; primary agate formation took place in Permian time (263.7 ± 2.2 Ma), whereas brecciation and secondary cementation by silica happened much later during tectonic activities in the Upper Oligocene/Miocene [172]; (**b**) agate of sedimentary origin from Rutba, Western desert (Iraque); (**c**) back side of the agate in (**b**) illustrating the typical cauliflower-like surface; scale bar is 2 cm.

In sedimentary rocks, agates are predominantly formed by silicification of concretions or pseudomorphism due to the replacement of soluble mineral species such as sulphates (e.g., anhydrite) or carbonates (calcite, dolomite) by silica [31,42,157,158,165,173,174]. Residues of plants and animals can also be silicified [79,80,175,176]. The silica originates from circulating pore water, which can precipitate under specific conditions. Geochemical data indicate a preferred silicification in shallow sediments, especially in areas with increased salinity (lagoons, marine shelf) under near-surface conditions and temperatures below 40 °C [177]. Sedimentary agates are often characterized by typical features such as a cauliflower-like surface, the frequent presence of quartzine besides chalcedony or the lack of horizontal banding (Figure 26b,c) [15].

5.2. *Temperature of Agate Formation*

One key point of scientific discussions concerning agate genesis is the temperature of formation. There is no doubt concerning the hydrothermal nature of most vein agates, and agate formation in sediments is running at relatively low temperatures under surface or diagenetic conditions. However, the arguments about the formation of agates in volcanic rocks are extremely widespread. There exist theories about an agate formation both at high temperatures (>400 °C) [1,15,56,178] and ideas about agate genesis at low temperatures (20–100 °C) in a sedimentary-diagenetic environment (e.g., [14,140,179,180]).

Investigations of active thermal vents such as in the Yellowstone Park or on Iceland, or the recent formation of agate-like structures on the seafloor [64] in so called "white smokers" point to a temperature range below 100 °C. However, recent observations concerning the formation of colloidal silicic acid in high-temperature hydrothermal fluids (300 up to >700 °C) indicate that agate formation at high temperatures in volcanic rocks seems to be possible [181]. Interestingly, experiments concerning the synthesis of chalcedony successfully run under elevated temperatures: White and Corwin [182] produced chalcedony during hydrothermal treatment of silica glass at 400 °C/340 bar, Flörke [183] synthesized chalcedony at temperatures above 400 °C, and Oehler [184] succeeded with syntheses at 100–300 °C and 3 kbar.

Modern analytical methods such as fluid inclusion studies, isotope or trace-element analyses provide useful information for the estimation of the temperatures of mineral-forming processes. In the case of fluid inclusion studies in agates, several limitations have to be taken into account. A direct determination of the temperature in agate (chalcedony) is mostly not possible because of the lack of measurable fluid inclusions. Data are only provided by fluid inclusion studies of the macrocrystalline quartz layers. Moreover, a heterogeneous trapping of inclusions has to be assumed for most of the investigated agates from acidic and basic volcanic rocks.

Taking into account all these facts, minimum homogenization temperatures for agates from volcanic rocks of some German occurrences scatter between ca. 95 °C and 186 °C (compare Table 3). The hydrothermal vein agate from Schlottwitz, Saxony (Germany) provided a temperature of 80 °C. These temperatures are in the same range as crystallisation temperatures calculated from inclusion studies of associated minerals in agates (e.g., calcite) [160].

An estimation of the temperature of formation of agates is also possible by using oxygen isotopes. Oxygen isotope data of agates from worldwide occurrences (Figure 18) are mainly in the range between +20‰ and +30‰ but show a scatter of $\delta^{18}O$ values over a broad range between +13‰ and +33‰. The estimated temperatures of agate formation using the fractionation curve of Matsuhisa et al. [139] result in a range between surface temperatures and ca. 230 °C. An exact calculation of temperatures is mostly not possible because of the unknown origin of the primary fluids, but the temperatures are in accordance with the temperatures obtained from fluid inclusion studies.

The results of both methods show relatively high formation temperatures for volcanic agates, which indicate that the mobilization and accumulation of silica started already during a late phase of or soon after the volcanic activity [37]. Considering temperatures obtained from the fluid inclusion homogenization, temperature calculations for different fluid compositions clearly indicate the mixing of dominant amounts of late magmatic (hydrothermal) fluids with smaller amounts of meteoric water [37]. The spatially resolved isotope analyses revealed that the scatter in $\delta^{18}O$-values may be as high as 10‰ even in adjacent chalcedony layers (compare Figure 19) confirming the idea of fluid mixing and/or isotope fractionation under non-equilibrium conditions.

Götze et al. [185] first tried to estimate the temperature of agate formation using the Al concentration in quartz ("Al-geothermometer" of Dennen et al. [186]). The temperatures were calculated from the amount of $[AlO_4]^0$ centers determined by EPR measurements [187]. The $[AlO_4]^0$ concentrations in macrocrystalline quartz of the agates provide reliable temperatures of formation between 65 and 200 °C, which are very close to the estimated temperatures using fluid inclusions or oxygen isotopes (Figure 27). In contrast, calculated temperatures for chalcedony of up to 655 °C as well as strong variations between different chalcedony bands within one agate sample indicate that these temperatures are invalid. It is assumed that the high temperatures are artifacts, which result from non-equilibrium crystallisation processes.

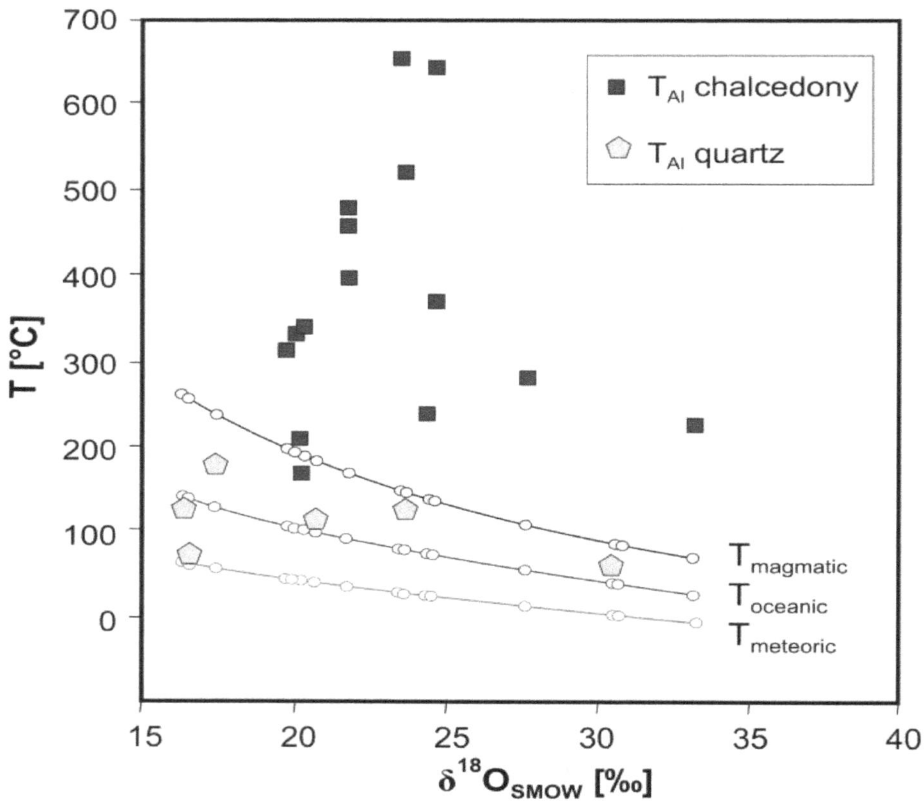

Figure 27. Temperatures of agate formation calculated from $[AlO_4]^0$ concentrations in chalcedony (T_{Al} chalcedony) and macrocrystalline quartz (T_{Al} quartz) compared to temperatures obtained for the same agate samples assuming oxygen isotope equilibration with meteoric ($T_{meteoric}$), oceanic ($T_{oceanic}$) or magmatic water ($T_{magmatic}$); data from Götze et al. [185].

The evaluation of the different methods applied for a calculation of the temperature of agate genesis illustrates that reliable results in most cases are only possible via the analysis of macrocrystalline quartz from the agates. Direct measurements of chalcedony bands provided only reasonable results for oxygen isotopes. In summary, it can be concluded that the temperature interval of agate formation ranges from surface conditions up to temperatures of ca. 230 °C. These temperatures are confirmed by the stability fields of associated paragenetic minerals, homogenization temperatures of fluid inclusions in paragenetic calcite, the low degree of maturity of carbonaceous matter found as cogenetic inclusions within some agates (e.g., [35,45,46,60]) as well as specific microstructural properties of agates. Results of Götze et al. [68] emphasized that the yellow CL, which can often be detected in agates, is a typical indication for low-temperature hydrothermal environment (mostly <250 °C) and is related to fast crystallisation processes from non-crystalline silica precursors.

The geochemical data indicate a mixing of heated meteoric water with magmatic fluids, which alter the volcanic host rocks and release SiO_2. There is no indication for a participation of supercritical fluids, although it cannot be excluded. Additionally, a formation of agate directly from an igneous melt is unlikely.

5.3. Origin and Supply of Silica

The SiO_2 sources for agate formation can clearly be related to the geological environment of the agates. Silica in hydrothermal vein agates and sedimentary agates originates from ascending hydrothermal fluids and silica-rich pore solutions, respectively. In contrast, SiO_2 in volcanic agates represents a product of late- or post-volcanic alteration/weathering of the volcanic host rocks [24,28,37,41,62,63,180].

The formation of agates in volcanic rocks especially occurs in marginal parts of the volcanic bodies nd zones with high activities of fluids and water. Geochemical data indicate that heated meteoric vater in interaction with magmatic fluids intensively reacts with the host rocks resulting in certain lteration processes ("autometasomatosis"). These alteration processes are reflected in the chemical omposition of the agates showing elevated concentrations of elements from the rock matrix (e.g., Al, 'e, K, Na, Ca; compare Table 1) and the occurrence of characteristic secondary minerals in agates see Table 4). In particular unstable volcanic glass and minerals (e.g., pyroxene, feldspar) are converted nto clay minerals, zeolites, and Fe-oxides/-hydroxides, accompanied by the release of enormous mounts of SiO_2 and alkali elements (Table 5). Although the principle processes in basic and acidic olcanics are similar, the specific rock composition and the chemical environment (T, Eh, pH, CO_2, SO_2, water, etc.) strongly influence the alteration processes [78,188–195].

Table 5. Mineral composition (in wt%) of a relatively fresh agate-bearing pitchstone from Chemnitz, Saxony (Germany) in comparison to the surrounding altered rock.

Phase	Pitchstone	Altered Pitchstone
Amorphous (glass)	63.7 ± 2.2	-
Plagioclase	16.1 ± 0.9	14.3 ± 1.0
K-feldspar	5.5 ± 0.5	8.7 ± 0.9
Quartz	1.7 ± 0.2	9.6 ± 0.7
Biotite	1.1 ± 0.9	-
Magnetite	0.8 ± 0.2	-
Ilmenite	0.5 ± 0.2	-
Montmorillonite	10.6 ± 1.9	62.0 ± 1.4
Klinoptilolite	-	5.4 ± 1.0

In mafic volcanic rocks, a greenish layer of sheet silicates (celadonite) can often be observed at he interface between agate geode and host rock. In addition, several minerals of clay and zeolite groups occur as secondary minerals in association with agates [78,196,197]. The process of the rock lecomposition and silica release in acidic volcanic rocks is mainly characterized by the formation of sheet silicates such as kaolinite, illite or montmorillonite. Holzhey [63] measured up to 18 wt% illite in he agate-bearing altered rhyolites of the Thuringian forest (Germany). Kryza [198] calculated that he metasomatic processes in 1kg rhyolite can mobilize up to 235 g SiO_2. Therefore, these processes significantly contribute to the release of silica for the agate formation.

Sufficient amounts of fluids (H_2O, but also CO_2 or F) have to be available to enable alteration and ransport processes of released silica to the agate cavities. The silica transport in aqueous solution is nainly limited by the extreme low solubility of SiO_2 in water. Several factors control silica solubility such as the pH (the solubility strongly increases at pH > 9), the temperature (100–140 mg/L at 20 °C; 300–380 mg/l at 90 °C) [199], and the silica phase which is in equilibrium with the solution. According o Landmesser [8], for instance, a silica concentration of 250 mg/L SiO_2 in the solution would require up to 0.5 m^3 water to flow in and out a cavity to form an agate of ca. 100 g. Since sufficient pathways or free liquids are lacking in most rocks, the silica supply and accumulation by diffusion processes via ntergranular pore space to the cavities is favored.

Due to the larger size (1–100 nm) and rather low diffusion rates of colloidal SiO_2 particles, the proposed main process of silica transport silica is via diffusion of monomeric silicic acid (H_4SiO_4) [8,200]. Recent investigations confirm this hypothesis, showing the release of polymeric and monomeric silicic acid during weathering processes of various silicate minerals. While monomeric silicic acid is the dominating species in a broad range of pH 1–9, polymeric forms including more than ten silicon atoms require higher pH values of above pH 10 [201]. Taking this into account and the fact that most weathering solutions show a pH value <9, the assumed transport mechanism for dissolved silica is via monomeric $Si(OH)_4$ (or oligomer).

The continuous supply of silicic acid results in the accumulation of silica and the filling of th cavities. Subsequent condensation processes lead to the formation of SiO_2 sols and amorphous silic In a first step, condensation processes cause the bridging of molecules of monomeric silicic acid an the development of Si–O–Si siloxane bonds [100].

$$\equiv\text{Si-OH} + \text{HO-Si}\equiv \quad \rightarrow \quad \equiv\text{Si-O-Si}\equiv + H_2O$$

As a result of this process, polymeric silicic acid is formed consisting of up to 10 Si atom which are linked via Si–O–Si bonds [201]. Further polycondensation results in the formation o discrete particles (sol). During succeeding reactions, the particles grow, form chains, and finall a three-dimensional network (gel; Figure 28). Even in the case of supersaturation, SiO_2 does no precipitate immediately because of the initial formation of di- and polysilicic acid (stable negative sol These sols are hydrophilic and hardly flocculate due to the water film. Factors which promote th flocculation and the transition from gel to a SiO_2 precipitate are the lowering of temperature and pH (e.g., mixing of fluids, contact with ground water, etc.), the presence of positively charged Al- an Fe-hydroxides from the residual solutions of the clay mineral formation, or electrolytic reactions [179 The formation of agates from such polydisperse colloidal systems can also explain the manifold an partially very complex agate structures.

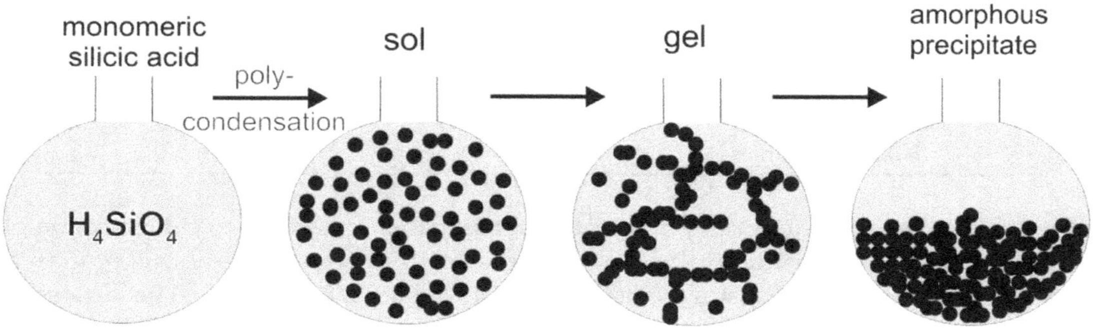

Figure 28. Schematic development of silica from monomeric silicic acid to an amorphous precipitate during the transport and accumulation processes.

Paragenetic minerals like e.g., calcite and fluorite and significant concentrations of element (e.g., B, Ge, U) in agates indicate that the composition of fluids which altering the volcanic rocks an also mobilize and transport the dissolved chemical compounds can vary significantly from "pure water. CO_2, for instance, is essential for the formation of calcite from a fluid phase. In additior chemical transport reactions (CTR) of gases and liquids are realized most probably by stable fluorid compounds like for example SiF_4, BF_3, GeF_4 or UO_2F_2 [34]. Si, B, Ge and U transport can be much better explained by CTR than exclusive transport by H_4SiO_4 in water. Actually, F concentrations of 2 up to 360 ppm have been found in agates of various occurrences [126]. In addition, REE distributior patterns measured in agates indicate that released HREE are preferably incorporated in carbonate and F-complexes during alteration and transport processes [134]. Pronounced Ce anomalies in certair agates also provide indications for a possible mixing of different fluids from contrasting source (e.g., magmatic volatiles, meteoric water) [55,135].

5.4. Formation of Agate Banding and Color

The results of the microstructural and geochemical investigations show that visible agate banding appears due to variations of silica phases, their respective crystal size and habit, and porosity The formation of spherulites from an amorphous (non-crystalline) silica precursor [202] and the following crystallisation characteristics of SiO_2 and self-organization of silica and impurities are the most important processes in the development of the characteristic agate microstructure [25,203,204

Assuming that the initial cavity was filled with a silica sol and/or amorphous silica, the silica's flocculation and crystallisation process must have been initiated by either changes in the physico-chemical conditions (e.g., T, pH) or reactions of positively charged compounds (e.g., Fe- or Al-hydroxides) with the negatively charged colloidal silica. The crystallisation mostly starts from the wall of the agate chamber with the formation of micro-granular or spherulitic quartz proceeding with fibrous chalcedony inwards to the center of the nodule and often ends up with macrocrystalline quartz (Figure 29). Indications for crystallisation processes of amorphous precursors under non-equilibrium conditions are provided by the results of CL and EPR studies, and are also documented by heterogeneous internal textures of macrocrystalline quartz crystals, which appear homogeneous in the optical microscope (Figure 29b,c).

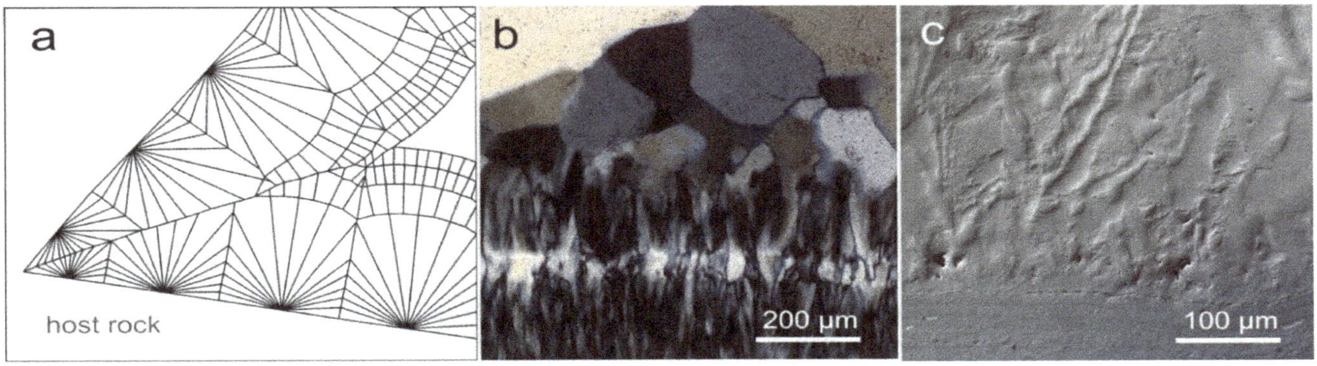

Figure 29. (**a**) Scheme of the formation of agate banding starting with separate spherulites and continuing inwards with fibrous chalcedony layers; (**b**) micrograph in transmitted light (crossed polars) of an agate from Los Indios (Cuba) showing a smooth transition from fibrous chalcedony to macrocrystalline quartz; (**c**) Nomarski DIC micrograph of the interface chalcedony/quartz revealing skeletal growth and irregular internal textures of macrocrystalline quartz, which point to rapid crystallisation under non-equilibrium conditions.

Wang and Merino [205], Heaney [23] and Merino et al. [27] published crystallisation models describing the formation of agate banding in the light of self-organizing processes. Therefore, the fibrous texture of chalcedony could be explained by the dynamics of silica and impurity (i.e., trace elements) transport in the surrounding environment. It is assumed that the impurities are initially precipitated together with the primary amorphous silica leading to a fine dissemination in the respective matrix. The first emerging chalcedony spherulites therefore may contain relatively high concentrations of trace elements. Because of the interaction of morphologically rather unstable crystallisation fronts of quartz and cation-enhanced quartz growth, a first generation of twisted chalcedony with high impurity content is formed. The ongoing silica release and lowering of trace elements during further crystallisation result in a residual medium of aqueous character with low silica concentration. Consequently, the advancing crystallisation front becomes more and more morphologically stable and macrocrystalline quartz evolves. The interface between chalcedony and quartz is often accompanied by a characteristic drop of both impurities and structural defects. This idea is supported by frequent inclusions of iron compounds and elevated contents of Al in the agates. For instance, Götze et al. [206] measured up to 0.55 wt% Al in early opal-CT of agates from Los Indios (Cuba) in contrast to 8–13 ppm Al in macrocrystalline quartz of the same sample.

Self-organization and Ostwald-ripening result in appropriate crystallisation sequences and the crystallinity of SiO_2 increases with proceeding maturity: amorphous SiO_2 → opal-CT/cristobalite → moganite → chalcedony (microcrystalline SiO_2) → macrocrystalline quartz (Figure 30). The succession of maturation can be explained by the decrease of the chemical potential and the Gibbs free energy ΔG according to the OSTWALD rule. During this process, the initial water content of up to 12% in amorphous silica decreases down to almost 0% in crystalline quartz [21]. The decrease in volume during ageing and dehydration of silica is compensated by fine dispersed pore space. The formation of completely filled nodules requires a continuous supply of silicic acid during the crystallisation and "self-organization" process, which may diffuse/penetrate through the existing layers of the agate. Alternating layers of chalcedony and macrocrystalline quartz indicate possible multi-step processes of silica supply and agate formation [51].

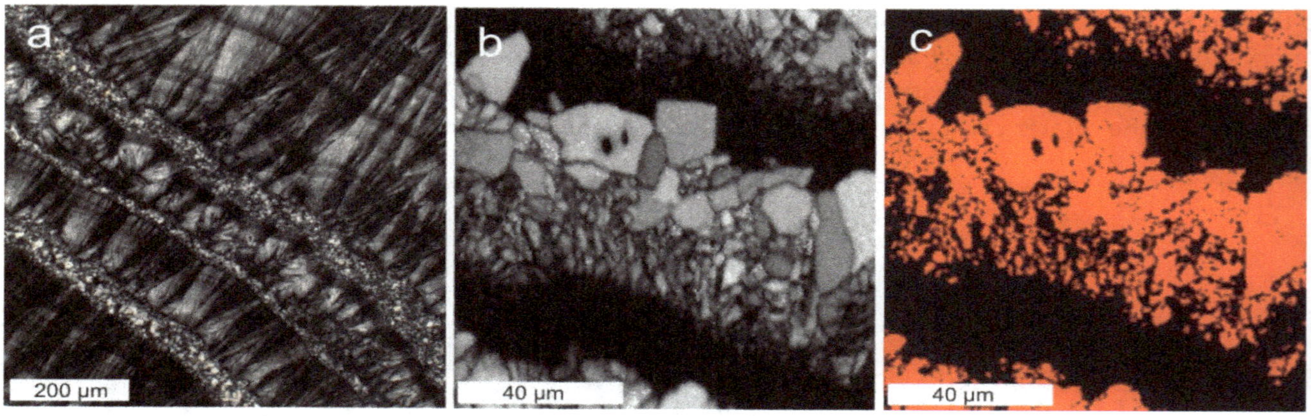

Figure 30. Micrographs showing the development of the band micro-structure in agates: (**a**) rhythmic banding of granular quartz and fibrous chalcedony in an agate from Los Indios (Cuba); (**b**) electron backscatter diffraction (EBSD) band contrast image revealing an increase in brightness with increasing SiO_2 crystallinity; (**c**) EBSD micrograph showing the distribution of α-quartz (red colour) in the agate bands; (**b**,**c**) demonstrate that the dark band areas consist of amorphous or strongly disordered silica.

It cannot be excluded that during ripening of amorphous silica certain layers/zones of different constitution (silica and/or water content, accumulation of trace elements, etc.) had already formed as precursors for the formation of a variable agate microstructure due to processes of diffusion and chemical differentiation (segregation; [57,207]). The formation of agate bands could then be related to a nonlinear reaction–diffusion–advection system, where fractal patterns are formed due to Belousov-Zhabotinsky (B-Z) chemically-oscillating reactions, similar to those in other minerals and materials [208–210]. Pabian and Zarins [24] proposed the B-Z reaction as a mechanism for agate genesis with an initial silica gel and cyclic chemical effects caused by later contact with alkaline water. However, apart from the continual need of circulating alkaline water, the mechanism fails as any gel would eventually collapse. Nevertheless, the agate formation from polydisperse colloidal systems allows the explanation of the wide spectrum of agate structures.

Up to now, only limited information concerning the timescale of the filling and crystallisation processes are available from technical processes or mathematical calculations. Oehler [184] crystallized silica gel to quartz in the laboratory in between 12 h and 180 days. However, these experiments run under high-pressure/high-temperature conditions, which might not apply for agate formation. Ortoleva et al. [25] calculated a duration of about 5000 a for the crystallisation and growth process of a spherical agate with a diameter of 3 cm. Other calculations indicate that the duration of the processes may range from some hundreds to some ten thousands of years [211]. Of course, the time will strongly be influenced by the specific conditions of formation.

The origin of color(s) and color banding in agates is closely connected with the formation of the agate microstructure. The most common colours in agate are white, grey and blue (Figure 31). These colours are caused by variations in crystallite size, microstructure (including water content) and porosity. SEM studies showed that white agate bands are not fibers but a relatively thick plate edge-like structure, whereas the clear areas are globular and mostly have high porosity [19,21]. These plate edges are able to totally reflect incident light while the non-white area allows some white light to be transmitted. The blue color is caused by the dispersion of light on the micro-particles (Tyndall effect). In particular, blue agate bands are relatively sensible for aging processes. Sometimes they lose their concise banding during several years (Figure 31b). This might be related to the loss of water, since storage in water can at least partly reactivate the blue color (Figure 31c).

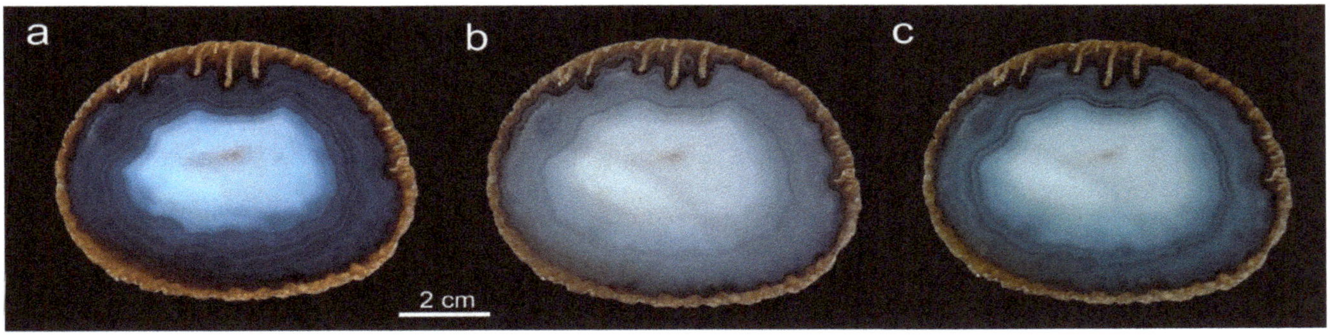

Figure 31. Aging of agate accompanied by loss of distinct banding and blue color demonstrated by an agate sample from Soledade, Rio Grande do Sul (Brazil): (**a**) freshly cut and polished agate with deep blue color and distinct banding; (**b**) the same agate after 3 years at room temperature and under natural light; (**c**) the agate after 6 weeks storage in water.

Although amethyst and smoky quartz can frequently be observed as macrocrystalline incrustations in the agate center, color centers related to point defects in chalcedony are rare. Up to now there is only evidence for the origin of violet color of chalcedony in "Grape agate" from Pantai Ngalo, Sulawesi (Indonesia) due to the amethyst color center [212]. In most cases, the colorful bands are correlated to the distribution and fixation of impurities in the silica matrix during the crystallisation process and often show conformity to the structural banding. Liesegang [11] tried to explain the formation of color banding by diffusion processes of metal ions in silica gels (Liesegang rings), and he suggested that also the rhythmic crystallisation of the agates is caused and influenced by these impurities. The appearance of Liesegang rings in a gel is indeed very similar to that what we see in a natural agate. However, the formation of differently colored bands and the distribution of color pigments in agates are mainly governed by the self-organizational processes of the silica matrix during crystallisation.

The most frequent color pigments are iron phases; different chemical compounds of manganese were also detected. Certain amorphous and crystalline Fe-oxides and Fe-hydroxides (in particular, hematite and goethite) are mostly responsible for the frequent red and also yellow coloration of agates. Different color shades are caused by different kinds of iron compounds, varying amounts or grain size of the color pigments. The mixture of different compounds also results in varying shades of colors.

The presence of iron oxides within the SiO_2 matrix is a result of high iron activity in the initial silica-bearing, mineralizing medium. The accumulation of fine-dispersed pigments along the banding of the chalcedony layers, parallel to the spherulitic structures, is probably a result of "self-purification" of the crystallizing SiO_2 phases. The incorporation of iron into the quartz structure is limited and therefore iron is rejected from the emerging quartz crystals. Consequently, it is transported with the crystallisation front and accumulated/adjusted along the banding (Figure 32).

If distinct banding in agates is lacking, color pigments can be distributed randomly in the silica matrix resulting in irregular color distribution (Figures 32 and 33a). Secondary effects due to alteration, weathering or infiltration of impurities along cracks can also change the primary colour banding in agates. In these cases, leaching or oxidation processes result in the formation of coloration or decoloration zones along the rims of the nodules or along secondary cracks (Figure 33b,c). Such secondary color distribution discordant to the agate banding is often called "chromatography" effect.

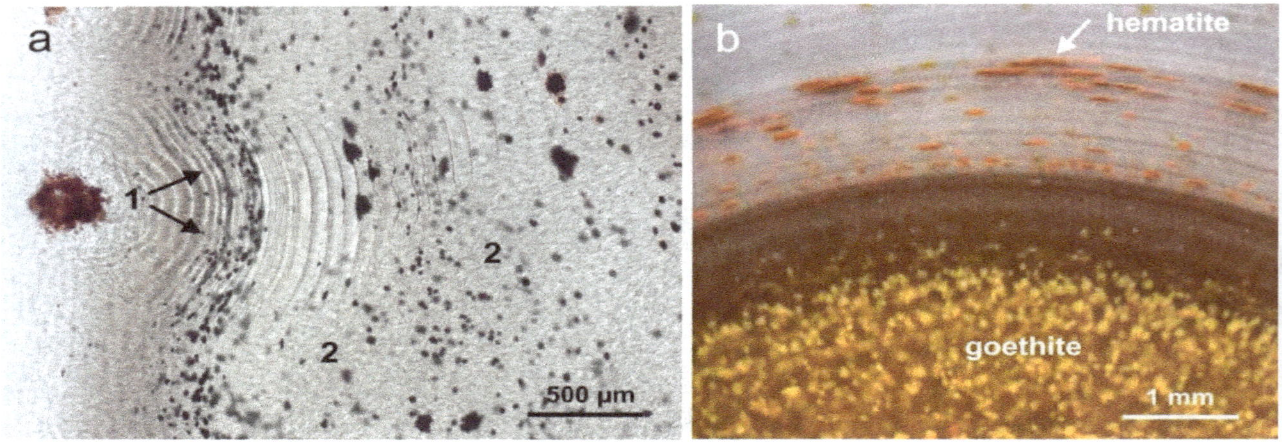

Figure 32. Incorporation of iron compounds as color pigments in agate: (**a**) micrograph in transmitted light of an agate from Hausdorf, Saxony (Germany) with iron oxides aligned along the banding (1) and random distribution in areas without banding (2); (**b**) agate from Pujiang, province Zhejiang (China) showing yellow color due to incorporation of goethite and red bands caused by hematite inclusions; note that hematite particles are flattened and aligned along the banding.

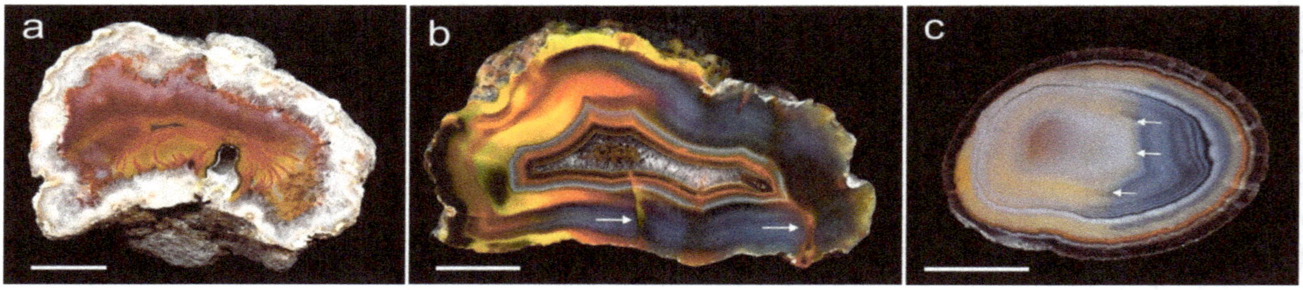

Figure 33. (**a**) Agate from Sierra del Chachahuen, Mendoza (Argentina) without distinct banding and irregular color distribution; (**b**) secondary discoloration (left side and along cracks, see arrows) in an agate from San Rafael, Mendoza (Argentina); (**c**) agate from Xuanhua, province Hebei (China) with effects of secondary bleaching and discoloration; scale bar is 2 cm.

5.5. Microstructural Peculiarities and Possible Biosignatures in Agates

Some agates may show structural peculiarities, which deviate from the common microstructures. One of the widely discussed aspects is the presence of so called "infiltration channels" ("escape tubes" or "flow channel"; Figure 34a), which predominantly occur in agates of mafic volcanic rocks. The shape and name of these structural features suggest that they represent channels of silica supply during agate formation [211]. Although in some specific cases the real impression of silica inflow exists (Figure 34b), a closer examination shows that in most agates these structures may have formed by deformation of non-solidified silica in a plastic state. Microscopic studies revealed that the primary orientation of chalcedony "fibers" changes at the interface and the banding is deformed. It is possible that this deformation is caused by increasing pressure during the formation process and central parts of the banding are squeezed outwards. Sometimes, agates even show multiple deformation features (Figure 34c).

Figure 34. Structural peculiarities in agates: (**a**) agate from Moctezuma, Chihuahua (Mexico) showing a so called "infiltration channel" ("escape tube"); (**b**) micrograph of the marginal part of an agate from Zschaitz, Saxony (Germany) resembling the inflow channel of silica into the agate, which cross-cuts the outermost chalcedony layer; (**c**) agate from Agate Creek, Queensland (Australia) with multiple deformation features; (**d**) cavity from the Dekkan Trap basalt, India completely filled with quartz stalactites (SFF); (**e**) green filamentous structure in a moss agate from Jalna, India; (**f**) Fe-rich agate of probably biogenic origin from the Mátra Mts., Hungary (inset) with filamentous pseudomorphs in the microcrystalline silica matrix; (**g**) agate from Kardzali (Bulgaria) with plume structures; (**h**) details of the reddish-brown plume structures in chalcedony matrix; (**i**) bundle-like structures resembling the shape of natural sheaf of wheat in an agate from the Darrell Friend Ranch near Ashwood, Oregon (USA) (sample courtesy G. Schiecke, photo A. Gabert); the unusual shape is probably caused by calcite traced in silica gel; (**k**) agate from Berschweiler near Idar-Oberstein (Germany) with inclusions of spiral forms probably developed from an amorphous silica precursor by self-organisation growth.

The formation of stalactitic fabrics in agates is also a rare anomaly in agates (Figure 34d). It is known only from a few deposits worldwide such as in the USA, Brazil, Uruguay, Germany, Poland, Russia or India. The Baker Egg Mine in New Mexico (USA) is famous for the high amount (up to 10%) of geodes with stalactites [43]. It is assumed that the stalactites have been formed simultaneously with the wall layers from a liquid that had completely filled the cavity. Recent studies on such pseudostalactitic fabrics (sub-surface filamentous fabrics—SFF; [213]) in cavities of the Deccan Trap basalts in India (Figure 34d) revealed that the growth of the SFF is not limited to vertical directions and might be initiated by microbial activities [195]. Residues of organic material (fungal chitin) in the innermost filamentous core as well as the morphometric characteristics of certain filamentous fabrics indicate a biological contribution during an early state of SFF formation. The internal structure of SFF is mostly made up of a core of clay minerals (in particular smectite), celadonite and zeolites followed by concentric layers of chalcedony. Similar results were obtained for moss agates (Figure 34e) which also show clear indications of biosignatures [33,195,213,214].

Other examples worldwide show that biosignatures and indications of biomineralization processes can be detected in agates and agate-like silica mineralization [33]. Several studies on ancient and modern hydrothermal systems revealed that microorganisms are closely connected with both terrestrial and deep-sea hydrothermal vents [215]. Almost all of the microfossils associated with hydrothermal systems are filamentous. These microorganisms are thermophilic (i.e., the organisms grow best at temperatures > 45 °C, sometimes >100 °C) and often obtain their energy from the oxidation of inorganic compounds ("chemolithoautotroph") (e.g., [216–219].

Müller et al. [220] found Fe-rich cryptocrystalline silica masses with agate-like structures in the southern part of the Mátra Mts. (Hungary), in the neighbourhood of the Gyöngyöstarján and Gyönggyösoroszi area (Figure 34f). These agate-like aggregates in fractures and cavities of an andesite consist mainly of quartz, moganite, opal-CT, hematite, goethite (together with calcite, nontronite, celadonite) and show characteristic signatures of microbial activities (Figure 34g).

Because of the general lack of residues of organic materials within agates, the detection of biosignatures is mostly based on morphological criteria indicating former activities of living organisms. However, recent studies have proved that certain chemical pathways exist, that produce inorganic mineral matter resembling living forms [221–225]. The processes resulting in such crystal architectures include a couple of different chemical interactions including the complexation of ions, selective adsorption onto growing crystal surfaces, or temporary stabilization of an amorphous precursor phase [226]. These results emphasized that morphological criteria alone cannot evidence the biogenic origin of mineral matter.

In conclusion, unusual mineral forms and structures in agates can also be related to such inorganic chemical processes, as in the case of "plume agates" (Figure 34g,h). Such plume structures represent the well-known phenomenon of "silica gardens"—i.e., plant-like structures that formed due to the addition of certain metal salts to sodium silicate solutions. Such aggregations of inorganic self-assembly originate from a combination of osmosis, chemical reactions and buoyancy [222,224]. Moreover, carbonates of certain alkaline earth elements can develop highly unusual shapes during crystallisation in a silica-rich environment such as in the case of calcite in silica gel [226]. Laboratory experiments resulted in finger-like structures or bundles resembling the shape of natural sheaf of wheat (Figure 34i). Further studies revealed that these spectacular aggregations consist of numerous calcite rhombohedra that were self-assembled along their c-axis and form laminar aggregates with rhythmically banded patterns [227].

A new kind of agates with unusual features was recently found in volcanic rocks of certain occurrences in the Saar-Nahe region (Germany). These agates include spirals of several tens of μm up to several hundreds of μm in size within the silica matrix (Figure 34k). Because of the specific shape, the conspicuous creations first led to a controversial discussion concerning a possible biogenic origin (microfossils such as foraminifera or snails) or an inorganic formation [228]. However, detailed nvestigations revealed that the spirals probably developed from an amorphous silica precursor by elf-organization growth and polymerization initiated by a screw dislocation (dislocation growth; [39]). 'he data indicate that agate formation runs through several structural states of SiO_2 with amorphous ilica as the first solid phase, and the specific conditions can produce manifold spectacular forms naking agates a unique product of nature.

. Conclusions

The present paper provides a comprehensive compilation of mineralogical and geochemical data f agates from worldwide occurrences in different host rocks. Based on these data in combination vith field observations and results of laboratory experiments an attempt is made to develop a model oncerning the formation of agates. The presented results illustrate that the processes of agate ormation are complex and often involve multiple steps. Moreover, agate formation in volcanic ocks, in sedimentary host rocks or the formation of hydrothermal vein agate can run under different onditions and involve divergent processes.

The majority of agates can be related to mafic (basalts, andesites) and acidic (rhyolites, rhyodacites) olcanic host rocks. Many field observations and analytical results indicate that volcanic agates form n a late stage of volcanism or after solidification of the volcanic host rocks. Most of the silica necessary or agate formation is supplied by alteration processes of the surrounding volcanic rocks by heated neteoric water, hydrothermal/magmatic fluids and other volatiles such as CO_2 or F-complexes. In vein gates, SiO_2 derives from ascending hydrothermal-magmatic solutions, whereas silica in sedimentary gates is mostly accumulated by SiO_2-rich pore solutions.

The transport of silica in the form of monomeric silicic acid H_4SiO_4 or oligomers is predominantly ealized by diffusion processes to cavities in the host rocks. In basic volcanic rocks, vesicular cavities orm from included gases and liquids during the solidification of the lava, whereas in acidic volcanics ithophysae (high-temperature crystallisation domains) are formed above T_g in the cooling melt as iost for the agates. Vein agates can be formed in fissures and veins within different types of host rocks, vhere a free movement of silica-bearing mineralizing fluids through a system of cracks is possible. n sedimentary rocks, formation of agates is often accompanied by replacement of pre-existing soluble ninerals (sulfates, carbonates) by silica.

The accumulation and condensation of silicic acid result in the formation of silica sols and morphous silica as precursors for the development of the typical agate structures. The process of crystallisation often starts with the spherulitic growth of chalcedony or micro-granular silica ontinuing into chalcedony fibres. Macrocrystalline quartz crystallizes when the SiO_2 concentration n the mineralizing fluid is low. The abundance of lattice defects and elevated concentrations of ertain trace elements in agates as well as heterogeneous internal textures and sector zoning in nacrocrystalline quartz crystals provide clear indications for crystallisation under non-equilibrium onditions. The formation of the typical agate microstructure is most likely governed by processes of self-organization. Depending on the specific physico-chemical conditions (e.g., T, pH, SiO_2 oncentration, trace-element contents), these crystallisation processes may proceed in a very complex vay and result in characteristic structures, SiO_2 phases, colors and inclusions. Oxygen isotopes, Al oncentrations and fluid inclusion studies provided a temperature range for agate formation between a. 20 and 230 °C.

During agate genesis not only silica minerals are formed, but also numerous other mineral phases. The spectrum of paragenically formed minerals can significantly vary in dependence on the genetic ype of agate. These differences can be related to the different types of host rocks and differences in the onditions of mineralization. The most common minerals associated with agates are clay minerals, e-oxides/-hydroxides, minerals of the zeolite group and carbonates (especially calcite).

The complex processes of agate formation, often running under non-equilibrium conditions, resul in a wide variety of peculiarities such as spectacular internal textures, "moss agates", "plume agates or stalactitic aggregations. Moreover, there are compelling indications for microbial activities that ar sometimes involved in agate creation, at least in a very early state of formation. These specific feature make agates a unique and spectacular product of nature. Although a lot of basic data concerning th formation of agates are now established, several open questions remain.

Author Contributions: J.G., R.M., and Y.P. conducted different analytical measurements, evaluated th mineralogical and geochemical data, and provided appropriate parts of the manuscript. J.G. compiled and wrot the final version of the manuscript. All authors have read and agreed to the published version of the manuscrip

Acknowledgments: The present publication would not have been possible without the help and promotion c numerous colleagues, students and friends over the last decades. Collaboration during sampling and scientifi discussions with Terry Moxon (Doncaster, UK), Gerhard Holzhey (Erfurt, Germany), Johann Zenz (Gloggnit: Austria), Klaus Schäfer (Vollmersdorf, Germany) successfully supported the preparation of the manuscrip Ulf Kempe (Freiberg, Germany), Reinhard Kleeberg (Freiberg, Germany), Harry Berek (Freiberg, Germany Lutz Nasdala (Vienna, Austria) and Axel Müller (Oslo, Norway) are gratefully acknowledged for their analytica suppport and the fruitful discussions. We thank Galina Palyanova (Novosibirsk, Russia) for the invitation to writ this review article for the special issue "Agates: Types, Mineralogy, Deposits, Host Rocks, Ages and Genesis and the helpful comments. Reviews of Magdalena Dumańska-Słowik (Kraków, Poland), Terry Moxon (Auckle Doncaster, UK) and two anonymous reviewers improved the quality of the manuscript significantly.

References

1. Blankenburg, H.-J. *Achat*; VEB Deutscher Verlag für Grundstoffindustrie: Leipzig, Germany, 1988; 203p.
2. Zenz, J. *Achate/Agate*; Bode-Verlag: Haltern, Germany, 2005; 656p.
3. Zenz, J. *Achate/Agates II*; Bode-Verlag: Haltern, Germany, 2009; 656p.
4. Zenz, J. *Achate/Agates III*; Bode-Verlag: Haltern, Germany, 2011; 656p.
5. Brückmann, U.F.B. *Abhandlung von Edelsteinen*; Waisenhaus-Buchhandlung: Braunschweig, Germany, 177
6. Collini, C. *Tagebuch Einer Reise, Welches Verschiedene Mineralogische Beobachtungen, Besonders Über Die Achat und Den. Basalt Enthält*; C.F. Schwan: Mannheim, Germany, 1776; 582p.
7. Noeggerath, M. On the porosity and colouring of agates, chalcedonies, etc. *Edinb. New Philos. J.* **1850**, *5* 166–172.
8. Landmesser, M. Das Problem der Achatgenese. *Mitt. Pollichia* **1984**, *72*, 5–137.
9. Daubrée, A. *Synthetische Studien zur Experimental-Geologie*; Vieweg-Verlag: Braunschweig, Germany, 1880; 122
10. Nacken, R. Über die Nachbildung von Chalzedon-Mandeln. *Nat. und Volk.* **1948**, *78*, 2–8.
11. Liesegang, R.E. Die Entstehung der Achate. *Zent. für Mineral.* **1910**, *11*, 593–597.
12. Liesegang, R.E. *Die Achate*; Verlag von Theodor Steinkopff: Dresden/Leipzig, Germany, 1915; 122p.
13. Flörke, O.W.; Köhler-Herbertz, B.; Langer, K.; Tönges, I. Water in Microcrystalline Quartz of Volcanic Origir Agates. *Contrib. Mineral. Petrol.* **1982**, *80*, 324–333. [CrossRef]
14. Fallick, A.E.; Jocelyn, J.; Donnelly, T.; Guy, M.; Behan, C. Origin of agates in volcanic rocks from Scotlan *Nature* **1985**, *313*, 672–674. [CrossRef]
15. Godovikov, A.A.; Ripinen, O.I.; Motorin, S.G. *Agaty*; Nedra: Moskva, Russia, 1987; 368p.
16. Macpherson, H.-G. *Agates*; The Natural History Museum and the National Museums of Scotland: Londor UK, 1989; 72p.
17. Moxon, T. On the origin of agate with particular reference to fortification agate found in the Midland Valle Scotland. *Chem. der Erde* **1991**, *51*, 251–260.
18. Moxon, T. *Agate: Microstructure and Possible Origin*; Terra Publications: Doncaster, UK, 1996; 106p.
19. Moxon, T. Agates: A study of ageing. *Eur. J. Mineral.* **2002**, *14*, 1109–1118. [CrossRef]
20. Moxon, T. *Studies on Agate—Microscopy, Spectroscopy, Growth, High Temperature and Possible Origir* Terra Publications: Doncaster, UK, 2009; 102p.
21. Moxon, T. A re-examination of water in agate and its bearing on the agate genesis enigma. *Mineral. Ma* **2017**, *81*, 1223–1244. [CrossRef]
22. Holzhey, G. Vorkommen und Genese der Achate und Paragensemineralien in Rhyolithkugeln au

Rotliegendvulkaniten des Thüringer Waldes. Ph.D. Thesis, TU Bergakademie, Freiberg, Germany, 1993; 132p.
23. Heany, P.J. A proposed mechanism for the growth of chalcedony. *Contrib. Mineral. Petrol.* **1993**, *115*, 66–74. [CrossRef]
24. Pabian, R.K.; Zarins, A. *Banded Agates—Origins and Inclusions*; Educational Circular No. 12; University of Nebraska: Lincoln, RI, USA, 1994; 32p.
25. Ortoleva, P.; Chen, Y.; Chen, W. *Agates, Geodes, Concretions and Orbicules: Self-Organized Zoning and Morphology*; Kruhl, J.H., Ed.; Springer: Heidelberg, Germany, 1994; pp. 283–305.
26. Heaney, P.J.; Davis, A.M. Observation and origin of self-organized textures in agates. *Science* **1995**, *269*, 1562–1565. [CrossRef]
27. Merino, E.; Wang, Y.; Deloule, E. Genesis of agates in flood basalts: Twisting of chalcedony fibers and trace-element geochemistry. *Am. J. Sci.* **1995**, *295*, 1156–1176. [CrossRef]
28. Cross, B.L. *The Agates of Northern Mexico*; Burgess Publishing Division: Edina, MI, USA, 1996; 130p.
29. Götze, J.; Nasdala, L.; Kleeberg, R.; Wenzel, M. Occurrence and distribution of "moganite" in agate/chalcedony: A combined micro-Raman, Rietveld, and cathodoluminescence study. *Contrib. Mineral. Petrol.* **1998**, *133*, 96–105. [CrossRef]
30. Götze, J.; Tichomirowa, M.; Fuchs, H.; Pilot, J.; Sharp, Z. Geochemistry of agates: A trace element and stable isotope study. *Chem. Geol.* **2001**, *175*, 523–541. [CrossRef]
31. Götze, J.; Möckel, R.; Kempe, U.; Kapitonov, I.; Vennemann, T. Origin and characteristics of agates in sedimentary rocks from the Dryhead area, Montana/USA. *Mineral. Mag.* **2009**, *73*, 673–690. [CrossRef]
32. Götze, J.; Martins, M.S.; Czarnobay, J.C. Achate aus Brasilien. *Veröff. Mus. für Nat. Chemnitz* **2010**, *33*, 63–78.
33. Götze, J.; Müller, A.; Polgári, M.; Pál-Molnár, E. Biosignaturen in Achat/Chalcedon–die Rolle von Mikroorganismen bei der Bildung von SiO_2. *Mineralienwelt* **2011**, *22*, 90–96.
34. Götze, J.; Schrön, W.; Möckel, R.; Heide, K. The role of fluids in the formation of agate. *Geochemistry* **2012**, *72*, 283–286. [CrossRef]
35. Götze, J.; Nasdala, L.; Kempe, U.; Libowitzky, E.; Rericha, A.; Vennemann, T. Origin of black colouration in onyx agate from Mali. *Mineral. Mag.* **2012**, *76*, 115–127. [CrossRef]
36. Götze, J.; Gaft, M.; Möckel, R. Uranium and uranyl luminescence in agate/chalcedony. *Mineral. Mag.* **2015**, *79*, 983–993. [CrossRef]
37. Götze, J.; Möckel, R.; Vennemann, T.; Müller, A. Origin and geochemistry of agates from Permian volcanic rocks of the Sub-Erzgebirge basin (Saxony, Germany). *Chem. Geol.* **2016**, *428*, 77–91. [CrossRef]
38. Götze, J.; Möckel, R.; Eulitz, B. "Karbonat-Achat" von Krásný Dvoreček. *Mineralienwelt* **2018**, *4*, 82–87.
39. Götze, J.; Berek, H.; Schäfer, K. Micro-structural phenomena in agate/chalcedony: Spiral growth. *Mineral. Mag.* **2019**, *83*, 281–291. [CrossRef]
40. Moxon, T.; Rios, S. Moganite and water content as a function of age in agate: An XRD and thermogravimetric study. *Eur. J. Mineral.* **2004**, *4*, 693–706. [CrossRef]
41. Petránek, J. Entstehung von gravitations und adhäsionsgebänderten Achaten in Raum und Zeit und in Abhängigkeit vom Klima. *Der Aufschluss* **2006**, *57*, 129–150.
42. Petránek, J. Sedimentäre Achate. *Der Aufschluss* **2009**, *60*, 291–302.
43. Colburn, R.P. *The Formation of Thundereggs*; Geode Kid Productions: Deming, NM, USA, 2008.
44. Dumańska-Słowik, M.; Natkaniec-Nowak, L.; Kotarba, M.J.; Sikorska, M.; Rzymełka, J.A.; Łoboda, A.; Gaweł, A. Mineralogical and geochemical characterization of the "bituminous" agates from Nowy Kościół (Lower Silesia, Poland). *Neues Jahrbuch für Mineralogie Abhandlungen* **2008**, *184*, 255–268.
45. Dumańska-Słowik, M.; Natkaniec-Nowak, L.; Weselucha-Birczyńska, A.; Gaweł, A.; Lankosz, M.; Wróbel, P. Agates from Sidi Rahal, in the Atlas Mountains of Morocco: Gemmological characteristics and proposed origin. *Gems Gemol.* **2013**, *49*, 148–159.
46. Dumańska-Słowik, M.; Powolny, T.; Sikorska-Jaworowska, M.; Gaweł, A.; Kogut, L.; Poloński, K. Characteristics and origin of agates from Płóczki Górne (Lower Silesia, Poland): A combined microscopic, micro-Raman, and cathodoluminescence study. *Spectrochim. Acta Part. A Mol. Biomol. Spectrosc.* **2018**, *192*, 6–15.
47. Clark, R. *South. Dakota's State Gemstone—Fairburn Agate*; Silverwind Agates: Appleton, CT, USA, 2009; 130p.
48. Lyashenko, E.A. Agates of Russia. *Mineral. Alm.* **2010**, *15*, 6–27.
49. Götze, J. Agate–Fascination between Legend and Science. In *Agates III*; Zenz, J., Ed.; Bode-Verlag: Haltern, Germany, 2011; pp. 19–133.
50. Hurst, J.T. *Dryhead Agate*; Agate Treasures-Schatzkammerachate Publishing: Boulder, CO, USA, 2012; 90p.

51. French, M.W.; Worden, R.H.; Lee, D.R. Electron backscatter diffraction investigation of length-fast chalcedon in agate: Implications for agate genesis and growth mechanisms. *Geofluids* **2013**, *13*, 32–44. [CrossRef]
52. Richter, S.; Götze, J.; Niemeyer, H.; Möckel, R. Mineralogical investigation of agates from Cordón de Lil Chile. *J. Andean Geol.* **2015**, *42*, 386–396.
53. Ottens, B.; Götze, J. *Achatwelt China*; extralapis 51; Christian Weise Verlag: München, Germany, 2016; 110p
54. Natkaniec-Nowak, L.; Dumańska-Słowik, M.; Pršek, J.; Lankosz, M.; Wróbel, P.; Gaweł, A.; Kowalczyk, J Kocemba, J. Agates from Kerrouchen (The Atlas Mountains, Morocco). *Minerals* **2016**, *6*, 77. [CrossRef]
55. Powolny, T.; Dumańska-Słowik, M.; Sikorska-Jaworowska, M.; Wójcik-Bania, M. Agate mineralization i spilitized Permian volcanics from "Borówno" quarry (Lower Silesia, Poland)—Microtextural, mineralogica and geochemical constraints. *Ore Geol. Rev.* **2019**, *114*, 103–130. [CrossRef]
56. Kigai, I.N. The genesis of agates and amethyst geodes. *Can. Mineral.* **2019**, *57*, 867–883. [CrossRef]
57. Howard, C.B.; Rabinovitch, A. A new model of agate geode formation based on a combination o morphological features and silica sol-gel experiments. *Eur. J. Mineral.* **2017**, *30*, 97–106. [CrossRef]
58. Gliozzo, E.; Cairncross, B.; Vennemann, T. A geochemical and micro-textural comparison of basalt-hoste chalcedony from the Jurassic Drakensberg and Neoarchean Ventersdorp Supergroup (Vaal River alluvia gravels), South Africa. *Int. J. Earth Sci.* **2019**, *108*, 1857–1877. [CrossRef]
59. Zhang, X.; Ji, L.; He, X. Gemological characteristics and origin of the Zhanguohong agate from Beipiac Liaoning province, China: A combined microscopic, X-ray diffraction, and Raman spectroscopic stud *Minerals* **2020**, *10*, 401. [CrossRef]
60. Pršek, J.; Dumańska-Słowik, M.; Powolny, T.; Natkaniec-Nowak, L.; Toboła, T.; Zych, D.; Skrepnicka, D Agates from Western Atlas (Morocco)—Constraints from mineralogical and microtextural characteristic *Minerals* **2020**, *10*, 198. [CrossRef]
61. Moxon, T.; Carpenter, M.A. Crystallite growth kinetics in nanocrystalline quartz (agate and chalcedony *Mineral. Mag.* **2009**, *73*, 551–568. [CrossRef]
62. Walger, E. Das Vorkommen von Uruguay-Achaten bei Flonheim in Rheinhessen, seine tektonische Auswertung und seine Bedeutung für die Frage nach der Achatbildung. *Jahresber. Mitt. Oberrh. Geol. Ve* **1954**, *36*, 20–31. [CrossRef]
63. Holzhey, G. Herkunft und Akkumulation des SiO_2 in Rhyolithkugeln aus Rotliegendvulkaniten des Thüringe Waldes. *Geowiss. Mitt. Thüringen* **1995**, *3*, 31–59.
64. Hopkinson, L.; Roberts, S.; Herrington, R.; Wilkinson, J. Self-organization of submarine hydrotherma siliceous deposits: Evidence from the TAG hydrothermal mound, 26°N Mid-Atlantic Ridge. *Geology* **199** *26*, 347–350. [CrossRef]
65. Taut, T.; Kleeberg, R.; Bergmann, J. Seifert software: The new Seifert Rietveld program BGMN and it application to quantitative phase analysis. *Mater. Struct.* **1998**, *5*, 57–66.
66. Neuser, R.D.; Bruhn, F.; Götze, J.; Habermann, D.; Richter, D.K. Kathodolumineszenz: Methodik un Anwendung. *Zentralblatt für Geologie und Paläontologie Teil I* **1995**, *1*, 287–306.
67. Götze, J.; Kempe, U. A comparison of optical microscope (OM) and scanning electron microscope (SEM based cathodoluminescence (CL) imaging and spectroscopy applied to geosciences. *Mineral. Mag.* **2008**, *72* 909–924. [CrossRef]
68. Götze, J.; Pan, Y.; Stevens-Kalceff, M.; Kempe, U.; Müller, A. Origin and significance of the yellow cathodoluminescence (CL) of quartz. *Am. Mineral.* **2015**, *100*, 1469–1482. [CrossRef]
69. Möckel, R.; Götze, J.; Sergeev, S.A.; Kapitonov, I.N.; Adamskaya, E.V.; Goltsin, N.A.; Vennemann, T Trace-element analysis by laser ablation inductively coupled plasma mass spectrometry (LA-ICP-MS): A cas study for agates from Nowy Kościoł, Poland. *J. Sib. Federal Univ. Eng. Technol.* **2009**, *2*, 123–138.
70. Flem, B.; Müller, A. In situ analysis of trace elements in quartz using Laser ablation inductively couple plasma mass spectrometry. In *Quartz: Deposits, Mineralogy and Analytics*; Götze, J., Möckel, R., Eds. Springer Geology: Heidelberg, Germany, 2012; pp. 219–236.
71. Monecke, T.; Bombach, G.; Klemm, W.; Kempe, U.; Götze, J.; Wolf, D. Determination of trace elements i quartz standard UNS-SpS and in natural quartz by ICP-MS. *Geostand. Newsl.* **2000**, *24*, 73–81. [CrossRef]
72. Venneman, T.W.; Morlok, A.; von Engelhardt, W.; Kyser, K. Stable isotope composition of impact glasse from the Nördlinger Ries impact crater, Germany. *Geochim. Cosmochim. Acta* **2001**, *65*, 1325–1336. [CrossRef]
73. Richter-Feig, J.; Möckel, R.; Götze, J.; Heide, G. Investigation of fluids in chalcedony/quartz of agates using Thermogravimetry-Mass-Spectrometry. *Minerals* **2018**, *8*, 72. [CrossRef]

4. Spötl, C.; Vennemann, T.W. Continuous-flow isotope ratio mass spectrometric analysis of carbonate minerals. *Rapid Commun. Mass Spectrom.* **2003**, *17*, 1004–1006. [CrossRef]
5. Moxon, T.; Nelson, D.R.; Zhang, M. Agate recrystallization: Evidence from samples found in Archaean and Proterozoic host rocks, Western Australia. *Aust. J. Earth Sci.* **2008**, *53*, 235–248. [CrossRef]
6. Gilg, H.A.; Morteani, G.; Kostitsyn, Y.; Preinfalk, C.; Gatter, I.; Strieder, A.J. Genesis of amethyst geodes in basaltic rocks of the Serra Geral Formation (Ametista do Sul, Rio Grande do Sul, Brazil): A fluid inclusion, REE, oxygen, carbon, and Sr isotope study on basalt, quartz, and calcite. *Miner. Depos.* **2003**, *38*, 1009–1025. [CrossRef]
7. Gilg, H.A.; Krüger, Y.; Taubald, H.; van den Kerkhof, A.M.; Frenz, M.; Morteani, G. Mineralisation of amethyst-bearing geodes in Ametista do Sul (Brazil) from low-temperature sedimentary brines: Evidence from monophase liquid inclusions and stable isotopes. *Miner. Depos.* **2014**, *49*, 861–877. [CrossRef]
8. Ottens, B.; Götze, J.; Schuster, R.; Krenn, K.; Hauzenberger, C.; Zsolt, B.; Vennemann, T. Exceptional multi-stage mineralization of secondary minerals in cavities of flood basalts from the Deccan Volcanic Province, India. *Minerals* **2019**, *1019*, 351. [CrossRef]
9. Lund, E.H. Chalcedony and quartz crystals in silicified coral. *Am. Mineral.* **1960**, *45*, 1304–1307.
0. Rössler, R. *Der Versteinerte Wald Von Chemnitz*; Museum für Naturkunde Chemnitz: Chemnitz, Germany, 2001; 252p.
1. Rich, P.V.; Rich, T.H.; Fenton, M.A.; Fenton, C.L. *The Fossil Book—A Record of Prehistoric Life*; Dover Publications: Mineola, NY, USA, 2020; 740p.
2. Graetsch, H. Structural characteristics of opaline and microcrystalline silica minerals. *Silica Rev. Mineral.* **1994**, *29*, 209–232.
3. Brewster, D. Über die Ursachen der Farben des irisierenden Achats. *Ann. Phys. und Chem.* **1844**, *137*, 134–138. [CrossRef]
4. Braitsch, O. Über die natürlichen Faser- und Aggregationstypen beim SiO_2, ihre Verwachsungsformen, Richtungsstatistik und Doppelbrechung. *Heidelb. Beiträge Mineral. Petrogr.* **1957**, *5*, 331–372.
5. Lange, P.; Blankenburg, H.-J.; Schrön, W. Rasterelektronenmikroskopische Untersuchungen an Vulkanitachaten. *Z. Geol. Wiss.* **1984**, *12*, 667–681.
6. Miehe, G.; Graetsch, H.; Flörke, O.W. Crystal structure and growth fabric of length-fast chalcedony. *Phys. Chem. Miner.* **1984**, *10*, 197–199. [CrossRef]
7. Folk, R.L.; Pittman, J.S. Length-slow chalcedony; A new testament for vanished evaporates. *J. Sediment. Petrol.* **1971**, *41*, 1045–1058.
8. Flörke, O.W.; Jones, J.B.; Schmincke, H.-U. A new microcrystalline silica from Gran Canaria. *Z. Krist.* **1976**, *143*, 156–165. [CrossRef]
9. Heaney, P.J.; Post, J.E. The widespread distribution of a novel silica polymorph in microcrystalline quartz varieties. *Science* **1992**, *255*, 441–443. [CrossRef]
0. Heaney, P.J.; Veblen, D.R.; Post, J.E. Structural disparities between chalcedony and macrocrystalline quartz. *Am. Mineral.* **1994**, *79*, 452–460.
1. Kingma, K.J.; Hemley, R.J. Raman spectroscopic study of microcrystalline silica. *Am. Mineral.* **1994**, *79*, 269–273.
2. Nasdala, L.; Götze, J.; Gaft, M.; Hanchar, J.; Krbetschek, M. Luminescence techniques in Earth Sciences. In *EMU Notes in Mineralogy*; Beran, A., Libowitzky, E., Eds.; Eötvös University Press: Budapest, Hungary, 2004; Volume 6, pp. 1–49.
3. Correns, C.W.; Nagelschmidt, G. Über Faserbau und optische Eigenschaften von Chalcedon. *Z. für Krist. A* **1933**, *85*, 199–213.
4. Hoffmann, S. *Untersuchungen über den Opalgehalt der Achate*; Schweizerbart: Stuttgart, Germany, 1942; pp. 238–276.
5. Flörke, O.W. Untersuchungen an amorphem und mikrokristallinem SiO_2. *Chem. der Erde* **1962**, *22*, 91–110.
6. Jones, B.; Renaut, R.W. Microstructural changes accompanying the opal-A to opal-CT transition: New evidence from the siliceous sinters of Geysir, Haukadalur, Iceland. *Sedimentology* **2007**, *54*, 921–948. [CrossRef]
7. Moxon, T.; Reed, S.J.B. Agate and chalcedony from igneous and sedimentary hosts aged from 13 to 3480 Ma: A cathodoluminescence study. *Mineral. Mag.* **2006**, *70*, 485–498. [CrossRef]
8. Moxon, T.; Reed, S.J.B.; Zhang, M. Metamorphic effects on agate found near the Shap granite, Cumbria,

England: As demonstrated by petrography, X-ray diffraction and spectroscopic methods. *Mineral. Ma* **2007**, *71*, 461–476. [CrossRef]
99. Iler, R.K. *The Chemistry of Silica: Solubility, Polymerization, Colloid and Surface Properties and Biochemistr* John Wiley & Sons: New York, NY, USA, 1979; p. 866.
100. Holzhey, G. Mikrokristalline SiO_2-Mineralisationen in rhyolithischen Rotliegendvulkaniten des Thüring Waldes (Deutschland) und ihre Genese. *Chem. der Erde* **1999**, *59*, 183–205.
101. Tennyson, C. "Struktur" und Farbenspiel des Edelopals. *Lapis* **1979**, *4*, 13–15.
102. Dong, G.; Morrison, G.; Jaireth, S. Quartz textures in epithermal veins, Queensland—Classification, origi and implication. *Econ. Geol.* **1995**, *90*, 1841–1856. [CrossRef]
103. Weil, J.A. A review of electron spin spectroscopy and its application to the study of paramagnetic defects i crystalline quartz. *Phys. Chem. Miner.* **1984**, *10*, 149–165. [CrossRef]
104. Weil, J.A. A review of the EPR spectroscopy of the point defects in α-quartz: The decade 1982–1992. In *Physic and Chemistry of SiO_2 and the Si-SiO Interface 2*; Helms, C.R., Deal, B.E., Eds.; Plenum Press: New York, N USA, 1993; pp. 131–144.
105. Stevens-Kalceff, M.A. Cathodoluminescence microcharacterization of point defects in a-quartz. *Mineral. Ma* **2009**, *73*, 585–606. [CrossRef]
106. Götze, J. Chemistry, textures and physical properties of quartz—Geological interpretation and technic application. *Mineral. Mag.* **2009**, *73*, 645–671. [CrossRef]
107. Götze, J.; Plötze, M.; Fuchs, H.; Habermann, D. Defect structure and luminescence behaviour c agate—Results of electron paramagnetic resonance (EPR) and cathodoluminescence (CL) studie *Mineral. Mag.* **1999**, *63*, 149–163. [CrossRef]
108. Mashkovtsev, R.I.; Li, Z.; Mao, M.; Pan, Y. ^{73}Ge, ^{17}O and ^{29}Si hyperfine interactions of the Ge E'_1 center i crystalline SiO_2. *J. Magn. Reson.* **2013**, *233*, 7–16. [CrossRef]
109. SivaRamaiah, G.; Lin, J.R.; Pan, Y. Electron paramagnetic resonance spectroscopy of Fe3+ ions in amethys Thermodynamic potentials and magnetic susceptibility. *Phys. Chem. Miner.* **2011**, *38*, 159–167. [CrossRef]
110. Pan, Y.; Hu, B. Radiation-induced defects in quartz. IV. Thermal properties and implications. *Phys. Chem. Mine* **2009**, *36*, 421–430. [CrossRef]
111. Botis, S.; Nokhrin, S.M.; Pan, Y.; Xu, Y.; Bonli, T. Natural radiation-induced damage in quartz. I. Correlation between cathodoluminescence colors and paramagnetic defects. *Can. Mineral.* **2005**, *43*, 1565–158 [CrossRef]
112. Nilges, M.J.; Pan, Y.; Mashkovtsev, R.I. Radiation-induced defects in quartz. I. Single-crystal W-band EP study of an electron irradiated quartz. *Phys. Chem. Miner.* **2008**, *35*, 103–115. [CrossRef]
113. Nilges, M.J.; Pan, Y.; Mashkovtsev, R.I. Radiation-induced defects in quartz. III. EPR, ENDOR and ESEE characterization of a peroxy radical. *Phys. Chem. Miner.* **2009**, *36*, 63–71. [CrossRef]
114. Pan, Y.; Nilges, M.J.; Mashkovtsev, R.I. Radiation-induced defects in quartz. II. W-band single-crystal EP study of natural citrine. *Phys. Chem. Miner.* **2008**, *35*, 387–397. [CrossRef]
115. Pan, Y.; Nilges, M.J.; Mashkovtsev, R.I. Multifrequency single-crystal EPR characterization and DFT modelin of new peroxy radicals in quartz. *Mineral. Mag.* **2009**, *73*, 517–535. [CrossRef]
116. Walsby, C.J.; Lees, N.S.; Claridge, R.F.C.; Weil, J.A. The magnetic properties of oxygen-hole aluminum centre in crystalline SiO_2. VI: A stable AlO_4/Li centre. *Can. J. Phys.* **2003**, *81*, 583–598. [CrossRef]
117. Götze, J. Application of cathodoluminescence (CL) microscopy and spectroscopy in geoscience *Microsc. Microanal.* **2012**, *18*, 1270–1284. [CrossRef]
118. Götze, J.; Plötze, M.; Habermann, D. Cathodoluminescence (CL) of quartz: Origin, spectral characteristic and practical applications. *Mineral. Petrol.* **2001**, *71*, 225–250. [CrossRef]
119. Götze, J.; Hanchar, J. *Atlas of Cathodoluminescence (CL) Microtextures*; GAC Miscellaneous Publication No. 1 Geological Association of Canada: St. John's, NL, Canada, 2018; 248p.
120. Ramseyer, K.; Baumann, J.; Matter, A.; Mullis, J. Cathodoluminescence colours of alpha-quartz. *Mineral. Ma* **1988**, *52*, 669–677. [CrossRef]
121. Siegel, G.H.; Marrone, M.J. Photoluminescence in as-drawn and irradiated silica optical fibers: An assessme of the role of non-bridging oxygen defect centres. *J. Non-Cryst. Solids* **1981**, *45*, 235–247. [CrossRef]
122. Alonso, P.J.; Halliburton, L.E.; Kohnke, E.E.; Bossoli, R.B. X-ray induced luminescence in crystalline SiO_2 *J. Appl. Phys.* **1983**, *54*, 5369–5375. [CrossRef]
123. Luff, B.J.; Townsend, P.D. Cathodoluminescence of synthetic quartz. *J. Phys. Condens. Matter* **1990**, 8089–8097. [CrossRef]

124. Perny, B.; Eberhardt, P.; Ramseyer, K.; Mullis, J. Microdistribution of aluminium, lithium and sodium in quartz: Possible causes and correlation with short-lived cathodoluminescence. *Am. Mineral.* **1992**, *77*, 534–544.

125. Götze, J.; Plötze, M.; Graupner, T.; Hallbauer, D.K.; Bray, C. Trace element incorporation into quartz: A combined study by ICP-MS, electron spin resonance, cathodoluminescence, capillary ion analysis and gas chromatography. *Geochim. Cosmochim. Acta* **2004**, *68*, 3741–3759. [CrossRef]

126. Blankenburg, H.-J.; Schrön, W. Zum Spurenelementchemismus der Vulkanitachate. *Chem. Erde* **1982**, *41*, 121–135.

127. Mason, B. Cosmochemistry, Part. I. Meteorites. In *Data of Geochemistry*; Fleischer, M., Ed.; U.S. Geological Survey Professional Paper: Reston, VA, USA, 1979.

128. Walenzak, Z. Geochemistry of minor elements dispersed in quartz (Ge, Al, Ga, Fe, Ti, Li and Be). *Arch. Mineral.* **1969**, *28*, 189–335.

129. Konstatinov, W.M. Uranium bearing lithophysae in acidic extrusive rocks. *Izvestiya Akademii Nauk SSSR Seria Geologika* **1968**, *7*, 43–49. (In Russian)

130. Zielinski, R.A. Uranium mobility during interaction of rhyolitic obsidian, perlite and felsite with alkaline carbonate solution: T = 120 °C, P = 210 kg/cm^2. *Chem. Geol.* **1979**, *27*, 47–63. [CrossRef]

131. Pan, Y.; Li, D.; Feng, R.; Wiens, E.; Chen, N.; Götze, J.; Lin, J. Uranyl binding mechanism in microcrystalline silicas: A potential missing link for uranium mineralization by direct uranyl co-precipitation and environmental implications. *Geochim. Cosmochim. Acta* **2021**, *292*, 518–531. [CrossRef]

132. Peppard, D.F.; Mason, G.W.; Lewey, S. A tetrad effect in the liquid–liquid extraction ordering of lanthanide(III). *J. Inorg. Nucl. Chem.* **1969**, *31*, 2271–2272. [CrossRef]

133. Monecke, T.; Dulski, P.; Kempe, U. Origin of convex tetrads in rare earth element patterns of hydrothermally altered siliceous igneous rocks from the Zinnwald Sn–W deposit, Germany. *Geochem. Cosmochim. Acta* **2007**, *71*, 335–353. [CrossRef]

134. Wood, S.A. The aqueous geochemistry of the rare-earth elements and yttrium. 2. Theoretical predictions of speciation in hydrothermal solutions to 350 °C at saturation water vapor pressure. *Chem. Geol.* **1990**, *88*, 99–125. [CrossRef]

135. Kempe, U.; Götze, J.; Belyatsky, B.V.; Plötze, M. Ce anomalies in monazite, fluorite and agate from Permian volcanics of the Saxothuringian (Germany). *J. Czech. Geol. Soc.* **1998**, *42*, 38.

136. Barsanov, G.P.; Plyusnina, I.I.; Jakovleva, M.E. Specific features of the chemical composition, physical properties and the structure of chalcedony. In *New Data of Minerals*; Nauka: Moscow, Russia, 1981; Volume 28, pp. 3–33. (In Russian)

137. Hoefs, J. *Stable Isotope Geochemistry*, 4th ed.; Springer: Heidelberg, Germany, 1997; 201p.

138. Ingerson, E.; Weshow, R.L. Oxygen isotope fractionation in the system quartz–water. *Geochem. Int.* **1965**, *2*, 691–707.

139. Matsuhisa, Y.; Goldsmith, J.R.; Clayton, R.N. Oxygen isotopic fractionation in the system quarz-albite–anorthite-water. *Geochim. Cosmochim. Acta* **1979**, *43*, 1131–1140. [CrossRef]

140. Harris, C. Oxygen-isotope zonation of agates from Karoo volcanics of the Skeleton Coast, Namibia. *Am. Mineral.* **1989**, *74*, 476–481.

141. Strauch, G.; Nitzsche, H.-M.; Holzhey, G. Isotopenuntersuchungen an Rhyolithen und Achatbildungen. *Neues Jahrb. für Mineral. Abh.* **1994**, *165*, 103–104.

142. Blankenburg, H.J.; Pilot, J.; Werner, C.D. Erste Ergebnisse der Sauerstoffisotopenuntersuchungen an Vulkanitachaten und ihre genetische Interpretation. *Chem. der Erde* **1982**, *41*, 213–217.

143. Harris, C. Oxygen isotope geochemistry of a quartz-agate geode from north-western Namibia. *Commun. Geol. Surv. S.W. Afr. Namib.* **1988**, *4*, 43–44.

144. Holzhey, G. Contribution to petrochemical-mineralogical characterization of alteration processes within the marginal facies of rhyolitic volcanics of lower Permian Age, Thuringian Forest, Germany. *Chem. Erde* **2001**, *61*, 149–186.

145. Fallick, A.E.; Jocely, J.; Hamilton, P.J. Oxygen and hydrogen stable isotope systematics in Brazilian agates. In *Geochemistry and Mineral Formation in the Earth Surface*; Rodriguez-Clemente, R., Tardy, Y., Eds.; Editorial CSIC: Madrid, Spain, 1987; pp. 99–117.

146. Duarte, L.C.; Hartmann, L.A.; Ronchi, L.H.; Berner, Z.; Theye, T.; Massonne, H.J. Stable isotope and mineralogical investigation of the genesis of amethyst geodes in the Los Catalanes gemological district, Uruguay, southernmost Paraná volcanic province. *Miner. Depos.* **2011** *46*, 239–255. [CrossRef]

147. Rezaei-Kahkhaei, M.; Ansarifar, O.; Ghasemi, H. Geochemistry and oxygen stable isotopes of Reza Abad agates, SE Shahrood, Central Iran: An approach to temperature and formation process. *J. Econ. Geol.* **2019**, *11*, 3.
148. Kita, I.; Taguchi, S. Oxygen isotopic behaviour of precipitating silica from geothermal water. *Geochem.* **1986**, *20*, 153–157. [CrossRef]
149. Blankenburg, H.-J.; Thomas, R.; Klemm, W.; Leeder, O. Interpretation der Ergebnisse von Einschlußunersuchungen an den Quarzinkrustaten aus Vulkanitachaten. *Z. Geol. Wiss.* **1990**, *18*, 81–85.
150. Frondel, C. Structural hydroxyl in chalcedony (Type B quartz). *Am. Mineral.* **1982**, *67*, 1248–1257.
151. Graetsch, H.; Flörke, O.W.; Miehe, G. The nature of water in chalcedony and opal-C from Brazilian agate geodes. *Phys. Chem. Miner.* **1985**, *12*, 300–306. [CrossRef]
152. Thomas, R.; Blankenburg, H.-J. Erste Ergebnisse über Einschlussuntersuchungen an Quarzen aus Achatmandeln und –kugeln basischer und saurer Vulkanite. *Z. Geol. Wiss.* **1981**, *9*, 625–633.
153. Commin-Fischer, A.; Berger, G.; Polvė, M.; Dubois, M.; Sardini, P.; Beaufort, D.; Formoso, M. Petrography and chemistry of SiO_2 filling phases in the amethyst geodes from the Serra Geral Formation deposit Rio Grande do Sul, Brazil. *J. South. Am. Earth Sci.* **2010**, *29*, 751–760. [CrossRef]
154. Ramboz, C.; Pichavant, M.; Weisbrod, A. Fluid immiscibility in natural processes: Use and misuse of fluid inclusion data. *Chem. Geol.* **1982**, *37*, 29–48. [CrossRef]
155. Roedder, E. *Fluid Inclusions*; Reviews in Mineralogy 12; Mineralogical Society of America: Reston, VA, USA, 1984; 645p.
156. Hall, D.L.; Sterner, S.M.; Bodnar, R.J. Freezing point depression of NaCl–KCl–H_2O solutions. *Econ. Geol.* **1988**, *83*, 197–202. [CrossRef]
157. Tripp, R.B. The mineralogy of Warsaw Formation geodes. *Iowa Acad. Sci. Proc.* **1959**, *66*, 350–356.
158. Hayes, J.B. Geodes and concretions from the Mississippian Warsaw Formation. Keokuk region, Iowa, Illinois Missouri. *Sediment. Petrol.* **1964**, *34*, 123–133.
159. Holzhey, G. Die Paragenese von Mineralen in Rhyolithkugeln (Lithophysen) aus Rotliegendvulkaniten des Thüringer Waldes. *Semana* **2016**, *31*, 25–50.
160. Schmitt-Riegraf, C. Magmenentwicklung und spät- bis post-magmatische Alterationsprozesse permischer Vulkanite im Nordwesten der Nahe-Mulde. *Münstersche Forsch. zur Geol. und Paläontologie* **1996**, *80*, 1–251.
161. Walger, E. Zur Entstehung des Calcitachates. *Fortschr. Mineral.* **1961**, *39*, 360.
162. Landmesser, M. Calcitachat: Zur Deutung eines verblüffenden mineralogischen Phänomens. *Mainz. Nat. Arch.* **1996**, *34*, 9–43.
163. Blankenburg, H.-J.; Unterricker, S.; Eichler, B.; Starke, R.; Stolz, W.; Rösler, H.J. Natürliche Radioaktivität chemische und Phasenzusammensetzung der Eisenoxide aus Vulkanitachaten. *Chem. Erde* **1986**, *45*, 159–166
164. Rosemeyer, T. Copper-banded agates from the Kearsarge copper-bearing amygdaloidal lode Houghton county, Michigan. *Rocks Miner.* **2012**, *87*, 352–365. [CrossRef]
165. Moxon, T. Agate in thin section. *Rocks Miner.* **2014**, *89*, 328–339. [CrossRef]
166. Breitkreuz, C. Spherulites and lithophysae—200 years of investigation on hightemperature crystallization domains in silica-rich volcanic rocks. *Bull. Volcanol.* **2013**, *75*, 705–720. [CrossRef]
167. Sakka, S.; MacKenzie, J.D. Relationship between apparent glass transition temperature and liquidus temperature for inorganic glasses. *J. Non-Cryst. Solids* **1971**, *6*, 145–162. [CrossRef]
168. Lofgren, G. Spherulitic textures in glassy and crystalline rocks. *J. Geophys. Res.* **1971**, *76*, 5635–5648. [CrossRef]
169. Breitkreuz, C.; Götze, J.; Weißmantel, A. Mineralogical and geochemical investigation of megaspherulites from Argentina, Germany and USA. *Bull. Volcanol.* **2020**. accepted for publication.
170. Herrington, R.J.; Wilkinson, J.J. Colloidal gold and silica in mesothermal vein systems. *Geology* **1993**, *21* 539–542. [CrossRef]
171. Jebrag, M. Hydrothermal breccias in vein-type ore deposits: A review of mechanisms, morphology and size distribution. *Ore Geol. Rev.* **1997**, *12*, 111–134. [CrossRef]
172. Haake, R.; Fischer, J.; Reißmann, R. Über das Achat-Amethyst-Vorkommen von Schlottwitz im Osterzgebirge *Mineralienwelt* **1991**, *2*, 20–24.
173. Chowns, T.M.; Elkins, J.E. The origin of quartz geodes and cauliflower cherts through the silification of anhydrite nodules. *J. Sediment. Petrol.* **1974**, *44*, 885–903.

74. Tucker, M.E. Quartz replaced anhydrite nodules ("Bristol diamonds") from the Triassic of the Bristol District. *Geol. Mag.* **1976**, *113*, 569–574. [CrossRef]
75. Jacka, A.D. Replacement of fossils by length-slow chalcedony and associated dolomitization. *J. Sediment. Petrol.* **1974**, *44*, 421–427.
76. Landmesser, M. Zur Entstehung von Kieselhölzern. *ExtraLapis* **1994**, *7*, 49–80.
77. Milliken, K.L. The silicified evaporate syndrome: Two aspects of silicification history of former evaporate nodules from Southern Kentucky and Northern Tennessee. *J. Sediment. Petrol.* **1979**, *49*, 245–256.
78. Shaub, B.M. *The Origin of Agates, Thundereggs, and Other Nodular Structures*; Agate Publishing Company: Northampton, MA, USA, 1989; 105p.
79. Landmesser, M. Zur Geothermometrie und Theorie der Achate. *Mitt. Pollichia* **1992**, *79*, 159–201.
80. Harder, H. Agates-formation as a multi component colloid chemical precipitation at low temperatures. *Neues Jahrb. Mineral. Mon.* **1993**, *H.1*, 31–48.
81. Williamson, B.J.; Wilkinson, J.J.; Luckham, P.F.; Stanley, C.J. Formation of coagulated colloidal silica in high-temperature mineralizing fluids. *Mineral. Mag.* **2002**, *66*, 547–553. [CrossRef]
82. White, J.F.; Corwin, J.F. Synthesis and origin of chalcedony. *Am. Mineral.* **1961**, *46*, 112–119.
83. Flörke, O.W. Transport and deposition of SiO_2 with H_2O under supercritical conditions. *Krist. Tech.* **1972**, *7*, 159–166. [CrossRef]
84. Oehler, J.H. Hydrothermal crystallization of silica gel. *Geol. Soc. America Bull.* **1976**, *87*, 1143–1152. [CrossRef]
85. Götze, J.; Plötze, M.; Tichomirowa, M.; Fuchs, H.; Pilot, J. Aluminium in quartz as an indicator of the temperature of formation of agate. *Mineral. Mag.* **2001**, *65*, 407–413. [CrossRef]
86. Dennen, W.H.; Blackburn, W.H.; Quesada, A. Aluminum in quartz as a geothermometer. *Contrib. Mineral. Petrol.* **1970**, *27*, 332–342. [CrossRef]
87. Agel, A.; Petrov, I. Im Quarzgitter substituiertes Aluminium als Indikator für dessen Bildungstemperatur. *Eur. J. Miner.* **1990**, *2*, 144.
88. Correns, C.W. The experimental weathering of silicates. *Clay Miner. Bull.* **1961**, *4*, 249–265. [CrossRef]
89. Wirsching, U. Experimente zum Einfluß des Gesteinsglas-Chemismus auf die Zeolithbildung durch hydrothermale Umwandlung. *Contrib. Mineral. Petrol.* **1975**, *49*, 117–124. [CrossRef]
90. Seyfried, W.E.; Bischo, J.L. Low temperature basalt alteration by seawater: An experimental study at 70 °C and 150 °C. *Geochim. Cosmochim. Acta* **1979**, *43*, 1937–1947. [CrossRef]
91. Giggenbach, W.F. Mass transfer in hydrothermal alteration systems—A conceptual approach. *Geochim. Cosmochim. Acta* **1984**, *48*, 2693–2711. [CrossRef]
92. Duplay, J.; Paquet, H.; Kossovskaya, A.; Tard, Y. Estimation de la température de formation des paragenèses saponite-céladonite et glauconite-nontronite dans les altérations sous-marines de basalte, par la méthode des corrélations entre éléments au sein de populations monominérales. *CR Acad. Sci. Paris* **1989**, *309*, 53–58.
93. Klammer, D. Mass change during extreme acid-sulphate hydrothermal alteration of a Tertiary latite, Styria, Austria. *Chem. Geol.* **1997**, *141*, 33–48. [CrossRef]
94. Polgári, M.; Hein, J.R.; Németh, T.; Pál-Molnár, E.; Vigh, T. Celadonite and smectite formation in the Úrkút Mn-carbonate ore deposit (Hungary). *Sediment. Geol.* **2013**, *294*, 157–163. [CrossRef]
95. Götze, J.; Hofmann, B.; Machałowski, T.; Tsurkan, M.V.; Jesionowski, T.; Ehrlich, H.; Kleeberg, R.; Ottens, B. Biosignatures in subsurface filamentous fabrics (SFF) from the Deccan Volcanic Province, India. *Minerals* **2020**, *10*, 540. [CrossRef]
96. Sukheswala, R.N.; Avasia, R.K.; Gangopadhyay, M. Zeolites and associated secondary minerals in the Deccan Traps of western India. *Mineral. Mag.* **1974**, *39*, 658–671. [CrossRef]
97. Mattioli, M.; Cenni, M.; Passaglia, E. Secondary mineral assemblages as indicators of multistage alteration processes in basaltic lava flows: Evidence from the Lessini Mountains, Veneto Volcanic Province, Northern Italy. *Period. Di Mineral.* **2016**, *85*, 1–24.
98. Kryza, R. Bilans chemiczny dla stref mineralizacji agatowej w wulkanitach rejonu Nowego Kośioła (Góry Kaczawskie). *Arch. Mineral.* **1983**, *39*, 125–132.
99. Krauskopf, K.B. Dissolution and precipitation of silica at low temperatures. *Geochim. Cosmochim. Acta* **1956**, *10*, 1–26. [CrossRef]
100. Landmesser, M. Mobility by metastability: Silica transport and accumulation at low temperatures. *Chem. Erde* **1995**, *55*, 149–176.

201. Dietzel, M. Dissolution of silicates and the stability of polysilicic acid. *Geochim. Et Cosmochim. Acta* **2000**, *6* 3275–3281. [CrossRef]
202. Gránásy, L.; Pusztai, T.; Tegze, G.; Warren, J.A.; Douglas, J.F. Growth and form of spherulites. *Phys. Re* **2005**, *72*, 11605–11619. [CrossRef]
203. Krug, H.-J.; Jacob, K.-H. Genese und Fragmentierung rhytmischer Bänderungen durch Selbstorganisatio *Z. Dtsch. Geol. Ges.* **1994**, *144*, 452–460.
204. Bryxina, N.A.; Sheplev, V.S. Auto-oscillation in agate crystallization. *Math. Geol.* **1999**, *31*, 297–309. [CrossRe
205. Wang, Y.; Merino, E. Self-organizational origin of agates: Banding, fiber twisting, composition, and dynami crystallizationmodel. *Geochim. Cosmochim. Acta* **1990**, *54*, 1627–1638. [CrossRef]
206. Götze, J.; Stanek, K.; Orozco, G. Auf Achatsuche in Kuba. *Mineralienwelt* **2020**, *3*, 56–63.
207. Goldbaum, J.; Howard, C.; Rabinovitch, A. Spatial Chirp of agate bands. *Minerals* **2019**, *9*, 634. [CrossRef]
208. Mikhailov, A.S.; Showalter, K. Control of waves, patterns and turbulence in chemical systems. *Phys. Re* **2006**, *425*, 79–194. [CrossRef]
209. Kumar, D.K.; Steed, J.W. Supramolecular gel phase crystallization: Orthogonal self-assembly unde non-equilibrium conditions. *Chem. Soc. Revs.* **2014**, *43*, 2080–2088. [CrossRef]
210. Papineau, D. Chemically oscillation reactions in the formation of botryoidal malachite. *Am. Mineral.* **202** *105*, 447–454. [CrossRef]
211. Walger, E.; Mattheß, G.; Von Seckendorff, V.; Liebau, F. The formation of agate structures: Models for silic transport, agate layer accretion, and for flow patterns and flow regimes in infiltration channels. *Neues Jahr Mineral. Abh.* **2009**, *186*, 113–152. [CrossRef]
212. Laurs, B.M.; Rossman, G.R. Grape-like "Manakarra" quartz from Sulawesi, Indonesia. *J. Gemmol.* **201** *36*, 101–102.
213. Hofmann, B.A. Subsurface filamentous fabrics. In *Encyclopedia of Geobiology*; Reitner, J., Thiel, V., Eds Springer: Berlin/Heidelberg, Germany, 2011; pp. 851–853.
214. Thewalt, U.; Dörfner, G. Wie kommt das Moos in den Achat und wie nicht? *Der Aufschluss* **2012**, *63*, 1–16.
215. Reysenbach, A.-L.; Cady, S.L. Microbiology of ancient and modern hydrothermal systems. *Trends Microbio* **2001**, *9*, 79–86. [CrossRef]
216. Jones, B.; Renault, R.W. Influence of thermophilic bacteria on calcite and silica precipitation in hot spring with water temperatures above 90 °C: Evidence from Kenya and New Zealand. *Can. J. Earth Sci.* **1996**, *3* 72–83. [CrossRef]
217. Konhauser, K.O.; Phoenix, V.R.; Bottrell, S.H.; Adams, D.G.; Head, I.M. Microbial-silica interaction i Icelandic hot spring sinter: Possible analogues for some Precambrian siliceous stromatolites. *Sedimentolog* **2001**, *48*, 415–433. [CrossRef]
218. Ferris, F.G.; Beveridge, T.J.; Fyfe, W.S. Iron-silica crystallite nucleation by bacteria in a geothermal sedimen *Nature* **1986**, *320*, 609–611. [CrossRef]
219. Parenteau, M.N.; Cady, S.L. Microbial biosignatures in iron-mineralized phototrophic mats at Chocolate Pot Hot Springs, Yellowstone National Park, United States. *PALAIOS* **2010**, *25*, 97–111. [CrossRef]
220. Müller, A.; Polgári, M.; Gucsik, A.; Pál-Molnár, E.; Koós, M.; Veres, M.; Götze, J.; Nagy, S.; Cserháti, C Németh, T.; et al. Cathodoluminescent features and Raman spectroscopy of Miocene hydrotherma biomineralization embedded in cryptocrystalline silica varieties, Central Europe, Hungary. In *Micro-Rama Spectroscopy and Luminescence Studies in the Earth and Planetary Sciences*; Gucsik, A., Ed.; American Institute o Physics: Houston, TX, USA, 2009; pp. 207–218.
221. Yang, H.; Coombs, N.; Ozin, G.A. Morphogenesis of shapes and surface patterns in mesoporous silica. *Natur* **1997**, *386*, 692–695. [CrossRef]
222. Glaab, F.; Kellermeier, M.; Kunz, W.; Morallon, E.; Garcia-Ruiz, J.M. Formation and evolution of chemica gradients and potential differences across self-assembling inorganic membranes. *Angew. Chem.* **2012**, *12* 4393–4397. [CrossRef]
223. Moxon, T.; Petrone, C.M.; Reed, S.J.B. Characterization and genesis of horizontal banding in Brazilian agat An X-ray diffraction, thermogravimetric and electron microprobe study. *Mineral. Mag.* **2013**, *77*, 227–24 [CrossRef]
224. Haudin, F.; Cartwright, J.H.E.; Braua, F.; De Wit, A. Spiral precipitation patterns in confined chemical garden *Proc. Natl. Acad. Sci. USA* **2014**, *111*, 17363–17367. [CrossRef]

125. Noorduin, W.L.; Grinthal, A.; Mahadevan, L.; Aizenberg, J. Rationally designed complex, hierarchical microarchitectures. *Science* **2013**, *340*, 832–837. [CrossRef]
126. Kellermeier, M.; Cölfen, H.; García-Ruiz, J.M. Silica Biomorphs: Complex biomimetic hybrid materials from "sand and chalk". *Eur. J. Inorg. Chem.* **2012**, *32*, 5123–5144. [CrossRef]
127. Dominguez-Bella, S.; Garcia-Ruiz, J.M. Textures in induced morphology crystal aggregates of $CaCO_3$: Sheaf of wheat morphologies. *J. Cryst. Growth* **1986**, *79*, 236–240. [CrossRef]
128. Lenz, G.; Schäfer, K. Spiralen im Achat—Ein biologisches oder ein mineralogisches Phänomen? *Lapis* **2008**, *1*, 21–24.

2

Chemometric Modeling of Trace Element Data for Origin Determination of Demantoid Garnets

Stefan G. Bindereif [1], Felix Rüll [1], Stephan Schwarzinger [1] and Clemens Schwarzinger [2,*]

[1] North Bavarian NMR-Centre, University of Bayreuth, 95447 Bayreuth, Germany; stefan.bindereif@uni-bayreuth.de (S.G.B.); Felix.Ruell@uni-bayreuth.de (F.R.); s.schwarzinger@uni-bayreuth.de (S.S.)

[2] Institute for Chemical Technology of Organic Materials, Johannes Kepler Univeristy Linz, 4040 Linz, Austri

* Correspondence: Clemens.schwarzinger@jku.at

Abstract: The determination of country of origin poses a common problem in the appraisal o gemstones and is in many cases still based on the observation of inclusions and growth features of a gem, whereas chemical analysis is only done by major labs. We have used Laser Ablation Inductivel Coupled Plasma Mass Spectrometry to analyze the trace element profiles of demantoid garnet from six different countries and identified a set of six elements, which are magnesium, aluminum titanium, vanadium, chromium, and manganese, that are necessary to assign the mining region with a good certainty. By using the logarithms of the trace element concentrations and subjecting them to chemometric modeling, we were able to separate the demantoids originating from Russia Pakistan, Namibia, Iran, and Madagascar very well, leaving only Italy with some uncertainty. Result are presented for an "all origins" model as well as pair-wise comparison of two locations at a time which lead to even better results.

Keywords: multivariate statistics; chemometrics; origin determination; garnets; demantoid LA-ICP-MS

1. Introduction

The correct determination of origin of gemstones has gained more and more importance recentl The desire to know where a gem comes from is not only scientifically driven but to a very large exten by the trade. For a scientist it might be hard to understand that for example two blue sapphires, which have the same size, weight, color, clarity and cut are appraised differently only because they hail from different sources. The trade, however, introduces factors such as rarity and history into the appraisa process and thus the blue sapphire originating from e.g., Madagascar (currently producing) will be less valuable than the one that has been found in the famed Kashmir region (found in 1881 after a landslide and now exhausted) [1,2]. Additionally, ethical issues associated with gemstone productior are becoming increasingly important. Hence, it will be necessary in certain cases to prove that gems do not come from conflict areas as those might be banned from trade in some countries, such as jade and rubies originating from Burma which were banned in the United States from 2008 to 2016.

In classical gemology, mainly optical characteristics have been used to determine origins, and above all, inclusions. Tiny crystals which grew in the host gem material are dependent on the geologica setting and thus provide a good tool to eliminate certain sources. For instance, in sapphires it is possible to assign certain origins by identifying inclusions [3]. However, the fact that inclusions are not always present, especially in high-priced gemstones, renders this approach problematic. In the last decades to optimize the inclusion-based process, major labs have started to exploit more advanced technologies among them the determination of chemical composition. By determining major and minor elements onclusions about e.g., sapphire genesis can be drawn, or the origin of copper bearing (Paraiba-type) ourmalines be determined [4]. The analytical arsenal used today includes techniques such as X-ray luorescence (XRF), laser induced breakdown spectroscopy (LIBS), or laser ablation inductively coupled lasma mass spectrometry (LA-ICP-MS). Each of these technologies has its merits but also weaknesses: vhereas XRF gives robust values for major elements, the other two techniques focus on trace elements. Iaving the lowest detection limits, typically in the ppb range, LA-ICP-MS further exhibits an excellent inearity of up to eight orders of magnitude. By contrast, higher concentrations very often present problem for this technique. However, having access to both, major components as well as trace lements, is desirable for obtaining the overall chemical composition in a greater context.

In our study we have focused on the determination of geographic origin of demantoids, which is he well accepted trade name of the green variety of andradite ($Ca_3Fe_2(SiO_4)_3$) garnets. Demantoids ave first been found in the Russian Urals in the river Bobrovka near the village Elizavetinskoye, close o Niznij Tagil (or often just called Tagil) in the middle of the 19th century [5,6]. Further deposits in the Jrals were later found in Poldnevskoye (Poldnevaya), Korkodin, and Ufaley—all within a range of 00 km of Ekaterinburg. Only in the 1990s a second major deposit was found in the Erongo region of Namibia [7], which together with Russia represents most of the cut gems available in today's market Figure 1) [8]. Smaller deposits of gem quality material can also be found in Val Malenco in Italy [9], Antetezambato in Madagascar [10,11], Khuzdar, Baluchistan in Pakistan [12,13], as well as in Belqeys nd Kerman, both located in Iran [14,15]. Certainly, there are many more deposits of demantoid garnets on the globe which are not mentioned here as to this date they do not play a major role in the gem market.

Figure 1. Demantoid gemstones (0.75–3.23 ct) from Russia (top row) and Namibia (bottom row) represent the majority of today's market.

Geologically, these deposits can be separated into two distinct groups, namely in (1) serpentinite nosted demantoids, such as the ones from the Urals, Val Malenco and Baluchistan and (2) those originating from skarns, such as the sites in Namibia and Madagascar. First successful attempts n determining the origin of demantoids based on chemical fingerprinting have shown promising esults, although only simple characteristics such as the Mn/Ti ratio and Cr and/or V content have een used [8,16]. In this study we have carried out systematic quantitative trace-element analysis of demantoid garnets from six different origins using LA-ICP-MS and subsequent chemometric modeling or targeted classification without using inclusion-based assumptions.

2. Materials and Methods

Representative rough demantoid samples have been collected from Tubussis (Namibia), Ufale (Russia), Karkodin (Russia), Poldnevskoye (Russia), Bobrovka (Russia), Val Malenco (Italy), Baluchista (Pakistan), Kerman (Iran) and Antetezambato (Madagascar). Cut stones have either been cut by one o the authors (CS) or were taken from the collection of the Austrian Gemmological Society or the collectio of Gerhard Brandstetter. Italian demantoids were in part supplied by Thomas Engeli, other sample were taken from the collections of the authors (CS and StS (Stephan Schwarzinger)). In particular fo many samples originating from Russian deposits no further site of collection was provided. A total o 136 samples (403 individual points) have been analyzed, 39 from Russia (117 points), 25 from Namibi (73 points), 33 from Pakistan (98 points), 5 from Madagascar (14 points), 8 from Iran (23 points) an 26 from Italy (78 points).

Concentrations of ^{24}Mg, ^{27}Al, ^{47}Ti, ^{51}V, ^{52}Cr, and ^{55}Mn were measured by Lase ablation—inductively coupled plasma mass spectrometry (LA-ICP-MS) using a Cetac LSX-213 G2 laser ablation system (Teledyne Cetac Technologies, Omaha, NE, USA) connected to a Therm Fisher Scientific X-Series II ICP MS system (Thermo Fisher Scientific, Waltham, MA, USA). ^{43}C ^{28}Si, and ^{56}Fe were also measured and utilized to monitor the smoothness of the Laser ablatio process. Other elements (^{7}Li, ^{59}Co, ^{60}Ni, ^{63}Cu, ^{64}Zn, ^{71}Ga, ^{74}Ge, and ^{88}Sr) were measured as wel for representative samples but did not show any significant concentrations. Tuning and calibratio were carried out using NIST 610, 612, and 614 glass standards; reference values were taken fron Jochum et al. [17]. The laser ablation was operated with a 100 μm spot size, a laser fluence of 9.2 J cm^{-} and a repetition rate of 10 Hz. 130–230 shots were made on each spot and the ablated product transferred to the ICP-MS with a flow rate of 500 mL.min^{-1} He gas. Each sample was measured a three different spots except for two pieces from Madagascar that have been sampled 5 times each.

Concentration profiles were further analyzed with the publicly available software AI(OMICS) (publicly available in-house written software), which provides a MatLab-based workflow for robus chemometric modelling [18]. Specifically, a table with sample IDs, meta data on the geographi origin of the sample as class information, and the respective element concentration profile for eacl sample were loaded in AI(OMICS)^{n}, which was operated in MatLab (version R2018b, The Mathwork Natick, MA, USA). Different strategies were applied, either considering all classes or considering jus pair-wise comparison of classes. As a first step, native element concentrations were subjected to th data expansion module of AI(OMICS)^{n}, which automatically generates, e.g., combinations of pair-wis ratios of parameters, logarithms of single parameters as well as parameter combinations for improvec screening of non-linear events contributing variance between groups. To calculate logarithms ir this context, concentrations below the limit of detection (LOD) were set to a fixed value of 0.1 ppn since negative and zero values cannot be logarithmized. The resulting variables were then subjectec to a Kruskal-Wallis H-test in order to detect class-related variance and ranked according to thei relevance [19]. A maximum of 50 of the most relevant variables got selected for further dimensionalit reduction by principal component analysis (PCA). Subsequently, linear discriminant analysis (LDA was applied to the extracted principal components (PCs). The number of PCs to be taken into accoun for LDA was determined by a cut-off at a threshold of 95% explained variance. This is a precaution ir order to avoid overfitting and thus the consideration of random error (noise) rather than relationship between variables. Models were validated using the Monte-Carlo cross-validation (MCCV) method [20 Specifically, 100 random models were generated with 90% of the data set as training data and th remaining 10% as test set. The overall quality of the models, i.e., the correctness of data predictec during MCCV, was judged using the multivariate correlation coefficient R^2, the multivariate correlatior coefficient after cross validation Q^2, the accuracy (ACC), the Matthews correlation coefficient (MCC the prevalence threshold as well as the confusion matrix. In particular the MCC has been found to be very reliable metric for judging the quality of chemometric models. In contrast to the ACC, whicl constitutes the ratio of the sum of true positive and negative classifications and the total populatior MCC also takes into account the false positive and false negative classifications. It can also be appliec well in cases where group sizes differ significantly. The prevalence threshold is applied to test whether a particular class is represented with enough samples under consideration of the accuracy and precision of the respective model. To assess model robustness averaged class accuracies and their standard deviations were calculated based on the 100-fold MCCV. The importance of each element for the respective model was judged by calculating the variable importance in projection (VIP). Based on this information box plots were generated for these elements in R version 3.5.4 using the 'ggplot2' package [21–27].

3. Results and Discussion

Garnets can be divided into 6 groups, which are pyrope, almandine, spessartine (pyralspite-series) and uvarovite, grossular, and andradite (ugrandite-series). Very often garnets are not pure endmembers of a series but represent mixtures of the members of the above-mentioned series and therefore do not exhibit well defined, narrow ranges of element-concentrations. We have therefore decided not to use any internal standard for the ICP-MS experiments, rather, we have used three different NIST glasses which cover concentrations of about 1–400 ppm for most elements in question. A comparison of our data with the ones published by Adamo et al. [9] for Italian demantoids using an internal standard approach shows very good agreement given the fact that the chemistry within a single crystal can vary dramatically as has been shown for an andradite slice from Ufalay [8]. The actual values are Cr: 2–5468 vs. 0–1134 ppm, Ti: 7–649 vs. 3–545 ppm, and V: 12–86 vs. 1–102 ppm for 31 points analyzed by Adamo and 78 points in this study.

As a first result this study reveals that the samples analyzed here all represent rather pure end members of the andradite series except for the samples originating from Madagascar. This class of samples exhibits from low to comparatively high aluminum concentrations of 40 ppm up to 8.5% (most being above 0.2%) indicating presence of a certain grossular content for this site. Hence, within the group of locations analyzed here aluminum serves as a marker element for Madagascar. Another element of interest is chromium, where highest concentrations can be found in samples from Russia and Iran, as low as 100 ppm but mostly in the range of 0.1 to 4%. Namibia has generally the lowest amounts of chromium; in Pakistani, Madagasy and Italian samples concentrations from below detection limit up to 1000 ppm were found. Manganese is high in Namibian and Russian samples (280–500 ppm) whereas magnesium is found to have highest concentrations in Pakistani and in part Russian material (2000 ppm—1.1%) and lowest concentrations in Namibian samples (generally below 500 ppm). A table listing all concentrations can be found as Table S1 in the Supplementary Materials .

A first approach was using the separation criteria which have been successfully applied to differentiate Namibian from Russian demantoids in our previous work [8]. There, the ratio of manganese/titanium was plotted against the sum of chromium and vanadium, however, this two-dimensional approach soon starts to produce significant overlapping of several origins and was thus discarded.

In this work, a total of 17 elements was analyzed to discover marker elements or reveal internal dependencies of element concentrations with respect to the origin of the samples. For successful multivariate modeling it is important to identify variables significantly contributing the inter-group variance and to eliminate other variables to reduce the risk of overfitting the data, which here is tested for by applying extensive Monte-Carlo cross-validation. In addition, it has to be considered that linear multivariate model approaches may have difficulties with non-linear data structures. Therefore, linearization of data, e.g., by logarithmization, may significantly improve performance of linear modeling algorithms. It should, however, be noted that the degree to which this is relevant will always depend on the particular data structure of the data set investigated. In this study the optimal number of variables as well as the optimal pre-processing of these variables was determined in an iterative manner. Specifically, elements that were not detected at concentrations above LOD were eliminated first. This way, the elements Li, Co, Ni, Cu, Zn, Ga, Ge, and Sr (eight out of 17) were not present in significant concentrations in a manner systematically correlating with the geographic origin and were subsequently not considered further. Additionally, Ca, Si and Fe—which were primarily used for monitoring of the laser ablation process—were excluded from further analysis as no standards are available in the relevant concentration regime preventing that the system could not be calibrated accordingly and the amounts detected were found to be too high. Following this step the data expansion module of AI(OMICS)^n, which aims at automation of certain steps of data pre-processing, was applied for the remaining six elements Al, Cr, Mg, Mn, Ti, and V. Specifically, inverse concentrations, products and ratios of all element concentrations measured as well as the calculated dependencies are generated to allow a systematic screening. Especially ratios, e.g., plotted in 2D graphs, have been shown in the past to contain important information as compared to using pure concentrations [8]. Similarly as concentrations of the various elements span several orders of magnitude and are heavily skewed transformation into their logarithm—as pointed out above—may improve chemometric modeling based on linear approaches, including PCA, PCA-LDA, partial least squares-discriminant analysis (PLS-DA), orthogonal PLS-DA (OPLS-DA) and others. Here, the natural logarithm (base-*e* log) has been chosen for model building with the PCA-LDA approach [28,29]. For this specific operation it was necessary to convert measurements numbers smaller or equal to zero to a small value but not zero. Considering the concentrations detected for the individual elements a fixed value of 0.1 was chosen empirically, which—being in the order of the value of the smallest concentrations of all elements measured—prevents unnaturally small logarithms. Several rounds of data elimination were performed considering the ranks of variables especially in the Kruskal-Wallis *H*-test as well as in the VIP-scores in PCA-LDA models. In addition, model quality was judged by the R^2 and Q^2 values as a function of the number of principal components to exclude overfitting. The Matthews correlation coefficient (MCC), which in contrast to R^2 and Q^2 also takes into consideration the effect of true and false positive/negative predictions as well, is also considered as a relative quality criterion between different models. The prevalence threshold, which takes into account the quality of discriminating power under consideration of the size of the groups, was judged, too. All models were generated as PCA-LDA for optimal comparison. Finally, it was found that logarithmizing of the pure element concentration as single data transformation was sufficient to achieve over 95% explained variance with no more than 3 to 5 principal components for pair-wise comparisons of locations. As logarithmizing of variables is a common way of converting non-linear problems into linear ones, it is not unexpected that LDA performs better with the logarithms of the element concentrations. Variables, such as ratios of concentrations or their logarithms were excluded to reduce multi-collinearity, which may significantly impact modeling by overfitting. Of note, PCA-LDA itself was chosen because it is known to be less prone to overfitting due to multi-collinear variables. Additionally, in contrast to e.g., a PLS-DA (where dimensionality reduction is supervised while it is un-supervised in the PCA approach), it allows better assessment of samples that do not belong to groups used to generate models through judgment of the Mahalanobis distance [30] of a sample relative to the group center in multidimensional space [31]. Note, too, that LDA in principle requires normality of the data, which is often not given in real-life data sets, in particular when only small sample numbers are available. Using logarithms of element concentrations is an important step towards normality, which, however, is not met for every element/country combination. Nevertheless, the LDA classification has been shown in the past to perform very robust, despite the assumption of normality is violated [32,33]. If the data were perfectly normally distributed LDA would determine the optimal solution for each hyperplane. In absence of normality several solutions for a hyperplane may exist, each describing only one local optimum. This in turn would lead to an increased number of mismatches in the cross-validation, which was not observed here. We wish to note that there are many other classification algorithms which in principle may perform similarly provided proper verification. In this study, however, the focus is not to find the best performing statistical model but to gain insights into whether trace element concentration allows classification of the origin at all and which elements play a role in this approach.

Table 1 shows the confusion matrix for a representative model "All origins". Shown are the model specific benchmark parameters accuracy (ACC), Matthews correlation coefficient (MCC), false lassifications, correct classifications and the prevalence threshold (minimum calculated fraction/true raction). Numbers in parenthesis are percent fractions for countries and correct and false classifications, espectively. n(total) denotes the number of total element profiles, n(class) the number of element rofiles, and n (latent variables) the number of principal components used to achieve 95% explained ariance. In addition, cumulative R^2 and Q^2 as well as the averaged group accuracy (mean ± standard eviation) from the 100-fold MCCV are shown to allow judgement of the model robustness. Diagonal lements in the confusion matrix represent the number (fraction) of correctly assigned cases. Vertical ff-diagonal elements correspond to false negative predictions (i.e., wrong origin assigned; e.g., from he 73 true Namibian Demantoids two are mis-assigned to Russia and three to Italy (first row)), vhile horizontal off-diagonal elements are false negative results (i.e., wrong origin is assumed; e.g., ive samples predicted to be from Namibia are indeed from Russia (first line)). The asterisk indicates hat the prevalence threshold expected for the accuracy and precision of the model is slightly missed or the origin Iran.

Table 1. Confusion matrix showing prediction accuracy of the method (for details refer to the text).

Model All Origins	ACC = 0.84	MCC = 0.78	R^2_{cum} = 1	Q^2_{cum} = 0.85			
n(total) = 403	True Namibia	True Pakistan	True Madagascar	True Russia	True Iran	True Italy	prevalence threshold (calc/observed)
Predicted Namibia	68 (93.2%)	0	0	5 (4.3%)	0	0	0.12/0.18
Predicted Pakistan	0	91 (92.9%)	1 (7.1%)	7 (6.0%)	0	7 (9.0%)	0.12/0.18
Predicted Madagascar	0	0	13 (92.9%)	0	0	0	0.03/0.03
Predicted Russia	2 (2.7%)	1 (1.0%)	0	96 (82.1%)	3 (13.0%)	18 (23.1%)	0.24/0.29
Predicted Iran	0	1 (1.0%)	0	3 (2.6%)	17 (73.9%)	0	0.10/0.06 *
Predicted Italy	3 (4.1%)	5 (5.1%)	0	6 (5.1%)	3 (13.0%)	53 (67.9%)	0.22/0.19
n(class)	73	98	14	117	23	78	
false classifications (n (%))	65 (16.1%)						
correct classifications (n (%))	338 (83.9 %)						
n(latent variables)	6						
Averaged Group Accuracy from MC cross validation	0.92 ± 0.10	0.91 ± 0.10	0.94 ± 0.15	0.81 ± 0.12	0.72 ± 0.31	0.68 ± 0.18	

*: Prevalence threshold missed for location Iran—see text for details. Model performance (correctly assigned) indicated by background color: >90%: green; >80 %: blue; >70 % yellow; rest: red.

The model with all six origins shows that at least three origins can be predicted with a considerable igh accuracy and precision (Figure 2, Table 1). In particular, demantoids from Namibia and Pakistan an be classified very well. Specimen from Madagascar can also be classified well, but the low revalence leads to a higher degree of variability as evidenced by the larger standard deviation of he group accuracy resulting from MCCV. Although samples from Russian deposits can be predicted vith accuracies around 80% some sample are mis-assigned to all other origins, except for Madagascar. This indicates a larger variability in the composition of these demantoids. Samples from Iran and taly could also be assigned only with accuracies around 70%. However, in contrast to the wrongly lassified samples from Russia the samples from Iran and Italy are mis-assigned to only two other rigins. In particular demantoids from Italy exhibit a tendency to be predicted in a false-positive nanner as Russian samples. This leads to the conclusion that a significant fraction of Italian demantoid amples shares a similar trace-element profile with the majority of the Russian samples. It should be nentioned that—although the prevalence index for five of the origins is met or even exceeded and only losely missed for the origin Iran—the smaller number of samples available for Iran and Madagascar dds to uncertainty. The question here is whether natural variability of these very difficult to come y samples is already adequately represented. Especially the narrow distribution of the profiles for Madagascar contributes to excellent classification properties for this particular deposit, which appears o source from an increased grossular content in the samples studied here. The small total number of data points also contributes to comparable large standard deviation of the class accuracy achieved during the 100-fold Monte Carlo cross-validation.

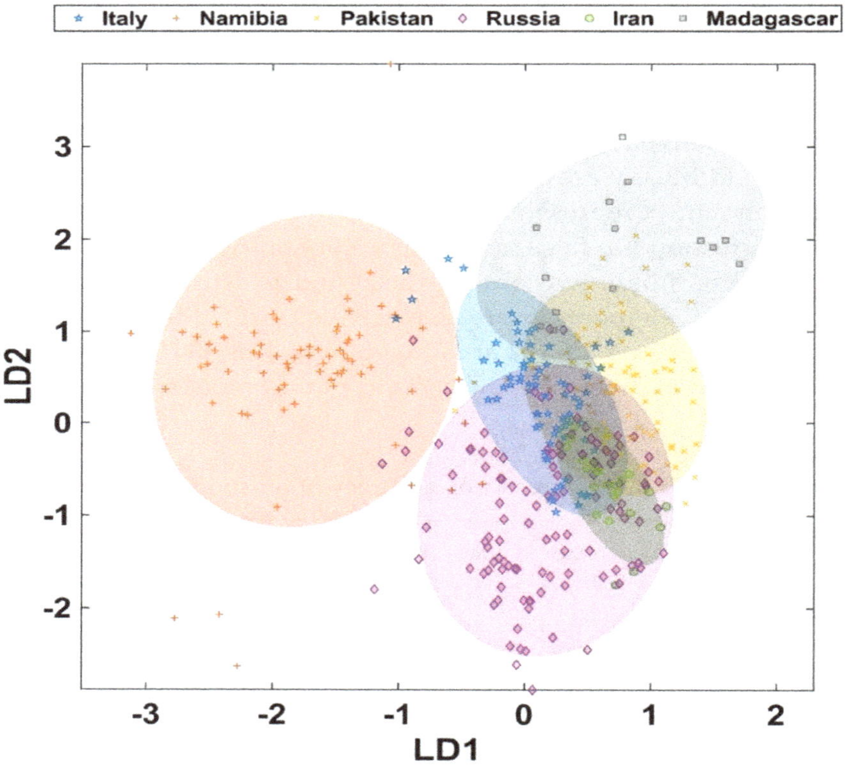

Figure 2. PCA-LDA score plot with all classes including the log-transformed concentrations of the six relevant elements Mg, Al, Ti, V, Cr and Mn. Each dot represents a single element profile from an individual measurement after MCCV. Axis shown represent the first two linear discriminants. Ellipses confer to the 95% confidence interval according to Hotellings T^2, which is the multivariate analog to the t-test [34]. Note that graphics represent a projection from 6-dimensional space, i.e., overlap in this projection does not necessarily mean samples cannot be separated (cf. confusion matrix in Table 1).

As already indicated above the number of samples has a critical influence on the model quality and should always be tested for using the prevalence threshold. In the above model the prevalence threshold for Iran was slightly missed as indicted by an asterisk in Table 2. This indicates that more samples will be needed to improve future models. However, at the same time this case represents an interesting test for what happens when locations with only small numbers of samples are included in modeling. While ideally all demantoid deposits in the world with all available specimens should be in a model this is not possible in reality, simply for sample availability, but also financial reasons associated with sample acquisition and analysis. As demantoids from Iran are present in the market we decided to include the available specimens into this model rather than to omit it and investigate the effects. While the model with six deposits does not allow satisfactory assignment of all samples in a single test the results from cross validation allow deepened insights into sample confusion. In fact testing all classes in one model is only one of several possible testing strategies. A different strategy is to test one deposit in a pair-wise manner against all other deposits (i.e., testing location X vs. (combined others)). However, in this case the group consisting of all but one location displays a very high variability preventing improved results for the particular classification problem studied here (data not shown). It should be noted that it is difficult to delineate a generally applicable approach, as this will be determined by the properties of the population investigated. For the present case another possibility is inspired by the directed acyclic graph (DAG) trees [35]. Here, models are constructed for only two deposits at a time, but for all possible combination of deposits. By doing so the model can take optimal advantage of the differences between a particular pair of groups, which may lead to improved classification performance.

Table 2. Confusion matrix for a representative model for pair-wise testing of the country of origin (15 individual models).

Model	ACC = 0.99	MCC = 0.94	
Italy vs. Madagascar	R^2_{cum} = 0.91	Q^2_{cum} = 0.89	
n (total) = 92	True Italy	True Madagascar	Prev. Threshold
Predicted Italy	78 (100%)	1 (7.1%)	0.24/0.85
Predicted Madagascar	0	13 (92.9%)	0/0.15
n (class)	78	14	
false classifications (*n* (%))	1 (1.1%)	Averaged Group Accuracy	
correct classifications (*n* (%))	91 (98.9%)	Italy	1.00±0.00
n (latent variables)	4	Madagascar	0.90±0.24

Model	ACC = 0.99	MCC = 0.94	
Namibia vs. Madagascar	R^2_{cum} = 0.94	Q^2_{cum} = 0.93	
n (total) = 87	True Namibia	True Madagascar	Prev. Threshold
Predicted Namibia	73 (100%)	1 (7.1%)	0.23/0.84
Predicted Madagascar	0	13 (92.9%)	0/0.16
n (class)	73	14	
false classifications (*n* (%))	1 (1.1%)	Averaged Group Accuracy	
correct classifications (*n* (%))	86 (98.9%)	Namibia	1.00±0.00
n (latent variables)	4	Madagascar	0.91±0.24

Model	ACC = 0.98	MCC = 0.92	
Pakistan vs. Madagascar	R^2_{cum} = 0.95	Q^2_{cum} = 0.35	
n (total) = 112	True Pakistan	True Madagascar	Prev. Threshold
Predicted Pakistan	97 (99.0%)	1 (7.1%)	0.18/0.875
Predicted Madagascar	1 (1.0%)	13 (92.3%)	0.10/0.125
n (class)	98	14	
false classifications (*n* (%))	2 (1.8%)	Averaged Group Accuracy	
correct classifications (*n* (%))	110 (98.2%)	Pakistan	0.99±0.03
n (latent variables)	5	Madagascar	0.95±0.19

Model	ACC = 1	MCC = 0.99	
Russia vs. Madagascar	R^2_{cum} = 0.87	Q^2_{cum} = 0.24	
n (total) = 131	True Russia	True Madagascar	Prev. Threshold
Predicted Russia	117 (100%)	0	0.05/0.89
Predicted Madagascar	0	14 (100%)	0/0.11
n (class)	117	14	
false classifications (*n* (%))	0	Averaged Group Accuracy	
correct classifications (*n* (%))	131 (100%)	Russia	1.00±0.00
n (latent variables)	4	Madagascar	0.998±0.03

Model	ACC = 1	MCC = 0.96	
Iran vs. Madagascar	R^2_{cum} = 0.88	Q^2_{cum} = 0.83	
n (total) = 23	True Iran	True Madagascar	Prev. Threshold
Predicted Iran	23	0	0.18/0.62
Predicted Madagascar	0	14	0/0.38
n (class)	23	14	
false classifications (*n* (%))	0	Averaged Group Accuracy	
correct classifications (*n* (%))	37 (100%)	Iran	1.00±0.00
n (latent variables)	3	Madagascar	0.95±0.16

Model	ACC = 1	MCC = 0.95	
Italy vs. Iran	R^2_{cum} = 0.75	Q^2_{cum} = 0.73	
n (total) = 101	True Italy	True Iran	Prev. Threshold
Predicted Italy	78 (100%)	0	0.07/0.77
Predicted Iran	0	23 (100%)	0.13/0.23
n (class)	78	23	
false classifications (*n* (%))	0	Averaged Group Accuracy	
correct classifications (*n* (%))	101 (100%)	Italy	0.99±0.03
n (latent variables)	3	Iran	0.99±0.11

Model	ACC = 1	MCC = 1	
Namibia vs. Iran	R^2_{cum} = 0.94	Q^2_{cum} = 0.92	
n (total) = 96	True Namibia	True Iran	Prev. Threshold
Predicted Namibia	73 (100%)	0	0/0.76
Predicted Iran	0	23 (100%)	0/0.24
n (class)	73	23	
false classifications (*n* (%))	0	Averaged Group Accuracy	
correct classifications (*n* (%))	96 (100%)	Namibia	1.00±0.00
n (latent variables)	4	Iran	1.00±0.00

Model	ACC = 0.99	MCC = 0.96	
Pakistan vs. Iran	R^2_{cum} = 0.87	Q^2_{cum} = 0.80	
n (total) = 121	True Pakistan	True Iran	Prev. Threshold
Predicted Pakistan	97 (99.0%)	0	0/0.81
Predicted Iran	1 (1.0%)	23 (100%)	0.11/0.19
n (class)	98	23	
false classifications (*n* (%))	1 (0.8%)	Averaged Group Accuracy	
correct classifications (*n* (%))	120 (99.2%)	Pakistan	0.99±0.04
n (latent variables)	4	Iran	1.00±0.00

Model	ACC = 0.93	MCC = 0.67	
Russia vs. Iran	R^2_{cum} = 0.92	Q^2_{cum} = 0.93	
n (total) = 140	True Russia	True Iran	Prev. Threshold
Predicted Russia	112 (95.7%)	5 (21.7%)	0.36/0.84
Predicted Iran	5 (4.3%)	18 (78.3%)	0.20/0.16
n (class)	117	23	
false classifications (*n* (%))	10 (7.1%)	Averaged Group Accuracy	
correct classifications (*n* (%))	130 (92.9%)	Russia	0.95±0.06
n (latent variables)	4	Iran	0.71±0.33

Model	ACC = 0.86	MCC = 0.68	
Italy vs. Russia	R^2_{cum} = 0.91	Q^2_{cum} = 0.87	
n (total) = 195	True Italy	True Russia	Prev. Threshold
Predicted Italy	58 (74.4%)	8 (6.8%)	0.24/0.40
Predicted Russia	20 (25.6%)	109 (93.2%)	0.35/0.60
n (class)	78	117	
false classifications (*n* (%))	28 (14.4%)	Averaged Group Accuracy	
correct classifications (*n* (%))	167 (85.6%)	Italy	0.73±0.15
n (latent variables)	4	Russia	0.93±0.11

Model	ACC = 0.95	MCC = 0.89	
Namibia vs. Russia	R^2_{cum} = 0.92	Q^2_{cum} = 0.92	
n (total) = 190	True Namibia	True Russia	Prev. Threshold
Predicted Namibia	69 (94.5%)	5 (4.3%)	0.19/0.38
Predicted Russia	4 (5.5%)	112 (95.7%)	0.18/0.62
n (class)	73	117	
false classifications (*n* (%))	9 (4.7%)	Averaged Group Accuracy	
correct classifications (*n* (%))	181 (95.3%)	Namibia	0.95±0.07
n (latent variables)	4	Russia	0.95±0.06

Model	ACC = 0.95	MCC = 0.88	
Pakistan vs. Russia	R^2_{cum} = 0.89	Q^2_{cum} = 0.89	
n(total) = 215	True Pakistan	True Russia	Prev. Threshold
Predicted Pakistan	95 (96.9%)	8 (6.8%)	0.21/0.46
Predicted Russia	3 (3.1%)	109 (93.2%)	0.18/0.54
n (class)	98	117	
false classifications (*n* (%))	11 (5.1%)	Averaged Group Accuracy	
correct classifications (*n* (%))	204 (94.5%)	Pakistan	0.96±0.06
n (latent variables)	4	Russia	0.93±0.08

Model	ACC = 0.94	MCC = 0.86	
Italy vs. Pakistan	R^2_{cum} = 0.88	Q^2_{cum} = 0.85	
n (total) = 165	True Italy	True Pakistan	Prev. Threshold
Predicted Italy	74 (94.9%)	7 (7.1%)	0.23/0.44
Predicted Pakistan	4 (5.1%)	91 (92.9%)	0.19/0.56
n (class)	78	98	
false classifications (*n* (%))	11 (6.3%)	Averaged Group Accuracy	
correct classifications (*n* (%))	165 (93.8%)	Italy	0.95±0.08
n (latent variables)	4	Pakistan	0.92±0.09

Model	ACC = 1	MCC = 1	
Namibia vs. Pakistan	R^2_{cum} = 0.94	Q^2_{cum} = 0.94	
n (total) = 171	True Namibia	True Pakistan	Prev. Threshold
Predicted Namibia	73 (100%)	0	0/0.43
Predicted Pakistan	0	98 (100%)	0/0.57
n (class)	73	98	
false classifications (*n* (%))	0	Averaged Group Accuracy	
correct classifications (*n* (%))	171 (100%)	Namibia	1.00±0.00
n (latent variables)	4	Pakistan	1.00±0.00

Model	ACC = 0.99	MCC = 0.96	
Italy vs. Namibia	R^2_{cum} = 0.95	Q^2_{cum} = 0.95	
n (total) = 151	True Italy	True Namibia	Prev. Threshold
Predicted Italy	78 (100%)	2 (2.7%)	0.17/0.52
Predicted Namibia	0	71 (97.3%)	0/0.48
n (class)	78	73	
false classifications (*n* (%))	2 (1.3%)	Averaged Group Accuracy	
correct classifications (*n* (%))	149 (98.7%)	Italy	1.00±0.00
n (latent variables)	4	Namibia	0.96±0.07

Models allowing invariably correct assignments are highlighted in red; models performing better than 99% and a standard deviation <5% are shown in blue.

Hence, in order to better understand these pair-wise relationships we further established 15 models comparing only two selected origins (Figure 3). This is particularly interesting as now for each case a different set of variables can be used to describe the inter-group variance in an optimized way. The approach also has the potential that separations between two particular deposits can be much clearer than in the multi-origin model. However, caution is advised as an unknown sample subjected to classification with only two classes will be assigned to one of the two available choices—regardless if it is indeed a member of one of the two. For instance, if a garnet from Madagascar is investigated for class membership in a pair-wise model trained with samples from Italy and Pakistan, it will be assigned to one of these classes.

From a purely mathematical point of view it is non-sense to use something in a model for which it is not trained for. However, in real-life this is a realistic scenario when testing an unknown sample especially if there is the suspicion of fraud. The unknown sample will be assigned to the class with the highest similarity. In case the sample does not belong to either of the groups tested for the Mahalanobis distance, which is a very useful metric for detecting multivariate outliers, for such a sample will likely be very large in comparison to samples that indeed belong to classes used for the training of the data set [36]. This is possible, for instance, through the QQ-plots (quantile-quantile) provided by the AI(OMICS)^{n} package, where Mahalanobis distance of all samples used for model building and the sample under investigation is plotted versus their respective Euclidean distance at a 95% confidence level ($\alpha = 5\%$). In general, if an unknown sample is initially run in the multi-origin data set, one can obtain a first estimate of the samples' country of origin. In a subsequent step this can be verified with the finer pair-wise models. In any case it is necessary to closely investigate the Mahalanobis distance for the sample in all models and to compare it with the typical confidence intervals obtained for true in-class samples.

Figure 3 clearly shows that even with only two depicted linear discriminators most origins can be clearly distinguished. In particular, Italy-Namibia, Italy-Iran, Italy-Madagascar, Namibia-Pakistan, Namibia-Iran, Namibia-Madagascar, Pakistan-Iran, and Iran-Madagascar can be separated essentially in full. For the remaining classification problems, it must not be forgotten that the graphs only represent 2D-projections of a multi-dimensional space and that the confusion matrix as well as additional metrics should be considered. Table 2 reveals that models for Madagascar typically exhibit a reasonable high accuracy which, however, varies considerably as a function of cross-validation. This can be explained to be a more frequent mis-assignment of only a few samples of the group, which due to the small group size accumulates to a large relative variation. Furthermore, Table 2 shows that the distinction of Italy vs. Russia and Russia vs. Iran works least as expressed by their low MCC of only 0.68 and 0.67, respectively. In both cases one group (Italy and Iran, respectively) shows a large number of mis-assigned samples. The first case is particularly interesting, as both groups are also present with comparable and larger sample numbers (Russia: 117, Italy: 78). While in this model Russian samples are mis-classified at a rate of only ~7% (8 out of 117), Italian samples are classified as being Russian demantoids at a rate of ~26% (20 out of 78). This means that the accuracy for both origins is different in this model, which has an impact on the interpretation of the results. Here, it is important to consider the testing hypothesis. This means that if the null hypothesis, H_0, is Russia (we assume the sample is from Russia) there is a 93% chance the result is correct, and the gem is indeed from Russia. More than 9 out of 10 tests will deliver a correct result. However, in the case H_0 is Italy (assuming the gem is from Italy) chances for the specimen being indeed an Italian demantoid are as low as 75% meaning one out of four tests will be mis-assigned. In such a case it is particularly important to critically evaluate all other models by attempting to rule out other solutions. Ultimately, there will always be a certain risk that a new sample is not well represented in the model. In such instances it is indicated to utilize additional approaches, such as analysis of inclusions (if present) to gain more confidence in the results obtained.

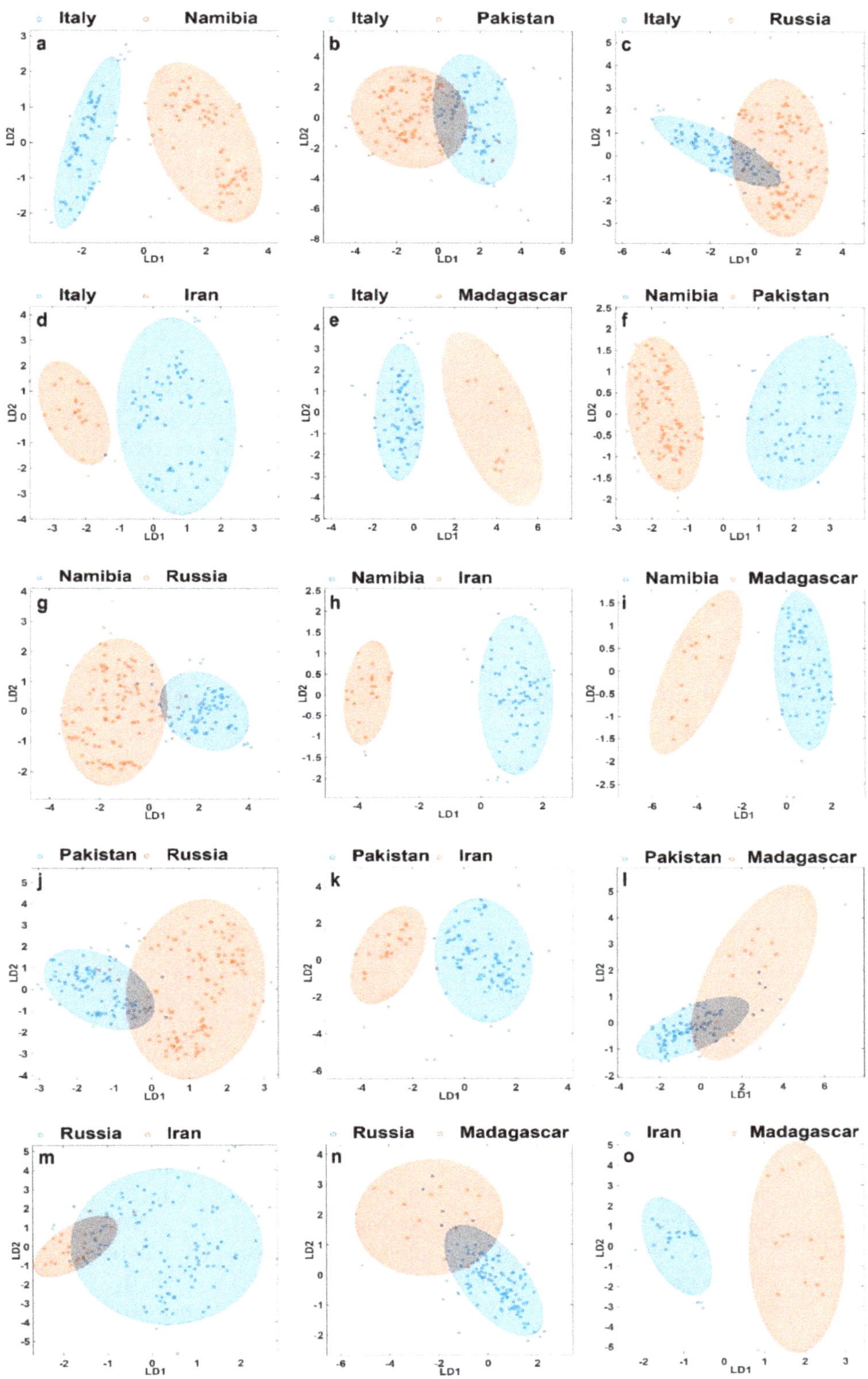

Figure 3. Representative PCA-LDAs for the pair-wise models (**a**) Italy-Namibia; (**b**) Italy-Pakistan; (**c**) Italy-Russia; (**d**) Italy-Iran; (**e**) Italy-Madagascar; (**f**) Namibia-Pakistan; (**g**) Namibia-Russia, (**h**) Namibia-Iran; (**i**) Namibia-Madagascar, (**j**) Pakistan-Russia, (**k**) Pakistan-Iran, (**l**) Pakistan-Madagascar, (**m**) Russia-Iran, (**n**) Russia-Madagascar, (**o**) Iran-Madagascar. Each dot represents a single element profile from an individual measurement after MC cross-validation. Axis shown represent the first two linear discriminants. Ellipses confer to the 95% confidence interval according to Hotellings T^2.

The confusion matrix for a representative model for pair-wise testing of the country of origin i shown in Table 2. Displayed are the model specific benchmark parameters accuracy (ACC), Matthew correlation coefficient (MCC), false classifications, correct classifications and the prevalence threshol (minimum calculated fraction/true fraction). Numbers in parenthesis are percent fractions for countrie and correct and false classifications, respectively. n(total) denotes the number of total element profile n(class) the number of element profiles, and n(latent variables) the number of principal component used to achieve 95% explained variance. In addition, cumulative R^2 and Q^2 as well as the average group accuracy (mean ± standard deviation) from the 100-fold MCCV are shown to allow judgemen of the model robustness. Diagonal elements in the confusion matrix represent the number (fractior of correctly assigned cases. Models highlighted in red and blue denote perfectly (exclusively correc classifications) respectively almost perfectly preforming models.

A closer inspection of the VIP scores, i.e., the natural logarithms of elements, of the individua elements provides insights into their importance for a particular model (Tables 3 and 4). Table provides the detailed results for all models generated in this study, with VIP1 having the highes impact on a model and VIP6 having the lowest. Overall, it can be seen that Al and Cr appear to pla a major role in the discriminant models, followed by Mg and Mn. However, it would be a wron conclusion to assume that V and Ti could be omitted from the analysis. Despite these two element playing a less important role in many of the generated models, they are essential components fo the discrimination of Russia-Madagascar, Pakistan-Iran, Italy-Iran, and Iran-Madagascar. Moreove valuable information about the importance of the elements for the within-class structure of the dat is gained as the VIP method takes into account all dimensions of the generated model. Hence, fo comprehensive analysis of the origin of a gem it is essential to record all six elements. In fact, it is th large range of concentrations found in natural gems which defines the origin of a specimen, whe enough elements are combined. This can be visualized best by boxplots of the natural logarithm of the individual elements as a function of the country of origin. In the best cases, an origin can b selected (or excluded) by simply plotting the concentration profile into the boxplot chart (Figure 4).

Table 3. Listing of the most important variables for each model ranked by their VIP-score.

Model	Type of Deposit	VIP1	VIP2	VIP3	VIP4	VIP5	VIP6
All origins		ln(Mg)	ln(Cr)	ln(Ti)	ln(Al)	ln(Mn)	ln(V)
Italy-Namibia	asbestos vs. skarn	ln(Al)	ln(Cr)	ln(Mn)	ln(Ti)	ln(Mg)	ln(V)
Italy-Pakistan	asbestos vs. asbestos	ln(Mn)	ln(Mg)	ln(Al)	ln(Cr)	ln(Ti)	ln(V)
Italy-Russia	asbestos vs. asbestos	ln(Mg)	ln(Al)	ln(V)	ln(Ti)	ln(Cr)	ln(Mn)
Italy-Iran	asbestos vs. asbestos	ln(Al)	ln(Ti)	ln(Cr)	ln(Mn)	ln(Mg)	ln(V)
Italy-Madagascar	asbestos vs. skarn	ln(Cr)	ln(Al)	ln(Mg)	ln(V)	ln(Ti)	ln(Mn)
Namibia-Pakistan	asbestos vs. skarn	ln(Al)	ln(Cr)	ln(Mg)	ln(Mn)	ln(V)	ln(Ti)
Namibia-Russia	asbestos vs. skarn	ln(Al)	ln(Mn)	ln(V)	ln(Ti)	ln(Cr)	ln(Mg)
Namibia-Iran	asbestos vs. skarn	ln(Mn)	ln(Mg)	ln(Cr)	ln(Ti)	ln(Al)	ln(V)
Namibia-Madagascar	skarn vs. skarn	ln(Cr)	ln(Mg)	ln(Mn)	ln(Al)	ln(Ti)	ln(V)
Pakistan-Russia	asbestos vs. asbestos	ln(Mg)	ln(Al)	ln(Cr)	ln(Ti)	ln(V)	ln(Mn)
Pakistan-Iran	asbestos vs. asbestos	ln(V)	ln(Cr)	ln(Al)	ln(Ti)	ln(Mg)	ln(Mn)
Pakistan-Madagascar	asbestos vs. skarn	ln(Cr)	ln(Mn)	ln(Mg)	ln(Al)	ln(Ti)	ln(V)
Russia-Iran	asbestos vs. asbestos	ln(Cr)	ln(Al)	ln(Mg)	ln(Mn)	ln(Ti)	ln(V)
Russia-Madagascar	asbestos vs. skarn	ln(V)	ln(Cr)	ln(Al)	ln(Mg)	ln(Ti)	ln(Mn)
Iran-Madagascar	asbestos vs. skarn	ln(Cr)	ln(V)	ln(Al)	ln(Ti)	ln(Mg)	ln(Mn)

Table 4. Scores of the individual elements as a function of their VIP-rank. Although it appears that Ti and V are overall of less importance Table 3 clearly indicates their important role for particular models.

Rank VIP	Al	Cr	Mg	Mn	Ti	V
1	4	5	3	2	0	2
2	4	5	3	2	1	1
3	4	3	4	2	1	2
4	3	1	1	3	7	1
5	1	2	4	1	6	2
6	0	0	1	6	1	8

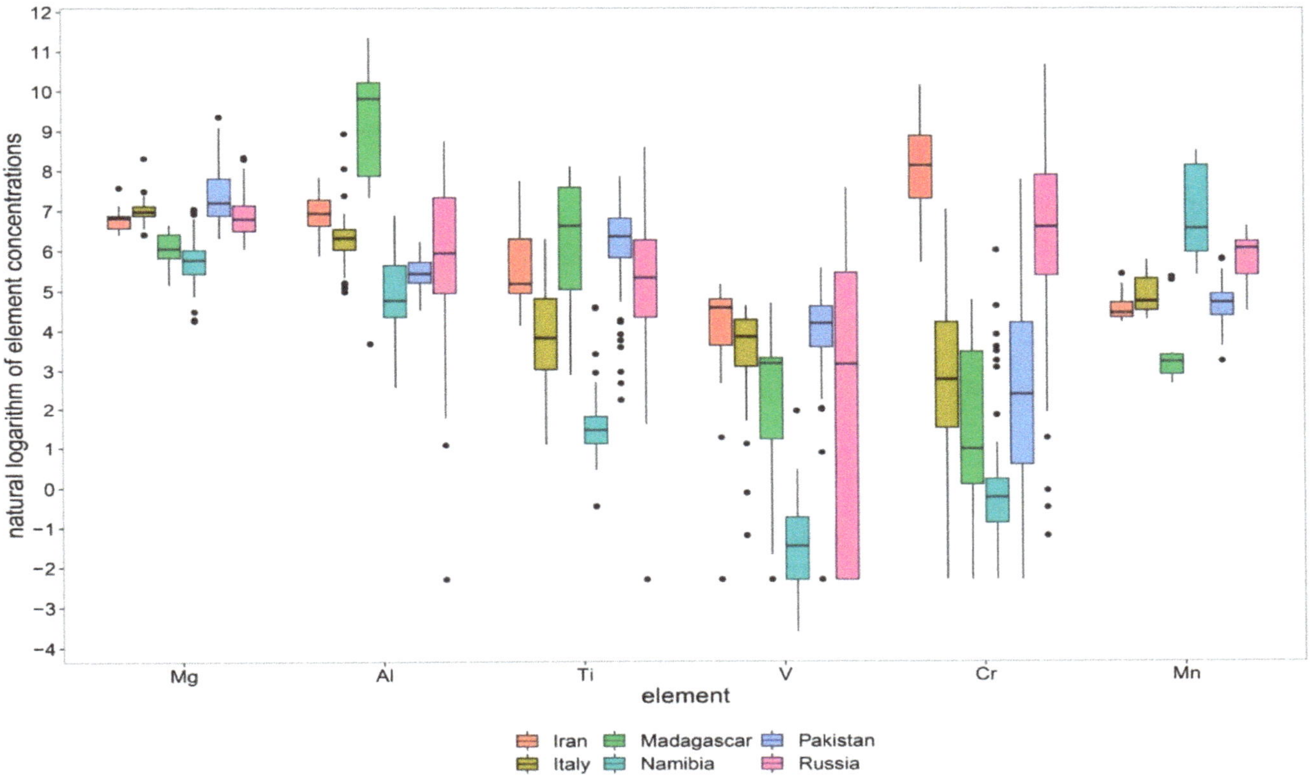

Figure 4. Boxplot showing the natural logarithm of all measured element concentrations. Samples are grouped and colored by origin. The graph displays the median, the first quartile, the third quartile as well as the maximum and minimum of each group. Outliers are represented as black dots.

4. Conclusions

We have successfully shown that the determination of country of origin is possible using a more quantitative approach than identification of inclusions, which is largely based on individual knowledge. By measuring minor and trace elements and applying modern chemometrics an excellent separation of five out of six countries of origin was achieved. The good separation with the complete dataset can even be improved when only a pair-wise comparison of two possible origins is used. This approach can be used when additional information, such as inclusions, can be used to rule out certain origins. Of course, the number of samples is essential for such studies and new samples will be added to the model as they become available. For the future it is planned to include even more elements into multivariate modelling. While it appears not rewarding to conduct large studies on elements that were shown to be not present in significant amounts in this and/or other studies, it seems promising to utilize major elements, such as Ca, Si, and Fe. As discussed for LA-ICP-MS proper calibration is difficult due to lacking standards and instrument limitations. This often prevents comparability of results obtained on different instruments. However, other technologies might be utilized in th future to obtain reproducible concentrations for these elements. Then, the data resulting from differen technologies can be fused and evaluated together. This approach has already been successfully applie to technologies as different as NMR-spectroscopy and multi-isotope ratio mass-spectrometry [37].

Still, any such approach is only as good as the samples that are available. In our study the numbe of samples from Madagascar and Iran was rather limited with 5 and 8, respectively, so in order t improve the models more samples have to be obtained from those localities. New localities ma be added as samples become available in significant numbers. Care has to be taken that selecte specimens properly reflect the natural variability at these locations. Nevertheless, when applyin such methods, the user has to be aware that likely not any sample population available will (a) eve represent all deposits in the world and (b) in significant numbers. Hence, interpretation shall alway be carried out having this in mind. When testing samples for fraud, analytical results often provid hints that something is not as advertised. It is advisable to back up results with independent method to get more evidence. Here, we have added one more tool to the analytical arsenal.

Author Contributions: Conceptualization, C.S. and S.S.; methodology, C.S. and S.S.; software, F.R. and S.S validation, S.G.B., F.R., S.S. and C.S.; formal analysis, S.G.B., F.R., S.S. and C.S.; investigation, S.G.B., F.R., S.S and C.S.; resources, C.S. and S.S.; writing—original draft preparation, C.S., S.S. and S.G.B.; writing—review an editing, C.S., S.S., S.G.B. and F.R.; visualization, S.G.B.; project administration, C.S. and S.S.; funding acquisitio C.S. and S.S. All authors have read and agreed to the published version of the manuscript.

Acknowledgments: The authors are grateful to Leopold Rößler (Austrian Gemmological Society), Gerhar Brandstetter, Ildar Latypov and Thomas Engeli for providing samples. Open Access Funding by the Universit of Linz.

References

1. World Sapphire Market Update 2020 Lotus Gemology. Available online: https://lotusgemolog com/index.php/library/articles/459-world-sapphire-market-update-2020-lotus-gemology (accessed o 28 September 2020).
2. Atkinson, D.; Kothavala, R.Z. Kashmir Sapphire. *Gems. Gemol.* **1983**, *19*, 64–76. [CrossRef]
3. Palke, A.C.; Saeseaw, S.; Renfro, N.D.; Sun, Z.; McClure, S.F. Geographic Origin Determination of Blu Sapphire. *Gems. Gemol.* **2019**, *55*, 536–579. [CrossRef]
4. Peretti, A.; Bieri, W.P.; Reusser, E.; Hametner, K.; Günther, D. Chemical Variations in Multicolore "Paraiba"-type Tourmalines from Brazil and Mozambique: Implications for Origin and Authenticit Determination. *Contrib. Gemol.* **2009**, *9*, 1–77.
5. Kolesar, P.; Tvrdý, J. *Zarenschätze*; Rainer Bode: Haltern, Germany, 2006; p. 204.
6. Burlakov, A.; Burlakov, E. Demantoid from Russia. *InColor* **2017**, *36*, 40–46.
7. Reif, S. Green Dragon Mine Demantoid from Namibia. *InColor* **2017**, *36*, 48–52.
8. Schwarzinger, C. Determination of demantoid garnet origin by chemical fingerprinting. *Monats. Chem.* **201** *50*, 907–912. [CrossRef]
9. Adamo, I.; Bocchio, R.; Diella, V.; Pavese, A.; Vignola, P.; Prosperi, L.; Palanza, V. Demantoid from Va Malenco Italy: Review and Update. *Gems. Gemol.* **2009**, *45*, 280–287. [CrossRef]
10. Pezzotta, F. Andradite from antetezambato, North Madagascar. *Miner. Rec.* **2010**, *41*, 209–229.
11. Pezzotta, F.; Adamo, I.; Diella, V. Demantoid and Topazolite from Antetezambato, Northern Madagasca Review and New Data. *Gems. Gemol.* **2011**, *47*, 2–14. [CrossRef]
12. Palke, A.C.; Pardieu, V. Demantoid from Baluchistan Province in Pakistan. *Gems. Gemol.* **2014**, *50*, 302–303
13. Adamo, I.; Bocchio, R.; Diella, V.; Caucia, F.; Schmetzer, K. Demantoid from Balochistan, Pakistan Gemmological and Mineralogical Characterization. *J. Gemmol.* **2015**, *34*, 428–433. [CrossRef]
14. Hairapetian, V.; Pelckmans, H. Emerging Mineral Finds from Northern Iran. *Rocks Miner.* **2017**, *6*, 540–54 [CrossRef]
15. Du Toit, G.; Mayerson, W.; Van Der Bogert, C.; Douman, M.; Befi, R.; Koivula, J.I.; Kiefert, L. Demantoid fro Iran. *Gems. Gemol.* **2006** *42*, 131.

6. Schwarzinger, C.; Engeli, T. LA-ICP-MS spectrometry and its use in gemology. *Ital. Gemol. Rev.* **2019**, *8*, 50–52.

7. Jochum, K.P.; Weis, U.; Stoll, B.; Kuzmin, D.; Yang, Q.; Raczek, I.; Jacob, D.E.; Stracke, A.; Birbaum, K.; Frick, D.A.; et al. Determination of reference values for NIST SRM 610–617 glasses following ISO guidelines. *Geostand. Geoanal. Res.* **2011**, *35*, 397–429. [CrossRef]

8. Rüll, F.; Schwarzinger, S. AI(OMICS)n: A MatLab Based Expert System for Chemometrics and Data Fusion. 2020. Available online: https://gitlab.com/ai-omics/ai-omics/ (accessed on 20 October 2020).

9. Kruskal, W.H.; Wallis, W.A. Use of ranks in one-criterion variance analysis. *J. Am. Stat. Assoc.* **1952**, *47*, 583–621. [CrossRef]

0. Xu, Q.; Liang, Y.Z.; Du, Y.-P. Monto Carlo cross-validation for selecting a model and estimating the prediction error in multivariate calibration. *J. Chemom.* **2004**, *18*, 112–120. [CrossRef]

1. Henningsson, M.; Sundbom, E.; Armelius, B.Å.; Erdberg, P. PLS model building: A multivariate approach to personality test data. *Scand. J. Psychol.* **2001**, *42*, 399–409. [CrossRef]

2. Johansson, E.; Eriksson, L.; Sandberg, M.; Wold, S. QSAR Model Validation. In *Molecular Modeling and Prediction of Bioactivity*; Springer: Boston, MA, USA, 2000. [CrossRef]

3. Wold, S.; Sjöström, M.; Eriksson, L. PLS-regression: A basic tool of chemometrics. *Chemom. Intell. Lab. Syst.* **2001**, *58*, 109–130. [CrossRef]

4. Matthews, B.W. Comparison of the predicted and observed secondary structure of T4 phage lysozyme. *Biochim. Biophys. Acta* **1975**, *405*, 442–451. [CrossRef]

5. Baldi, P.; Brunak, S.; Chauvin, Y.; Andersen, C.A.; Nielsen, H. Assessing the accuracy of prediction algorithms for classification: An overview. *Bioinformatics* **2000**, *16*, 412–424. [CrossRef] [PubMed]

6. Balayla, J. Prevalence threshold (φe) and the geometry of screening curves. *PLoS ONE* **2020**, *15*, e0240215. [CrossRef] [PubMed]

7. R Core Team. *R: A Language and Environment for Statistical Computing*; R Foundation for Statistical Computing: Vienna, Austria, 2020; Available online: https://www.R-project.org/ (accessed on 20 October 2020).

8. Huelin, S.R.; Longerich, H.P.; Wilton, D.H.; Fryer, B.J. The determination of trace elements in Fe–Mn oxide coatings on pebbles using LA-ICP-MS. *J. Geochem. Expl.* **2006**, *91*, 110–124. [CrossRef]

9. van den Berg, R.A.; Hoefsloot, H.C.; Westerhuis, J.A.; Smilde, A.K.; van der Werf, M.J. Centering, scaling, and transformations: Improving the biological information content of metabolomics data. *BMC Genom.* **2006**, *7*, 142. [CrossRef] [PubMed]

0. Kolb, P.; Bindereif, S.G.; Rüll, F.; Schwarzinger, S. unpublished data, to be published elsewhere.

1. Mahalanobis, P.C. On the Generalized Distance in Statistics. *Proc. Natl. Inst. Sci. India* **1936**, *2*, 49–55.

2. Moawed, S.A.; Osman, M.M. The robustness of binary logistic regression and linear discriminant analysis for the classification and differentiation between dairy cows and buffaloes. *Int. J. Stat. Appl.* **2017**, *7*, 304. [CrossRef]

3. Barragués, J.I.; Morais, A.; Guisasola, J. (Eds.) *Probability and Statistics: A didactic Introduction*; CRC Press: Boca Raton, FL, USA, 2014.

4. Hotelling, H. The generalization of Student's ratio. *Ann. Math. Statist.* **1931**, *2*, 360–378. [CrossRef]

5. Platt, J.C.; Cristianini, N.; Shawe-Taylor, J. Large Margin DAGs for Multiclass Classification. *Adv. Neural Inf. Process. Syst.* **2000**, *12*, 547.

6. Ghorbani, H. Mahlanobis distance and its application for detecting multivariate outliers. *Facta Univ. (NIS) Ser. Math. Inform.* **2019**, *34*, 583. [CrossRef]

7. Bindereif, S.G.; Brauer, F.; Schubert, J.-M.; Schwarzinger, S.; Gebauer, G. Complementary use of ^{1}H NMR and multi-element IRMS in association with chemometrics enables effective origin analysis of cocoa beans (*Theobroma cacao* L.). *Food Chem.* **2019**, *299*, 125105. [CrossRef]

3

Gem Corundum Deposits of Greece: Geology, Mineralogy and Genesis

Panagiotis Voudouris [1,*], Constantinos Mavrogonatos [1], Ian Graham [2], Gaston Giuliani [3], Vasilios Melfos [4], Stefanos Karampelas [5], Vilelmini Karantoni [4], Kandy Wang [2], Alexandre Tarantola [6], Khin Zaw [7], Sebastien Meffre [7], Stephan Klemme [8], Jasper Berndt [8], Stefanie Heidrich [9], Federica Zaccarini [10], Anthony Fallick [11], Maria Tsortanidis [1] and Andreas Lampridis [1]

1 Faculty of Geology and Geoenvironment, National and Kapodistrian University of Athens, 15784 Athens, Greece; kmavrogon@geol.uoa.gr (C.M.); m.tsortanidi@hotmail.com (M.T.); andreaslampridis18@gmail.com (A.L.)
2 School of Biological, Earth and Environmental Sciences, University of New South Wales, Sydney, NSW 205 Australia; i.graham@unsw.edu.au (I.G.); kandy.wang@student.unsw.edu.au (K.W.)
3 Université Paul Sabatier, GET/IRD et Université de Lorraine, C.R.P.G./C.N.R.S., 54501 Vandoeuvre, CEDE France; giuliani@crpg.cnrs-nancy.fr
4 Faculty of Geology, Aristotle University of Thessaloniki, 54124 Thessaloniki, Greece; melfosv@geo.auth.gr (V.M.); wilelminikar@gmail.com (V.K.)
5 Bahrain Institute for Pearls & Gemstones (DANAT), WTC East Tower, P.O. Box 17236 Manama, Bahrain; stefanos.karampelas@danat.bh
6 UMR GeoResources, Faculté des Sciences et Technologies, Université de Lorraine, 54506 Vandoeuvre-lès-Nancy, France; alexandre.tarantola@univ-lorraine.fr
7 CODES Centre of Ore Deposit and Earth Sciences, University of Tasmania, Tas 7001, Australia; khin.zaw@utas.edu.au (K.Z.); sebastien.meffre@utas.edu.au (S.M.)
8 Institut für Mineralogie, Westfälische-Wilhelms Universität Münster, 48149 Münster, Germany; stephan.klemme@uni-muenster.de (S.K.); jberndt@uni-muenster.de (J.B.)
9 Mineralogisch-Petrographisches Institut, Universität Hamburg, 20146 Hamburg, Germany; stefanie.heidrich@mineralogie.uni-hamburg.de
10 Department of Applied Geosciences and Geophysics, University of Leoben, 8700 Leoben, Austria; federica.zaccarini@unileoben.ac.at
11 Isotope Geosciences Unit, S.U.E.R.C., Glasgow G75 0QF, UK; anthony.fallick@glasgow.ac.uk
* Correspondence: voudouris@geol.uoa.gr

Abstract: Greece contains several gem corundum deposits set within diverse geological settings mostly within the Rhodope (Xanthi and Drama areas) and Attico-Cycladic (Naxos and Ikaria islands tectono-metamorphic units. In the Xanthi area, the sapphire (pink, blue to purple) deposits are stratiform, occurring within marble layers alternating with amphibolites. Deep red rubies in the Paranesti-Drama area are restricted to boudinaged lenses of Al-rich metapyroxenites alternating with amphibolites and gneisses. Both occurrences are oriented parallel to the ultra-high pressure/high pressure (UHP/HP) Nestos suture zone. On central Naxos Island, colored sapphires are associated with desilicated granite pegmatites intruding ultramafic lithologies (plumasites), occurring either within the pegmatites themselves or associated metasomatic reaction zones. In contrast, on southern Naxos and Ikaria Islands, blue sapphires occur in extensional fissures within Mesozoic metabauxites hosted in marbles. Mineral inclusions in corundums are in equilibrium and/or postdate corundum crystallization and comprise: spinel and pargasite (Paranesti), spinel, zircon (Xanthi), margarite, zircon, apatite, diaspore, phlogopite and chlorite (Naxos) and chloritoid, ilmenite, hematite, ulvospinel, rutile and zircon (Ikaria). The main chromophore elements within the Greek corundums show a wide range in concentration: the Fe contents vary from (average values) 1099 ppm in the blue sapphires of Xanthi, 424 ppm in the pink sapphires of Xanthi, 2654 ppm for Paranesti rubies, 4326 ppm for the Ikaria sapphires, 3706 for southern Naxos blue sapphires, 4777 for purple and 3301 for pink sapphire from Naxos plumasite, and finally 4677 to 1532 for blue to colorless sapphires from Naxos plumasites, respectively. The Ti concentrations (average values) are very low in rubies from Paranesti (41 ppm), with values of 2871 ppm and 509 in the blue and pink sapphires of Xanthi, respectively, of 1263 ppm for the Ikaria blue sapphires, and 520 ppm, 181 ppm in Naxos purple, pink sapphires, respectively. The blue to colorless sapphires from Naxos plumasites contain 1944 to 264 ppm Ti, respectively. The very high Ti contents of the Xanthi blue sapphires may reflect submicroscopic rutile inclusions. The Cr (average values) ranges from 4 to 691 ppm in the blue, purple and pink colored corundums from Naxos plumasite, is quite fixed (222 ppm) for Ikaria sapphires, ranges from 90 to 297 ppm in the blue and pink sapphires from Xanthi, reaches 9142 ppm in the corundums of Paranesti, with highest values of 15,347 ppm in deep red colored varieties. Each occurrence has both unique mineral assemblage and trace element chemistry (with variable Fe/Mg, Ga/Mg, Ga/Cr and Fe/Ti ratios). Additionally, oxygen isotope compositions confirm their geological typology, i.e., with, respectively $\delta^{18}O$ of $4.9 \pm 0.2‰$ for sapphire in plumasite, 20.5‰ for sapphire in marble and 1‰ for ruby in mafics. The fluid inclusions study evidenced water free CO_2 dominant fluids with traces of CH_4 or N_2, and low CO_2 densities (0.46 and 0.67 g/cm^3), which were probably trapped after the metamorphic peak. The Paranesti, Xanthi and central Naxos corundum deposits can be classified as metamorphic sensu stricto (s.s.) and metasomatic, respectively, those from southern Naxos and Ikaria display atypical magmatic signature indicating a hydrothermal origin. Greek corundums are characterized by wide color variation, homogeneity of the color hues, and transparency, and can be considered as potential gemstones.

Keywords: corundum megacrysts; ruby; sapphire; plumasite; metamorphic-metasomatic origin; Greece

---

1. Introduction

Rubies and sapphires, the two different varieties of the mineral corundum, are among the most popular gemstones used in jewelry. The color in corundum varies from brown, pink to pigeon-blood-red, orange, yellow, green, blue, violet etc. The main chromophore trace elements in corundum are Cr, Fe, Ti and V. Cr^{3+} produces pink and red, while Fe^{2+}–Ti^{4+} pairs produce blue. Distinction of primary gem quality-rubies and sapphires in magmatic and metamorphic is mainly based on the classification schemes presented by Garnier et al. [1], Giuliani et al. [2] and Simonet et al. [3]. Corundum deposits were mainly formed in three periods, the Pan-African orogeny (750–450 Ma), the Himalayan orogeny (45–5 Ma) and the Cenozoic continental rifting with related alkali basalt volcanism (65–1 Ma) [2,4–10]. Magmatic and metamorphic ruby and sapphire deposits related to the Pan-African orogeny are present in southern Madagascar, East Africa, South India and Sri Lanka; those related to the Himalayan orogeny mostly occur in marbles from Central and Southeast Asia. Alkali basalt-related sapphires occur in Australia, northern and central Madagascar, Nigeria, Cameroon, French Massif Central and southeast Asia.

The use of trace element content of gem corundums, in association with their oxygen isotopic signature, is an effective tool in characterizing and interpreting their origin, and can also be used for the exploration, classification and comparison of gem corundum deposits especially those of disputed origin (e.g., placers, xenoliths, etc.) [2,7,11–17].

The presence of corundum in Greece (Xanthi, Naxos and Ikaria islands) has been known of since several years ago, but mostly from unpublished company reports [18–20]. It was only during the 80s, when detailed mineralogical, petrological studies investigated the geological environment of formation for the emery-hosted corundum on Naxos, Samos and Ikaria island [21], and the marble-hosted corundum in Xanthi area [22,23]. Iliopoulos and Katagas [24] described metamorphic conditions of formation for corundum-bearing metabauxites from Ikaria island. Fieldwork during

1998–2010 by the first two authors resulted in additional findings of corundum megacrysts i Greece, mostly within the Rhodope (Xanthi and Drama areas) and Attico-Cycladic (Naxos and Ikari islands) tectono-metamorphic units [25–29]. According to these authors, both occurrences in th Rhodope are classified as metamorphic deposits related to meta-limestones (Xanthi deposit) and mafi granulites (Paranesti deposit), based on the scheme of Simonet et al. [3], and supported the theory o their formation during the high temperature-medium pressure retrograde metamorphic episode o carbonates and eclogitic amphibolites during the Cenozoic collision along the Nestos Suture Zone. Th plumasite-hosted corundum deposits at Naxos island and those hosted in metabauxites from Ikari island are classified as metasomatic and metamorphic, respectively [27,29]. Recently, Wang et al. [30 compiled multi-analytical geochemical, mineralogical and petrological studies on rubies from Paranes Drama area, discussing their origin and comparing the deposit to those in other mafic-ultramafi complexes especially that of Winza in Tanzania.

In addition to the above localities, corundum has been also reported from the Koryfes Hil prospect, which is a telescoped porphyry-epithermal system hosted within Tertiary granitoids in th Kassiteres-Sapes area, on the south of the Rhodope massif (Figure 1). Hydrothermal corundum occur within a transitional sericitic-sodic/potassic alteration of a quartz-feldspar porphyry, which form the root zone of an advanced argillic lithocap comprising diaspore, topaz, pyrophyllite and alunit supergroup minerals [31]. Hydrothermal corundum forms up to 3-mm-large aggregates of dee blue-colored crystals that are separated from quartz (and K-feldspar/albite) by a sericite rim. Electro microprobe analyses indicate minor amounts of Fe (up to 0.34 wt. % FeO), Mg (up to 0.02 wt. % MgO) Ti (up to 0.13 wt. % TiO_2) and Cr (up to 0.12 wt. % Cr_2O_3) substituting for Al in the structur The Koryfes Hill corundum occurrence belongs to those present in porphyry deposits elsewher and is attributed to the magmatic class of corundum deposits, according to the classification schem of Simonet et al. [3]. It was suggested by Voudouris [31] that the assemblage corundum-sericite a Koryfes Hill formed as a result of rapid cooling of ascending magmatic-hydrothermal solutions unde pressures between 0.6 and 0.3 kb. However, due to its small grain size, which probably rather preven its use as a gemstone this corundum occurrence will not be further discussed in the present paper.

The aim of this work is to present detailed information on the geology, mineralogy, geochemistr and fluid characteristics involved in the formation of rubies and sapphires in Greece, by expanding o the previous work of the Wang et al. [30], which was only focused on the Paranesti area. New lase ablation inductively coupled plasma mass spectrometry (LA-ICP-MS) and electron probe microanalysi (EPMA) data on corundum will help define their origin and compare them with other famou occurrences elsewhere.

2. Materials and Methods

For this study, the following 18 samples, collected by the first author, were investigated: Naxo samples (Nx1a, Nx1b, Nx2a, Nx2b, Nx3a, Nx3b, Nx4, Nx5a, Nx5b), Ikaria samples (Ik1a, Ik1b), Xanth samples (Go1a, Go1b, Go5a, Go5b) and Drama samples (Dr1a, Dr1b, Dr2), ranging in color fron colorless, to pink, purple, red and blue (Table 1). All occurrences are primary, which is rare for blu sapphire and relatively rare for ruby (S. Karampelas, pers. communication).

Thirty-five thin and ten thin-and-polished sections of corundum-bearing samples and host rock were studied by optical and a JEOL JSM 5600 scanning electron microscope (SEM) equipped wit back-scattered imaging capabilities, respectively, at the Department of Mineralogy and Petrology a the University of Athens (Athens, Greece). Quantitative analyses were carried out at the "Euger F. Stumpfl" Laboratory installed at the University of Leoben, Austria, using a Superprobe Jeol JX 8200 wavelength-dispersive electron microprobe (WDS), and at the Department of Mineralogy an Petrology, University of Hamburg (Hamburg, Germany) using a Cameca-SX 100 WDS. Analytica conditions were as follows: at Leoben, 20 kV accelerating voltage, 10 nA beam current, 1 μm bear diameter. Peak and backgrounds counting times were 20 and 10 s for major and 40 and 20 s for trac elements, respectively. The X-ray lines used were AlKα, SiKα, TiKα, GaKα, FeKα, CeLα, VKα, MgKα

CrKα, and CaLα. The standards used were: corundum for Al, wollastonite for Si and Ca, rutile for Ti, GaAs for Ga, ilmenite for Fe, synthetic V for V, chromite for Mg and Cr. At Hamburg, accelerating voltage of 20 kV, a beam current of 20 nA and counting time of 20 s. The X-ray lines used were: AlKα, SiKα, TiKα, GaKα, FeKα, CeLα, VKα, MgKα, CrKb, and CaLα. The standards used were: andradite and vanadinite (for Si, Ca and V), and synthetic Al_2O_3 (for Al), TiO_2 (for Ti), Fe_2O_3 (for Fe), Cr_2O_3 (for Cr), Ga_2O_3 (for Ga) and MgO (for Mg). Corrections were applied using the PAP online program [32].

Table 1. Sample description of the studied Greek corundum crystals.

Sample	Variety	Color	Host Rock	Locality
Dr1a, Dr1b, Dr2	Ruby	Red	Pargasite-schist	Paranesti, Drama
Go1a, Go1b	Sapphire	Blue-colorless	Marble	Gorgona, Xanthi
Go5a	Sapphire	Pink-purple	Marble	Gorgona, Xanthi
Go5b	Sapphire	Pink	Marble	Gorgona, Xanthi
Nx1a, Nx2a, Nx2b, Nx3a, Nx3b, Nx4	Sapphire	Blue-colorless	Plumasite	Naxos (Kinidaros)
Nx1b	Sapphire	Purple	Plumasite	Naxos (Kinidaros)
Nx5a	Sapphire	Blue	Metabauxite	Naxos (Kavalaris)
Nx5b	Sapphire	Pink	Plumasite	Naxos (Kinidaros)
Ik1a, Ik1b	Sapphire	Blue	Metabauxite	Ikaria

LA-ICP-MS analyses were conducted at the CODES ARC Centre of Excellence in Ore Deposits of the University of Tasmania, Australia, and the Institute of Mineralogy, University of Münster, Germany. Analytical conditions were as follows: The LA-ICP-MS analysis at the CODES was performed under standard procedures using a New Wave UP-213 Nd: YAG Q-switched Laser Ablation System coupled with an Agilent HP 4500 Quadrupole ICP-MS. The international standard NIST 612 was used as the primary standard to calculate concentrations and correct for ablation depth, and the basaltic glass BCR-2 was used as the secondary standard. Analyses were standardized on the theoretical Al content for corundum at Al 529,227 ppm, with an error in analytical precision of 2–3%. Corundum sample ablation at the University of Münster was done with a pulsed 193 nm ArF excimer laser (Analyte G2, Photon Machines). A repetition rate of 10 Hz and an energy of ~4 J/cm^2 were used throughout the entire session. The beam spot diameter was set to 35 μm. Trace element analysis has been carried out with an Element 2 mass spectrometer (ThermoFisher). Forward power was 1250 W and reflected power <1 W, gas flow rates were 1.2 L/m for He carrier gas, 0.9 L/m and 1.2 L/m for the Ar-auxiliary and sample gas, respectively. Argon cooling gas flow rate was set to 16 L/min. Before starting analysis, the system was tuned on a NIST 612 reference glass measuring ^{139}La, ^{232}Th and $^{232}Th^{16}O$ to get stable signals and high sensitivity, as well as low oxide rates ($^{232}Th^{16}O/^{232}Th < 0.1\%$) during ablation. A total of 30 elements were quantitatively analyzed using the NIST 612 glass as an external standard and ^{27}Al as internal standard, which had previously been determined by electron microprobe. Overall time of a single analysis was 60 s (20 s for background, 40 s for peak after switching laser on). Concentrations of measured elements were calculated using the Glitter software [33,34]. Standard reference glass BHVO-2G and BIR1-G were analyzed in order to monitor for precision and accuracy in the silicate phases in the course of this study. The obtained results match the published range of concentrations given in the GeoReM database (version 23).

Fluid inclusions studies were carried in a total of 14 double-polished sections prepared at the Department of Mineralogy, Petrology and Economic Geology of the Aristotle University of Thessaloniki (Thessaloniki, Greece). Routine heating and freezing runs were performed at a LINKAM THM-600/TMS 90 heating–freezing stage coupled to a Leitz SM-LUX-POL microscope. Calibration of the stage was achieved using organic standards with known melting points, and the H_2O phase transition from ice to liquid; the precision of the measurements, including reproducibility, was ±0.2 °C. The SoWat program [35] was used to process fluid inclusion data.

Oxygen isotope analyses were performed at the Scottish Universities Environmental Research Centre, Glasgow, Scotland. The oxygen isotope composition of corundum was obtained using a modification of the CO_2 laser fluorination system similar to that described by Sharp [36], which was similar applied by Giuliani et al. [14]. The method involves the complete reaction of ~1 mg of ground corundum. This powder is heated by a CO_2 laser, with ClF_3 as the fluorine reagent. The released oxygen is passed through an in-line Hg-diffusion pump before conversion to CO_2 on platinized graphite. The yield is then measured by a capacitance manometer. The gas-handling vacuum line is connected to the inlet system of a dedicated VG PRISM 3 dual-inlet isotope-ratio mass spectrometer. Duplicate analyses of corundum samples suggest that precision and accuracy are $\pm 0.1‰$ (1σ). All oxygen isotope ratios are shown in $\delta^{18}O$ (‰) relative to Vienna standard mean ocean water (VSMOW).

3. Geological Setting

3.1. Regional Geology

The Hellenide orogen formed as a result of the collision between the African and Eurasian plates above the north-dipping Hellenic subduction zone from the Late Jurassic to the present [37–39]. From north to south, it consists of three continental blocks (Rhodopes, Pelagonia, and Adria-External Hellenides) and two oceanic domains (Vardar and Pindos Suture Zones). In the Aegean region, continuous subduction of both oceanic and continental lithosphere beneath the Eurasian plate since the Early Cretaceous resulted in a series of magmatic arcs from the north (Rhodope massif) to the south (Active South Aegean Volcanic Arc) [40]. The geodynamic evolution of Hellenides is characterized by a collisional phase taking place during the Mesozoic, followed by a continuous southward slab retreat in a back-arc setting, started at the Eocene but still ongoing, that triggered large-scale extension concomitant with thrusting at the southern part of the Hellenic domain [39,41]. The gem corundum deposits in Greece are spotted in two tectono-metamorphic units of the greater Hellenides Orogen: the Rhodope- and the Attico-Cycladic Massifs [29,30,42].

3.1.1. Rhodope Massif

The Rhodope Massif is considered part of the European continental margin [41,43]. It is a heterogeneous crustal body composed in its eastern and central parts of two sub-domains (Figure 1): the Northern Rhodope Domain and the Southern Rhodope Core Complex (both separated by the Nestos thrust fault and the Nestos Suture Zone) [41]. The Northern Rhodope Domain consists of a Lower unit of high-grade basement including orthogneisses derived from Permo-Carboniferous protoliths; this Unit includes four metamorphic core complexes (the Arda, Biala Reka-Kechros, and Kesebir-Kardamos migmatitic domes) which are considered to be equivalent to orthogneisses in the Southern Rhodope Core Complex (SRCC) [41,44]. The latter also consists of Triassic marbles with intercalated amphibolitic and metapelitic rocks [44,45].

The upper tectonic unit of the Northern Rhodope Domain includes high-grade basement rocks that have both continental and oceanic affinities, and with protoliths ranging in age from Neoproterozoic through Ordovician, and Permo-Carboniferous to Early Cretaceous [45]. The rocks of the Intermediate unit experienced high- to ultra-high-pressure metamorphism with subsequent high-grade amphibolite-facies overprint [46]. The Rhodope Massif has undergone a polycyclic alpine orogeny including UHP-eclogitic metamorphism, followed by HP granulite-facies, amphibolite-facies and finally by greenschist facies metamorphic events starting from Jurassic (~200–150 Ma) and lasting up to the Oligocene [47,48]. In the Rhodope region, syn-orogenic exhumation of the metamorphic pile initiated in the early Eocene (~55 Ma) and core complex extension followed the Cretaceous syn-metamorphic SW-directed thrusting [41,43,49,50]. The metamorphic core complexes were progressively exhumed along several ductile- to brittle shear zones, active from ~42–12 Ma [49,51].

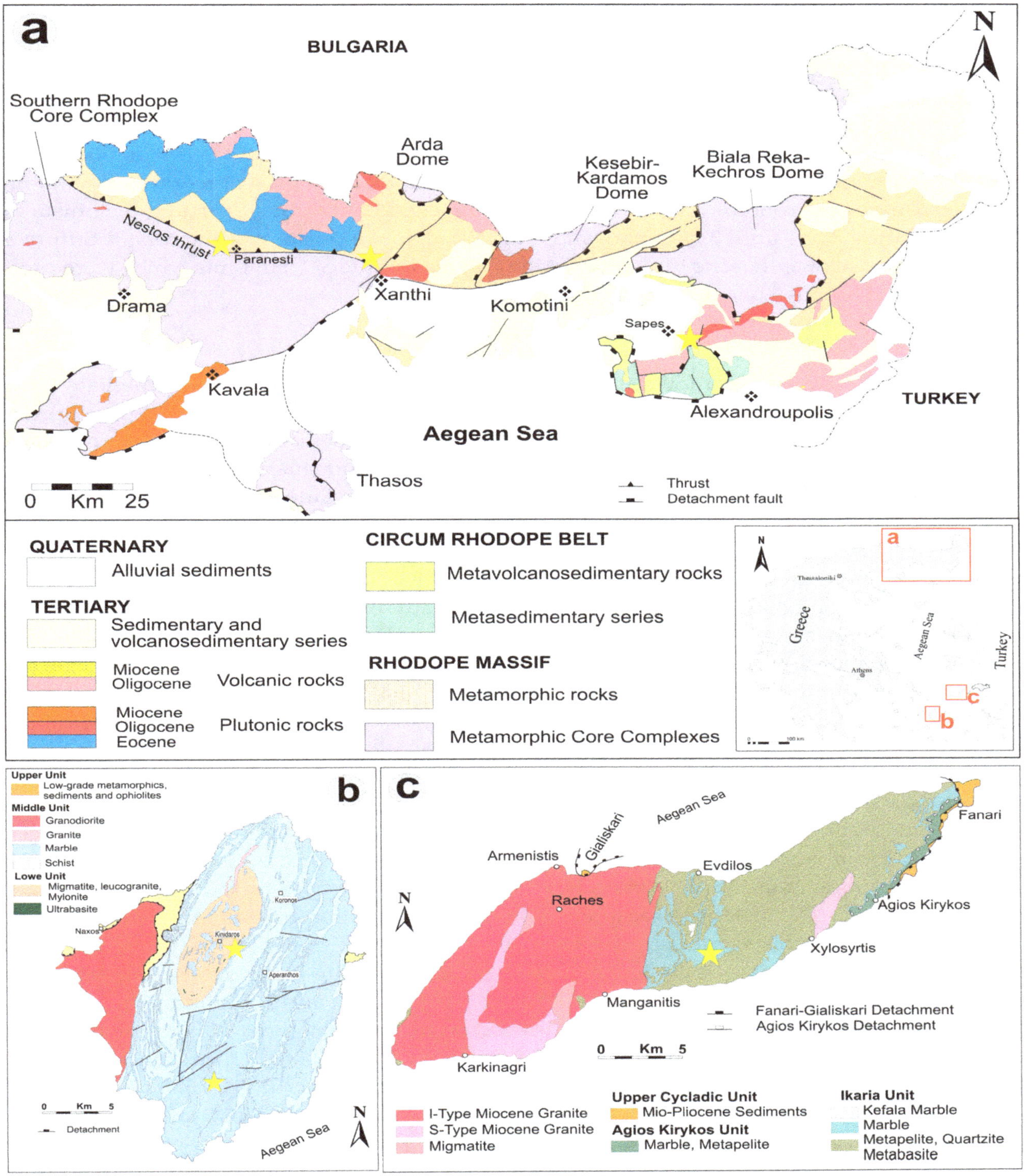

Figure 1. (**a**) Simplified geological map of eastern Rhodope massif, Greece showing the main tectonic zones and the distribution of Cenozoic igneous rocks (after Melfos and Voudouris [52] and references therein), showing location of corundums at Xanthi, Paranesti and Kassiteres/Sapes areas (stars); (**b**) Geological map of Naxos island showing locations of sapphires in plumasites at Kinidaros and meta-bauxites at Kavalaris Hill (stars) (after Duchêne et al. [53]; modified after Ottens and Voudouris [54]); (**c**) Geological map of Ikaria showing the location of blue sapphire in Atsida area (after Beaudoin et al. [55]; modified after Ottens and Voudouris [54]). Yellow stars mark the studied corundum occurrences.

Separating the upper and lower tectonic units in the Rhodope complex, the Nestos Sutur Zone a NW trending shear zone was extensively studied by Papanikolaou and Panagopoulos [56 Krenn et al. [57], Nagel et al. [58] and Turpaud and Reischmann [45] and is composed from th bottom to the top of the following lithological types: (a) a lower 1-km-thick highly sheared "mélange zone, consisting of amphibolites, garnet-kyanite schists, migmatites, orthogneisses and marbles; (b) 1-km-thick sequence of augen-gneisses; (c) two layers of marbles intercalated with amphibolites an mylonitic amphibolites; (d) a layer of ortho-gneisses, characterized by biotite-gneisses with highl migmatized base (Sidironero). Gautier et al. [48] suggest a continuous thrusting tectonism alon the Nestos Suture Zone until 33 Ma ago, and Nagel et al. [58] indicate that the Nestos Suture Zon constitutes the base of an Eocene thrusting wedge that also includes UHP units which were probabl merged with the Nestos Zone during the thrusting event.

3.1.2. Attico-Cycladic Massif

The Cyclades consist of a lowermost Pre-Alpidic Basement Unit, the intermediate Cycladi Blueschist Unit, and the uppermost Pelagonian Unit [59]. The Blueschist Unit represents polymetamorphic terrane which tectonically overlies the basement gneiss and consists of metamorphosed volcano-sedimentary sequence of clastic metasedimentary rocks, marbles, calc-schist and mafic and felsic meta-igneous rocks [38,39,60]. The Cyclades have been exhumed sinc the Eocene as metamorphic core complexes formed in low-, medium-, and/or high-temperatur environments [50]. An Eocene (~52–53 Ma) high-pressure eclogite- to blueschist-facies metamorphisn was followed by syn-orogenic exhumation of the blueschists in a cold retrograde path and then b early Oligocene amphibolite to greenschist facies, and finally, by high-temperature medium-pressur metamorphism and associated migmatites in the early Miocene [37,60–63]. The amphibolite t greenschist metamorphic event occurred during extension-related exhumation and was coeval wit back-arc extension at the Rhodopes in northern Greece [64–66]. Exhumation of the Cycladic rocks a metamorphic core complexes was accommodated during the Oligocene–Miocene by several ductil to brittle detachment systems. The extensional event also allowed for various granitoids (granite granodiorite, leucogranite) to be intruded throughout the Cyclades between 15 and 7 Ma [64].

On Naxos island, a migmatitic dome (pre-alpine basement), is overlain by the Blueschist uni containing alternating layers of marbles, schists and gneisses, and by the upper Pelagonian tectoni unit (Figure 1b). Blueschist relics at the SE part of the island indicate temperatures from 400 t 460 °C at minimum pressures of 7–9 kbar [21] for the high-pressure/low-temperature (HP/LT) Eocen metamorphic event. The metamorphic grade of the Oligocene–Miocene event increases from 400 °C at 6 kbar in marbles, schists, gneisses and amphibolites at the SE of the island, towards the core o the dome, where migmatites in the leucogneissic and amphibolite-facies sillimanitic schists rocks suggest temperatures up to 700 °C at 6–8 kbar [21,67–70]. Jansen and Schuiling [67] distinguishe six metamorphic zones (I–VI) at Naxos island based on mineral isograds. Maximum temperature are 420 °C for zone I, inferred by the presence of diaspore; 420 to 500 °C (appearance of biotite) fo zone II; 500 °C to 540–580 °C (disappearance of chloritoid) for zone III, 540–580 °C to 620–660 °C (kyanite-sillimanite transition) for zone IV, 620–660 °C to 660–690 °C (onset of melting) for zon V; and >690 °C for zone VI. The Naxos migmatitic core is surrounded by a discontinuous block o ultramafic (meta-peridotites) lenses, representing a thrust zone along which the metamorphic comple lies on top of the pre-alpine bedrock [71]. The peridotites underwent high-P metamorphism and then after cooling amphibolite-facies conditions (T~600 °C), were finally re-heated and metamorphose together with the country rocks. According to Jansen and Schuiling [67], pegmatites penetrating th ultramafic bodies in the sillimanite stability zone, are desilicated and composed of phlogopite, anorthite corundum, chlorite, zoisite, tourmaline and beryl. Anorthite crystals are composed almost entirel of anorthite (98% An), while margarite can be found in places. According to Andriessen et al. [72 and Katzir et al. [71], aplites and pegmatites are of Early Miocene age (19–20 Ma), and were forme as a result of the crystallization of an in situ anatectic liquid, that was produced during the high-T metamorphic event. They represent a system that channeled fluids through the leucogneissic core to the lower metamorphic unit and resulted in the metasomatization of peridotites and in situ recrystallization of the peridotitic blackwalls. Corundum-bearing emery deposits occur in the metamorphic zones I–V following the increase in metamorphic grade from SE towards NW to the migmatite core [21,73].

Ikaria Island (Figure 1c), similarly to Naxos, represents a Miocene migmatite-cored metamorphic core complex, where peak-metamorphic temperatures range from 390 °C in the upper parts of the structure down to 625 °C in the core of the dome in the vicinity of migmatites and S-type granite [55,74]. Three main tectonic units are distinguished and are, from bottom to top, the Ikaria, Agios Kirykos and Fanari units, limited by two major shear zones [55,75,76]. The non-metamorphic Fanari unit consists of Miocene to Pliocene sandstones, conglomerates and ophiolitic molasses. The intermediate Agios Kirykos unit (previously called the Messaria unit) consists of alternating marble and metapelite layers, metamorphosed in greenschist-facies conditions. Finally, the Ikaria unit is composed of amphibolite-facies (ca. 6–8 kbar and 600–650 °C) metasediments including micaschist and marble layers and minor metabasites. Peak-metamorphic conditions were retrieved from the basal parts of the succession. Two synkinematic granitic bodies intrude the Ikaria unit: (a) an I-type Bt-granite (Raches granite) in the western part of the island, with K-Ar ages of 22.7 Ma [64]; and (b) an S-type Bt-Ms-granite (Xilosirtis granite) in the southern part of the island, with Rb-Sr ages of 18.1 Ma [64]. Liati and Skarpelis [77], Iliopoulos and Katagas [24], and Iliopoulos [75] studied the Ikaria metabauxitic rocks hosted in marbles of the Ikaria unit, and recorded a Jurassic age for the formation of bauxitic deposits, and upper greenschist to lower amphibolite-facies conditions for their metamorphism.

3.2. Local Geology

3.2.1. Xanthi

The Gorgona-Stirigma corundum occurrence is stratiform and distributed in marble layers (Figure 2a), reaching up to 50 m width, alternating with eclogitic amphibolites and gneisses [22,27,42]. The chemical composition of the marbles varies widely, highly locally impure marbles, rich in alumino-silicate minerals, are characterized by a mineralogical assemblage composed of calcite, dolomite, corundum, spinel, pyrophyllite, muscovite, paragonite, amphibole, zoisite, margarite, chlorite, olivine, serpentine, phlogopite, rutile, titanite, anorthite, garnet and Ni-tourmaline [22,23,27,42]. Spinel occurs as a rim around corundum crystals. The color of spinel ranges from blue to green and brown. Liati [22] defined the following contact assemblages: calcite-anorthite-zoisite, calcite-corundum-spinel, calcite-corundum-zoisite and margarite-zoisite-anorthite-chlorite. The corundum crystals occur within phyllosilicate-rich micro-shear zones along marble layers. Sapphire is of a pink, orange, purple to blue color, usually of tabular or barrel-shaped euhedral form and reaches sizes of up to 4 cm (Figure 3a–c). In some cases, blue corundum alters to green spinel. Red corundum has also been reported in this area, mainly from zoisite-bearing amphibolites [23], but could not be identified in the present study.

3.2.2. Paranesti

The Paranesti corundum occurrence is stratiform, oriented parallel to the main regional foliation, and distributed mainly in boudin-like lenses of amphibole schist (Figure 2b) [27,30,42]. The outcrops on the surface are spotted west of Perivlepto village at one hillside and one roadside location [30]. In the first site, the corundum-bearing amphibole schist lenses are surrounded by a narrow clinochlore schist zone and occur together with boudins of corundum-kyanite-amphibole schists. The schists are intruded by white-colored quartz-feldspar-mica-garnet pegmatitic veins, which do not present characteristics of desilication (Figure 2c). In the second site, the lenses are hosted by amphibolites, intercalated with kyanite-bearing gneisses and kyanite-amphibole-chlorite schists. Again, the transition from amphibolitic lithology towards corundum mineralization is made through a clinochlore-rich schist zone. Ruby crystals from Paranesti, range in size from <1 mm to 50 mm size (average size 5–10 mm, and are of pale pink to deep red color [30]. Their morphology is mainly flat tabular with basal planes paralleling the orientation of the main regional foliation [30], and less commonly prismatic and barrel-shaped (Figure 3d,e). They occur together with pargasite, which is the main amphibole of the assemblage, forming a similar image with Tanzanian rubies which are surrounded by zoisite. The ruby crystals coexist locally with kyanite (Figure 3f).

Figure 2. Field photographs demonstrating the geological setting of the ruby and sapphire occurrences in Greece. (**a**) Marble layers alternating with eclogitic amphibolites at Gorgona/Xanthi; (**b**) Boudinaged hornblende schists, intercalated within eclogitic amphibolites, marbles and metapegmatites at Paranesti/Drama; (**c**) The main ruby working at Paranesti. The pegmatite body at the lower part of the photograph shows no desilification; (**d**,**e**) Ultrabasic rocks between the migmatite dome-like core and the overlying kyanite-sillimanite grade schists and marbles in the two plumasite localities east of Kinidaros, Naxos island; (**f**) Corundum-bearing plumasite at the contact to ultrabasites at the locality shown in Figure 2e; (**g**) Panoramic view of the metabauxite bearing marbles at Kavalaris Hill, southern Naxos island; (**h**) Panoramic view for the metabauxite-bearing marbles of Ikaria unit; (**i**) Corundum-bearing metabauxite lenses within the Ikaria unit marbles.

3.2.3. Naxos

The corundums studied from Naxos island occur in two localities, representing different geological environments.

(a) In the central part of the island and about 2 km east and southeast of Kinidaros, two corundum-bearing plumasites (desilicated pegmatites) were studied (Figure 2d–f) [27,29,42]. The meta-peridotites occurred initially as lenses of few to several tens of meters size and are surrounded by phlogopitite, which formed at the contact between the metaperidotites and the intruding pegmatites. Today, most of the plumasites and metaperidotites in these two localities occur as loose fragments mainly due to erosion and agricultural activities in the area.

Figure 3. Field photographs and handspecimens demonstrating gem corundum crystals of Greece: (**a**–**c**) Pink, blue, to purple sapphires within Xanthi marbles; (**d**,**e**) Deep red-colored ruby with both tabular and barrel-shaped crystals within pargasite schist from Paranesti/Drama; (**f**) Ruby associated (replacing) kyanite, Paranesti/Drama; (**g**) Barrel-shaped blue and colorless sapphires hosted within anorthite-rich desilicated pegmatite (plumasite) and associated with phlogopite and tourmaline (schorlite) from the first Kinidaros locality, Naxos island; (**h**) Purple to pink sapphire within plumasite the later rimmed by phlogopite at the left part, second Kinidaros locality, Naxos island; (**i**) Blue sapphire in fissures within metabauxite at Kavalaris Hill, southern Naxos island; (**j**,**k**) Blue sapphires and margarite as free growing crystals of probably hydrothermal origin within fissure of metabauxites, Ikaria island.

The plumasites are surrounded by an up to 20-cm-wide zone of phlogopite and chlorite (blackwal (Figure 3g). Within the plumasites, colorless to blue, purple and pink corundum may occur either a isolated crystals within the plagioclase matrix, and/or associated with tourmaline, and phlogopit (Figure 3g,h). Purple to pink sapphires have been also found entirely enclosed in phlogopitite in th blackwalls. The sapphire crystals are barrel-shaped, display macroscopic color zoning from colorles cores to blue rims or pink cores to purple rims and reach sizes up to 3 cm [27,42].

(b) In the southern part of the island, close to Kavalaris Hill (Figure 2g), corundu represents a single mineral rock termed "corundite" by Feenstra and Wunder [73], which wa formed by the dissociation of former diasporites, in meta-carstic bauxites during prograd regional metamorphism. This first corundum occurrence in metabauxites also recorded a corundum-in isograd (T~420–450 °C and P~6–7 kbar, [73]), separates diasporites with th assemblage diaspore-chloritoid-muscovite-paragonite-calcite-hematite-rutile from emery characterize by corundum-chloritoid-muscovite-paragonite-margarite-hematite-rutile. Corundum and chloritoi intergrowths occur as a rim in metabauxites, while the corundum crystals occur mainly in the contac between marbles and metabauxites. The corundum from this locality does not occur in well-shape crystals; it is, rather, massive, and of blue color (Figure 3i).

3.2.4. Ikaria

The metabauxite occurrence is located on mountain Atheras (Figure 2h) [20,75]. Alread Iliopoulos [75], reported the presence of corundum as mineralogical constituent of meta-bauxit (emery) lenses occurring within dolomitic and calcic marble, which lies on top of gneiss, both bein part of the Ikaria unit. In this area, four main layers of metabauxites, of up to 1 m width, wer detected (Figure 2i). However, the corundum megacrysts described by Voudouris et al. [27,42 represent a second, late, generation of corundum formation within the emeries, since they fill togethe with margarite extensional fissures and networks of veins discordant to the metabauxite foliatio (Figure 3j,k) [27,42]. The corundums are deep-blue in color, tabular to euhedral, reaching sizes up to cm and are accompanied by Fe-chlorite, hematite, rutile and diaspore.

4. Mineralogy

The Paranesti rubies are dull to transparent, exhibiting clear parting and polysynthetic twinnin The rubies are associated with edenitic hornblende and tremolite, and rimmed by margarite, muscovit chlorite and chromian spinel (Figure 4a,b and Figure 5a,b). The ruby crystals are surrounde by a margarite rim. Spinel appears as the main component of primary inclusions in the rubie The association of ruby in this deposit is pargasite, chlorite (mainly clinochlore and rarely nimite margarite, tremolite and/or monazite, while for the host rocks the assemblage is pargasite, anorthit clinozoisite, chlorite and monazite. The Paranesti rubies are rimmed by Cr-bearing spinel and als include ulvospinel. Corundum is in equilibrium with amphibole, while in a later, retrograde stag margarite-muscovite, Cr-bearing spinel, orthoclase, chlorite, smectite and vermiculite, rim and/o replace corundum. Cr-bearing spinel is replaced by chlorite.

The corundums from Xanthi are transparent, with very clear parting and fine cracks. Polysyntheti twinning is common. Corundum may be isolated within the carbonate matrix (Figure 4c) or rimme by amesite (an Al-rich serpentine member), margarite, titanite or by brown and green spinel. Blu sapphires are zoned, with alternating deep blue and colorless domains (Figure 4c). This zoning, o irregular color distribution in the sapphires, is attributed to different Fe and Ti contents in the crystal (see next paragraph). Ni-bearing brown tourmaline accompanying corundum in the paragenesi contains up to 4.4 wt. % NiO, much higher than the Ni content reported in tourmaline from Samos b Henry and Dutrow [78].

The sapphire crystals from Naxos present strong pleochroism with their color ranging fro dark blue to colorless (Figure 4d–f). They are transparent, with smaller or bigger inner fracture Hexagonal barrel-shaped corundum crystals exhibit blue color zoning. The blue color of corundum i bserved either as a blue core surrounded by a white rim or as a blue-zoned outer rim surrounding colorless core. Polysynthetic twinning is observed along the colorless parts of the crystals. apphires are associated with anorthite, phlogopite, zircon, margarite, muscovite, tourmaline and hlorite (Figures 4d–f and 5c–i). The sapphires are enveloped by an assemblage consisting of ligoclase/labradorite-orthoclase-muscovite and then by phlogopite, chlorite (Figure 5c–h). In sample Jx1, blue sapphire associated with phlogopite is rimmed by an intergrowth of oligoclase, anorthite, rthoclase and muscovite. Margarite, zircon, diaspore, phlogopite and chlorite are included in blue apphire; however, it is considered that margarite, diaspore and chlorite are retrograde minerals ostdating corundum formation (this study). In samples Nx4 and Nx4a, blue sapphire includes hlogopite, zircon and apatite and surrounded by anorthite, muscovite and tourmaline with minor mounts of monazite and chlorite. Phlogopite and anorthite are replaced by muscovite, margarite and hlorite. Tourmaline postdates anorthite and muscovite.

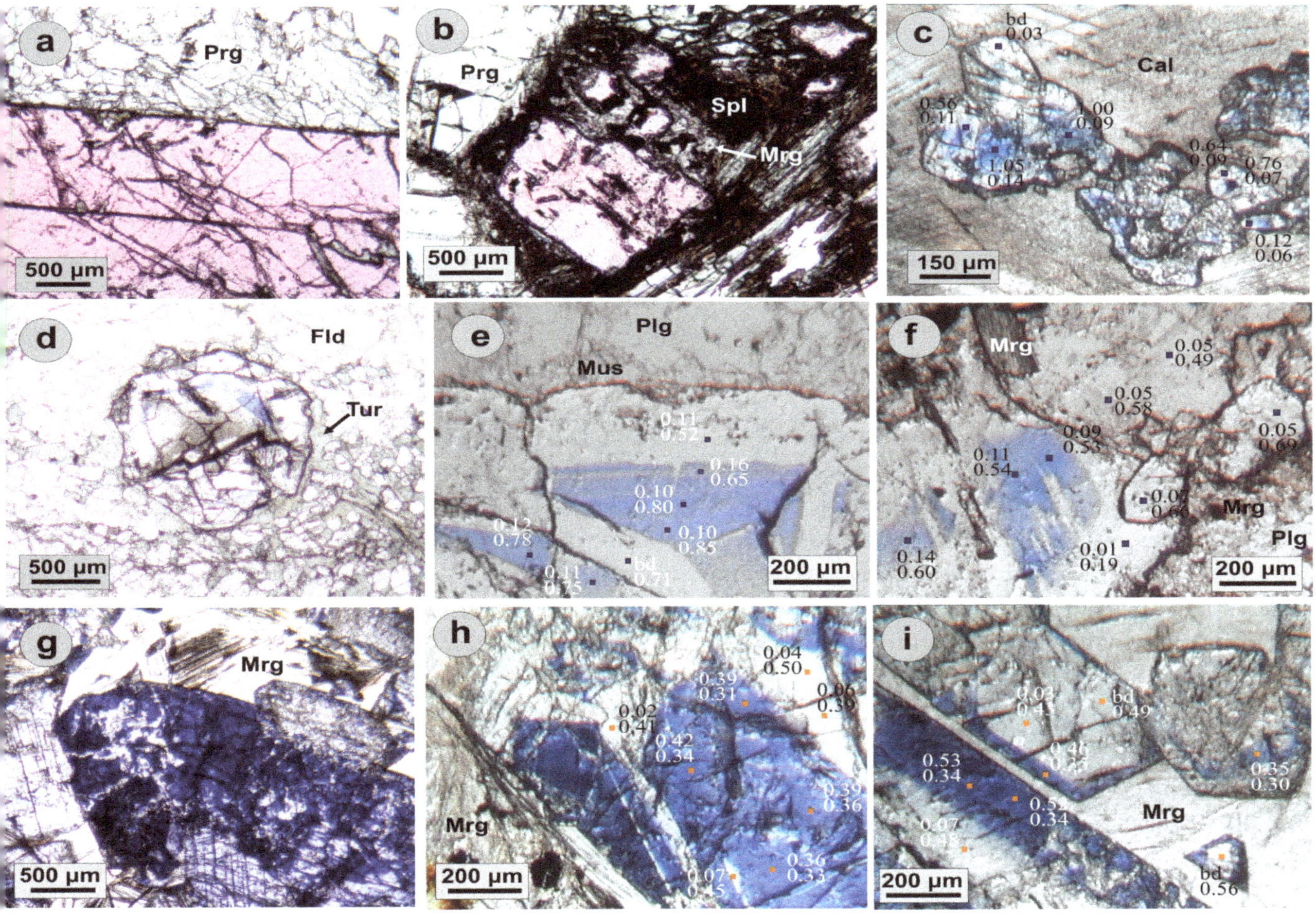

Figure 4. Microphotographs (plane polarized light) demonstrating mineralogical assemblages, color- and chemical zoning of corundum-bearing assemblages from Greece. Values in microphotographs (**c**,**e**,**f**,**h**,**i**) correspond to TiO_2 and Fe_2O_3 content in wt. % (upper and lower values, respectively) as obtained by electron probe microanalyses. (**a**,**b**) Ruby associated with pargasite (Prg) from Paranesti, Drama area. Note replacement of ruby by spinel (Spl) and margarite (Mrg) in Figure 4b (sample Dr1a); (**c**) blue to colorless sapphires included in calcite, Xanthi area (sample Go1b); (**d**,**e**) Zoned blue to colorless sapphires within anorthite (Plg) and rimmed by tourmaline (Tur) and muscovite (Mus), Kinidaros, Naxos island (sample Nx4); (**f**) Sapphire with blue core and colorless rim associated with margarite (Mrg) and anorthite (Plg), Kinidaros, Naxos island (sample Nx3); (**g**) Idiomorphic deep blue colored sapphire crystals surrounded by margarite, Ikaria island (sample Ik1b); (**h**,**i**) Zoned blue to colorless idiomorphic sapphires, surrounded by margarite, Ikaria island (sample Ik2).

On Ikaria Island, blue sapphires in the metabauxites are accompanied by the assemblag margarite + chloritoid + hematite/ilmenite + rutile (Figures 4g–i and 5j–l). Chloritoid fill fracture and also occurs as inclusions in margarite. The retrogression has continued to low temperature as suggested by the presence of diaspore replacing corundum (Figure 4j–l) and chlorite replacin chloritoid. Corundum forms colorless cores and deep blue rims (Figure 4h,i). While transparent, thei blue color is quite dark, with colorless spots appearing all over their surface. Corundum include ilmenite, hematite, ulvospinel, rutile and zircon. In addition, a Zn-bearing green colored spinel, whic forms solid solution between hercynite and gahnite, has been reported with chlorite and/or margarit by Iliopoulos [75] and Iliopoulos and Katagas [24].

In sample Ik1a, Ik1b, and Ik2 blue sapphire is surrounded by margarite, chlorite and green-spine Chlorite fills fractures and represents latest deposition. Chloritoid, ilmenite, ulvospinel, rutil Ti-hematite and zircon occur as inclusions in sapphire. Sapphire is associated with hematite, zirco and chloritoid and is replaced by diaspore and margarite.

5. Mineral Chemistry

5.1. EPMA

Electron probe microanalyses (EPMA) for the Greece corundums are presented in Table 2. EPM data from Paranesti/Drama rubies showed an extreme enrichment in Cr_2O_3, which reaches up t 2.65 wt. %, with the higher concentrations being consistent with intense red-colored grains. Beyon Cr_2O_3, the Paranesti rubies are characterized by moderate Fe_2O_3 vaues (up to 0.44 wt. %) and b traces of MgO, Ga_2O_3, SiO_2, TiO_2 and CaO, which range from below detection up to 0.17, 0.15, 0.0 0.05 and 0.02 wt. %, respectively. Finally, V_2O_3 content is mostly below detection, but in some case values up to 0.07 wt. % were detected.

Colored sapphires from Gorgona/Xanthi are characterized by elevated contents of TiO_2, Mg and Fe_2O_3. The blue color varieties show the highest TiO_2 content, which reaches up to 1.05 wt. % while MgO and Fe_2O_3 reach up to 0.27 and 0.22 wt. %, respectively. In the pink-colored grains, TiO and Fe_2O_3 contents are significantly lower (up to 0.17 and 0.06 wt. %, respectively) in comparison t the blue ones, but MgO content increases up to 0.36 wt. %. SiO_2 and CaO values are generally belo 0.08 wt. %, but in one analysis SiO_2 is 0.51 wt. %, probably indicating a fine-grained silicate inclusio V_2O_3 content tends to be higher in the pink varieties rather than the colorless to blue, with value ranging from below detection to 0.06 and 0.01 wt.%, respectively. Pink-colored grains contain mor than double Ga_2O_3 content (up to 0.17 wt. %) compared to the blue ones (up to 0.07 wt. %). Finall colorless grains diasplay very low TiO_2 content (from below detection up to 0.12 wt. %) and simila Fe_2O_3 contents to the blue domains.

In colored sapphires from the Naxos plumasite, Fe_2O_3 is the dominant impurity. In the blu vareties, the Fe_2O_3 content reaches up to 0.93 wt.%, and decreases significantly in the colorless (u to 0.69 wt. %), pink- (up to 0.50 wt. %) and the purple-colored grains (up to 0.34 wt. %). A simila decreasing trend is remarked for the Ga_2O_3 content, which reaches up to 0.18 wt. % in the white t blue sapphires, and decreases slightly in the pink and purple varieties, reaching up to 0.15 and 0.10 w %, respectively. On the other hand, TiO_2 and MgO seem both to follow a different trend: the highes values for both oxides characterize the purple-colored grains (up to 0.58 and 0.54 wt. %, respectivel Blue grains display intermediate values (up to 0.38 and 0.49 wt. %), followed by the pink grains (up t 0.27 and 0.35 wt. %), respectively. TiO_2 content in the colorless areas ranges from below detection t 0.12 wt. % similarly to the blue domains. The Cr_2O_3 content is up to 0.1 wt. % in the colorless to blu and purple varieties, and almost doubles in the case of pink sapphires (up to 0.18 wt. %). Traces o SiO_2, V_2O_3 and CaO were also detected in all samples (up to 0.13, 0.09 and 0.03 wt. %, respectively).

Blue sapphires from Ikaria island appear enriched in TiO_2 and Fe_2O_3 values, which reach up t 0.69 and 0.57 wt. %, respectively. SiO_2 shows a wide range from below detection to 0.44 wt. %, whil MgO content is constantly low (up to 0.05 wt. %). The Cr_2O_3 and Ga_2O_3 content are almost fixed eaching up to 0.14 wt. %. Traces of V_2O_3 and CaO were also detected (up to 0.04 and 0.02 wt. %, espectively). Colorless domains are characterized by very low TiO_2 values, which range from below letection up to 0.12 wt. %.

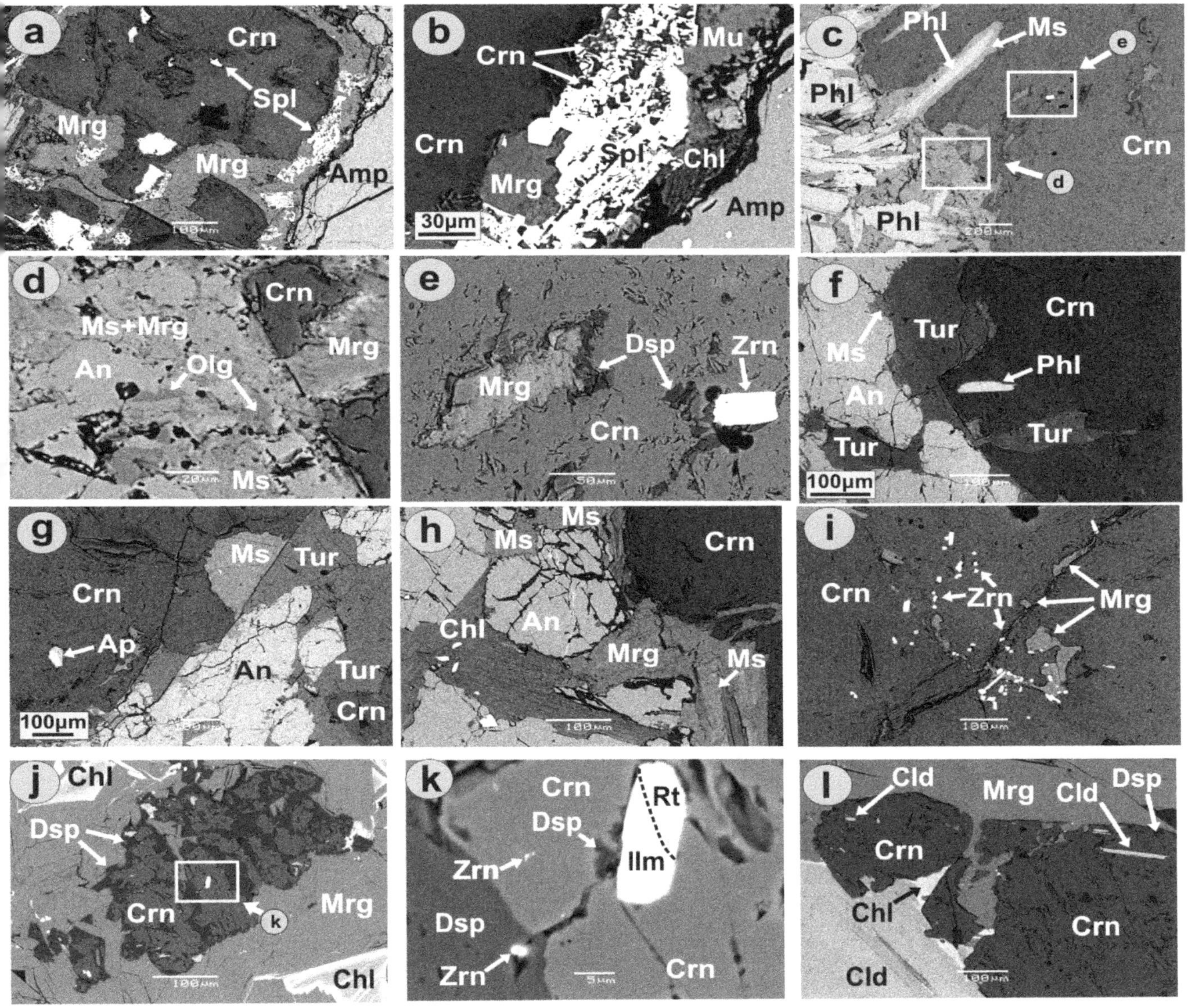

Figure 5. Microphotographs (SEM-BSE images) demonstrating mineralogical assemblages of corundum-bearing assemblages from Greece. (**a**,**b**) Ruby (Crn) including and rimmed by chromian spinel (Spl), margarite (Mrg), muscovite (Mu), chlorite (Chl) and pargasitic hornblende (Amp) (sample Dr1, Paranesti/Drama); (**c**,**d**) Blue sapphire (Crn) in association with anorthite (An) and phlogopite (Phl) replaced by muscovite (Mus), margarite (Mrg), and oligoclase (Olg) (Naxos island, sample Nx1); (**e**) Zircon (Zrn) is included in sapphire, diaspore (Dsp) and margarite (Mrg) fill fractures and are retrograde minerals. Naxos island, sample Nx1); (**f**,**g**) Blue sapphire including phlogopite and fluorite (Fl) is associated with anorthite (An) and replaced by tourmaline (Tur) and muscovite (Mus) (Naxos island, sample Nx4); (**h**) Blue sapphire and anorthite are rimmed by muscovite, margarite and chlorite (Chl) the late probably replacing phlogopite (Naxos island, sample Nx4a); (**i**) Blue sapphire including zircon is replaced by margarite (Naxos island, sample Nx4a); (**j**–**l**) Blue sapphire associated with margarite includes and replaces chloritoid (Cld) and are retrogressed to diaspore and chlorite, respectively. Zircon and Ilmenite (Ilm), rutile (Rt) are included in corundum (Ikaria island, sample Ik2).

Table 2. Chromophores and key trace elements electron probe microanalyses (EPMA) (wt. %) of the Greek corundum crystals.

Sample	Color	ppm	MgO	Al_2O_3	SiO_2	CaO	TiO_2	V_2O_3	Cr_2O_3	Fe_2O_3	Ga_2O_3	Total
Dr1a (n = 8)	red	aver	bd	96.96	0.03	0.01	0.01	0.01	1.82	0.39	0.02	99.23
		min	bd	96.52	0.03	bd	bd	bd	1.74	0.34	bd	98.80
		max	bd	97.39	0.04	0.02	0.01	0.01	1.99	0.43	0.09	99.66
Dr1b (n = 13)	red	aver	bd	96.98	0.04	0.01	0.01	0.01	2.14	0.37	0.02	99.55
		min	bd	96.13	0.02	bd	bd	bd	1.48	0.33	bd	98.95
		max	bd	97.58	0.09	0.01	0.01	0.01	2.65	0.44	0.07	100.10
Dr2 (n = 8)	red	aver	0.02	99.79	0.01	0.01	0.01	0.01	0.32	0.19	0.05	100.41
		min	bd	99.18	bd	bd	bd	bd	0.18	0.15	bd	99.78
		max	0.17	100.65	0.03	0.02	0.05	0.074	0.43	0.21	0.15	101.30
Go1a (n = 11)	blue-colorless	aver	0.07	98.02	0.09	0.01	0.59	0.01	0.04	0.11	0.01	98.94
		min	0.02	97.37	bd	bd	bd	bd	bd	0.07	bd	98.19
		max	0.27	98.61	0.51	0.01	0.94	0.01	0.08	0.14	0.04	99.36
Go1b (n = 15)	blue-colorless	aver	0.04	99.21	0.03	0.02	0.56	0.01	0.01	0.11	0.02	100.00
		min	bd	98.46	0.01	bd	bd	bd	bd	0.031	bd	99.70
		max	0.10	99.73	0.07	0.08	1.05	0.01	0.05	0.21	0.07	100.38
Go5b (n = 14)	pink	aver	0.05	100.93	0.01	0.01	0.06	0.01	0.03	0.04	0.05	101.18
		min	bd	100.22	bd	bd	bd	bd	bd	bd	bd	100.40
		max	0.36	101.58	0.03	0.02	0.17	0.06	0.06	0.06	0.17	101.99
Ik1a (n = 8)	blue	aver	bd	97.88	0.35	0.01	0.23	0.01	0.05	0.36	0.03	98.90
		min	bd	97.08	0.24	bd	bd	bd	bd	0.27	bd	98.16
		max	bd	98.45	0.44	0.016	0.53	0.04	0.13	0.50	0.12	99.36
Ik1b (n = 16)	blue	aver	0.01	99.13	0.04	0.01	0.25	0.01	0.03	0.43	0.03	99.91
		min	bd	98.44	0.02	bd	bd	bd	bd	0.14	bd	99.61
		max	0.01	99.68	0.06	0.01	0.68	0.04	0.10	0.57	0.14	100.21
Nx1a (n = 8)	blue-colorless	aver	0.01	98.25	0.09	0.01	0.22	bd	0.04	0.31	0.02	98.94
		min	bd	97.85	0.07	bd	bd	bd	bd	0.23	bd	98.77
		max	0.05	98.80	0.12	0.02	0.38	bd	0.12	0.45	0.08	99.42
Nx1b (n = 14)	purple	aver	0.12	99.80	0.02	0.01	0.27	0.01	0.01	0.27	0.04	100.54
		min	bd	98.08	bd	bd	0.08	bd	bd	0.20	bd	98.79
		max	0.54	100.57	0.07	0.02	0.58	0.06	0.04	0.34	0.11	101.28
Nx2b (n = 22)	blue-colorless	aver	0.10	100.26	0.02	0.01	0.07	0.01	0.01	0.45	0.05	100.98
		min	bd	93.73	bd	bd	bd	bd	bd	0.136	bd	99.21
		max	0.49	101.81	0.07	0.03	0.19	0.08	0.04	0.69	0.18	101.96
Nx3b (n = 15)	blue-colorless	aver	0.01	98.98	0.05	0.01	0.08	0.01	0.01	0.66	0.01	99.80
		min	bd	98.33	0.01	bd	0.01	bd	bd	0.19	bd	99.30
		max	0.02	99.47	0.13	0.01	0.16	0.01	0.04	0.93	0.05	100.23
Nx4 (n = 16)	blue-colorless	aver	0.01	98.09	0.09	0.01	0.10	0.01	0.01	0.72	0.01	99.02
		min	bd	97.69	0.05	bd	bd	bd	d	0.52	bd	98.67
		max	0.01	98.55	0.12	0.01	0.17	0.01	0.01	0.85	0.05	99.43
Nx5b (n= 12)	pink	aver	0.10	100.64	0.01	0.01	0.06	0.01	0.05	0.35	0.04	101.27
		min	bd	97.20	bd	bd	bd	bd	bd	0.16	db	98.91
		max	0.35	100.7	0.06	0.02	0.27	0.09	0.18	0.45	0.15	102.70

bd = below detection; aver = average; min = minimum; max = maximum.

5.2. LA-ICP-MS

The trace element results for averages, ranges and critical ratios are listed in Tables 3 and 4. Analyzed rubies from Paranesti/Drama are particularly enriched in Cr (up to 15,347 ppm), and less so in Fe (up to 4348 ppm), with the highest values for both elements coming from the most dark-colored grains. The Mg, Ga and V values are consistently low, (up to 31, 24 and 6 ppm, respectively). The low V contents result in consistent very low V/Cr ratios, typically <1. Calcium and Ti values display wide variations between 117–1444 ppm and 8–148 ppm, respectively, with higher values characterizing the most light-colored grains. Other trace elements are close or below detection limits.

Table 3. Chromophores and key trace elements (Mg and Fe) LA-ICP-MS analyses (ppm) of the Greek corundum crystals and their different chemical ratios.

Sample.	Color	ppm	Mg	Ti	V	Cr	Fe	Ga	V/Cr	V + Cr/Ga	Fe/Mg	Ga/Mg	Fe/Ti	Cr/Ga
Dr1a ($n = 8$)	red	aver	15	17	2.4	9814	2020	14	0.0003	718	162	1.11	133	718
		min	8	8	2	8326	1799	12	0.0002	602	86	0.53	75	602
		max	25	27	2.7	12,116	2181	16	0.0003	879	247	1.68	225	879
Dr1b ($n = 13$)	red	aver	17	39	3.79	14,322	3795	14	0.0003	998	295	1.13	109	998
		min	9	22	3.01	11,625	3352	12	0.0002	750	122	0.46	68	750
		max	31	58	5.71	15,347	4348	16	0.0004	1252	440	1.72	187	1252
Dr2 ($n = 8$)	red	aver	17	67	2.4	3289	2149	20	0.0008	162	157	1.48	58	161
		min	13	27	2.1	2431	1918	19	0.0006	122	91	0.95	16	122
		max	24	148	2.9	4056	2438	24	0.0011	190	209	2	83	189
Go1a ($n = 11$)	blue-colorless	aver	380	5007	90	97	1096	89	1.89	2.98	3.16	0.26	0.26	1.91
		min	227	1915	40	35	554	29	0.23	1.15	1.32	.07	0.10	0.29
		max	601	6462	159	251	1339	119	4.15	7.73	5.73	0.49	0.68	6.30
Go1b ($n = 15$)	blue-colorless	aver	264	3503	66	70	910	89	1.19	1.59	4.33	0.46	0.35	0.86
		min	78	1007	23	35	363	64	0.26	0.84	1.66	0.14	0.12	0.30
		max	536	6361	207	187	1309	121	2.96	3.71	9.25	0.94	1.20	2.92
Go5a ($n = 8$)	pink-purple	aver	28	102	20	104	1291	47	0.37	2.88	60	2.46	24	2.46
		min	13	30	5	4	962	9	0.04	0.34	21	0.43	4.98	0.43
		max	49	205	60	298	1782	100	1.36	8.13	130	7.45	56	7.45
Go5b ($n = 14$)	pink	aver	35	509	65	297	424	78	0.30	4.67	14	2.54	1.45	3.83
		min	17	39	47	105	294	72	0.04	2.13	8.17	4.42	0.51	1.27
		max	53	810	79	1082	494	87	0.68	14.68	24	6.57	9.19	14.08
Ik1a ($n = 8$)	blue	aver	13	1135	120	222	3237	85	0.57	4.04	445	12	3.10	2.65
		min	5	653	70	168	2710	72	0.22	3.30	62	1.50	1.93	1.97
		max	54	1942	154	313	3627	93	0.84	5.29	711	19	4.73	4.32
Ik1b ($n = 16$)	blue	aver	16	1390	111	224	5415	94	0.49	3.54	348	6.07	8.49	2.38
		min	13	267	31	103	3903	68	0.20	2.13	207	3.54	0.93	1.52
		max	24	4508	164	279	7324	114	0.67	5.24	491	9.76	20.3	3.21
Nx1a ($n = 8$)	blue-colorless	aver	47	368	16	691	1532	46	0.02	15	35	1.06	7.32	15
		min	32	105	14	504	1377	42	0.02	11	24	0.70	1.78	10
		max	71	943	19	851	1804	53	0.04	19	43	1.37	14	18

Table 3. *Cont.*

Sample.	Color	ppm	Mg	Ti	V	Cr	Fe	Ga	V/Cr	V + Cr/Ga	Fe/Mg	Ga/Mg	Fe/Ti	Cr/Ga
Nx1b (n = 14)	purple	aver	64	520	22	43	4677	63	0.52	1.03	82	1.12	12	0.48
		min	39	124	15	36	3324	55	0.34	0.82	27	0.37	4.16	0.67
		max	199	848	31	48	6670	76	0.86	1.14	123	1.57	44	0.83
Nx2a (n = 8)	blue-colorless	aver	45	489	24	3.6	3210	49	14	0.55	84	1.30	16	0.07
		min	24	40	19	0.9	2852	47	1.83	0.44	42	0.67	4.15	0.02
		max	82	761	36	11	3474	53	42	0.68	127	1.99	86	0.22
Nx2b (n = 22)	blue-colorless	aver	63	560	26	51	4241	60	0.50	2.23	94	1.23	2.14	1.65
		min	16	89	13	31	1758	48	0.09	0.91	24	0.31	0.97	0.51
		max	208	966	42	73	6678	72	1.01	6.01	217	3.01	7.29	4.42
Nx3a (n = 8)	blue-colorless	aver	50	264	23	115	3189	52	1.11	2.68	68	1.12	94	2.24
		min	35	10	17	5	2843	47	0.10	0.51	42	0.68	5.57	0.08
		max	79	598	30	247	3922	59	5.10	4.99	88	1.37	316	4.46
Nx3b (n = 15)	blue-colorless	aver	105	1944	28	71	2773	62	0.50	2.23	30	0.68	1.86	1.65
		min	98	454	25	23	2220	53	0.09	0.91	16	0.34	0.97	0.41
		max	198	3222	34	246	3317	78	1.01	6.01	50	0.93	7.29	4.42
Nx4 (n = 16)	blue-colorless	aver	51	377	21	45	3828	83	5.98	0.87	81	1.85	41	0.57
		min	16	18	9	1	1812	50	0.11	0.18	41	0.96	3.47	0.01
		max	117	846	32	111	6361	184	30	1.83	126	4.61	189	1.38
Nx5a (n = 8)	blue	aver	5	462	52	262	3706	87	0.20	3.64	1069	24	9.17	3.04
		min	2	238	43	227	3301	84	0.13	3.16	241	6.28	4.32	2.64
		max	14	784	65	339	4268	90	0.24	4.48	2309	47	18	3.97
Nx5b (n = 12)	pink	aver	52	181	20	428	3096	45	0.05	9.95	77	1.08	24	9.52
		min	17	60	14	274	2138	37	0.03	6	34	0.36	7.13	5.61
		max	133	348	43	548	4716	52	0.10	13	208	2.07	59	12

Table 4. Other trace elements LA-ICP-MS analyses (ppm) of the Greek corundum crystals.

Sample	ppm	Be	Na	Si	Ca	Mn	Ni	Zn	Rb	Sr	Zr	Nb	Mo	Sn	Ba	Ta	W	Pb	Th	U
Dr1a (*n* = 8)	aver	bd	4.5	294	175	0.19	0.26	0.04	0.05	0.13	0.02	0.03	0.01	0.18	0.45	0.01	0.04	0.02	0.01	0.01
	min	bd	1.0	185	117	0.14	0.10	0.03	0.00	0.00	0.02	0.02	0.01	0.07	0.01	0.01	0.02	0.01	0.01	0.01
	max	bd	20	520	282	0.20	0.47	0.08	0.37	0.59	0.03	0.04	0.01	0.47	3.15	0.01	0.12	0.02	0.01	0.01
Dr1b (*n* = 13)	aver	bd	236	1807	780	bd	57	bd	bd	bd	bd	0.07	0.36	bd	9.80	0.05	0.17	bd	0.02	0.01
	min	bd	152	bd	352	bd	32	bd	bd	bd	bd	0.03	0.09	bd	0.39	0.01	0.11	bd	0.01	0.01
	max	bd	323	2099	1202	bd	309	bd	bd	bd	bd	0.11	1.27	bd	47	0.14	0.29	bd	0.02	0.01
Dr2 (*n* = 8)	aver	bd	bd	bd	1414	bd	bd	bd	bd	bd	bd	0.15	bd	bd	bd	0.02	0.48	bd	0.01	0.01
	min	bd	bd	bd	1384	bd	bd	bd	bd	bd	bd	0.04	bd	bd	bd	0.02	0.23	bd	0.01	0.01
	max	bd	bd	bd	1444	bd	bd	bd	bd	bd	bd	0.21	bd	bd	bd	0.02	0.73	bd	0.01	0.01
Go1a (*n* = 11)	aver	bd	bd	bd	1748	bd	bd	bd	bd	bd	bd	28	bd	38	0.62	1.60	68	0.73	6.98	0.32
	min	bd	bd	bd	1482	bd	bd	bd	bd	bd	bd	0.46	bd	11	0.55	0.09	1.11	0.70	2.14	0.03
	max	bd	bd	bd	2014	bd	bd	bd	bd	bd	bd	90	1.46	79	0.69	4.01	311	0.76	21	0.72
Go1b (*n* = 15)	aver	bd	492	bd	2265	bd	bd	bd	bd	bd	bd	8.76	0.22	16	0.52	1.64	44	5.43	10	0.10
	min	bd	434	bd	1859	bd	bd	bd	bd	bd	bd	0.29	0.11	2.14	0.23	0.11	0.48	2.23	1.07	0.01
	max	bd	524	bd	2491	bd	bd	bd	bd	bd	bd	55	0.34	97	0.74	11	498	8.63	53	0.26
Go5a (*n* = 8)	aver	0.2	1.3	224	136	0.20	0.75	0.22	0.01	0.04	0.02	0.11	0.01	27	0.02	0.02	0.16	0.06	0.40	0.01
	min	0.1	bd	14	88	0.20	0.04	0.03	0.01	0.01	0.01	0.01	0.01	0.09	0.01	0.00	0.03	0.01	0.01	0.01
	max	0.5	3.3	369	217	0.20	2.04	0.44	0.01	0.24	0.05	0.52	0.01	213	0.06	0.04	0.46	0.34	1.17	0.03
Go5b (*n* = 14)	aver	bd	bd	bd	bd	bd	bd	bd	bd	bd	bd	0.16	0.45	bd	0.87	0.07	bd	bd	0.03	0.02
	min	bd	bd	bd	bd	bd	bd	bd	bd	bd	bd	0.16	0.33	bd	0.57	0.06	bd	bd	0.01	0.01
	max	bd	bd	bd	bd	bd	bd	bd	bd	bd	bd	0.16	0.61	bd	1.17	0.08	bd	bd	0.06	0.05
Ik1a (*n* = 8)	aver	0.45	116	2479	1700	0.44	0.56	2.25	0.16	1.45	0.07	2.56	0.01	4.97	0.35	0.96	1.19	2.95	1.74	0.12
	min	0.10	24	353	312	0.23	0.37	0.94	0.02	0.30	0.04	1.59	0.01	3.23	0.11	0.34	0.79	1.40	0.85	0.04
	max	0.70	313	6492	4158	1.05	0.81	4.30	0.48	3.39	0.11	4.03	0.01	7.66	0.71	1.73	1.89	4.85	2.45	0.23
Ik1b (*n* = 16)	aver	bd	bd	bd	bd	bd	bd	bd	bd	bd	bd	7.92	0.34	6.88	1.74	0.39	4.64	bd	4.34	0.04
	min	bd	bd	bd	bd	bd	bd	bd	bd	bd	bd	0.40	0.11	2.83	1.13	0.10	0.74	bd	0.04	0.01
	max	bd	bd	bd	bd	bd	bd	bd	bd	bd	bd	37	0.84	23	2.34	1.06	12	bd	31	0.22
Nx1a (*n* = 8)	aver	2.08	5.30	543	14	0.20	1.10	0.03	0.03	0.66	0.05	1.16	0.01	32	0.39	15	2.74	0.02	0.40	0.01
	min	0.18	1.00	399	14	0.20	0.79	0.03	0.00	0.01	0.00	0.11	0.01	5.61	0.01	3.82	0.16	0.02	0.01	0.01
	max	11.03	34	1110	14	0.20	1.65	0.03	0.18	4.75	0.15	3.89	0.01	108	3.01	46	8.74	0.02	2.45	0.02
Nx1b (*n* = 14)	aver	2.91	205	2303	1551	bd	bd	bd	bd	bd	bd	7.77	0.24	1040	0.11	0.39	43	bd	0.72	0.01
	min	1.31	141	2069	1351	bd	bd	bd	bd	bd	bd	0.41	0.12	166	0.10	0.16	4.04	bd	0.02	0.01
	max	4.50	253	2537	1874	bd	bd	bd	bd	bd	bd	19	0.49	1712	0.12	0.65	147	bd	2.57	0.02
Nx2a (*n* = 8)	aver	1.21	36	804	120	0.21	0.71	0.03	0.20	1.73	1.23	9.87	0.01	566	0.05	1.58	105	0.02	0.13	0.01
	min	0.10	1	466	14	0.11	0.35	0.03	0.01	0.01	0.76	0.23	0.01	197	0.01	0.18	0.63	0.02	0.01	0.01
	max	3.62	271	2209	648	0.35	1.39	0.03	0.04	11	2.25	36	0.01	887	0.17	7.14	480	0.02	0.48	0.03

Table 4. *Cont.*

Sample	ppm	Be	Na	Si	Ca	Mn	Ni	Zn	Rb	Sr	Zr	Nb	Mo	Sn	Ba	Ta	W	Pb	Th	U
Nx2b (n = 22)	aver	3.53	112	1998	2118	bd	bd	bd	0.92	0.88	bd	12	0.33	608	bd	3.12	147	0.53	1.48	0.08
	min	2.71	90	1848	1844	bd	bd	bd	0.66	0.68	bd	0.11	0.09	5	bd	0.18	1.08	0.51	0.01	0.01
	max	5.64	141	2252	2418	bd	bd	bd	1.25	1.15	bd	117	0.68	1842	bd	18	877	0.56	13	0.49
Nx3a (n = 8)	aver	2.79	1.58	501	14	0.20	1.06	0.03	0.02	0.22	0.83	7.71	0.01	340	0.27	2.16	100	0.02	0.07	0.01
	min	0.25	1.00	412	14	0.20	0.59	0.03	0.01	0.01	0.01	0.98	0.01	4.95	0.01	0.53	11	0.02	0.01	0.00
	max	17	5.65	693	14	0.20	1.67	0.03	0.11	1.56	1.64	36	0.01	574	2.05	6.02	424	0.02	0.21	0.01
Nx3b (n = 15)	aver	4.25	bd	bd	bd	bd	bd	bd	bd	bd	bd	9.87	0.23	560	0.17	4.85	49	bd	3.36	0.02
	min	3.28	bd	bd	bd	bd	bd	bd	bd	bd	bd	0.16	0.12	198	0.12	0.34	0.49	bd	0.04	0.01
	max	5.82	bd	bd	bd	bd	bd	bd	bd	bd	bd	48	0.34	947	0.22	13	245	bd	20	0.08
Nx4 (n = 16)	aver	7.05	1.69	500	14	0.23	0.83	0.03	0.01	0.02	1.80	112	0.01	244	0.06	111	34	0.03	0.83	0.06
	min	0.10	1.00	267	14	0.15	0.28	0.03	0.00	0.00	0.01	1.16	0.01	9.19	0.01	0.31	1.75	0.02	0.01	0.01
	max	38	12	920	14	0.61	1.92	0.03	0.17	0.12	7.56	769	0.01	684	0.38	549	157	0.10	4.69	0.55
Nx5a (n = 8)	aver	1.12	1.12	372	163	16	1.17	0.79	0.01	0.04	0.06	1.53	0.01	3.64	0.66	0.23	6.70	1.80	7.40	0.02
	min	0.10	1.00	328	98	0.20	0.70	0.22	0.01	0.01	0.02	1.14	0.01	2.31	0.07	0.11	3.80	0.18	4.20	0.01
	max	1.86	1.99	415	239	110	1.89	2.25	0.01	0.08	0.10	1.91	0.01	5.45	2.75	0.35	9.45	7.72	10	0.03
Nx5b (n = 12)	aver	bd	bd	bd	bd	bd	bd	bd	bd	bd	bd	1.49	0.63	4.30	bd	12	6.28	0.67	0.02	0.01
	min	bd	bd	bd	bd	bd	bd	bd	bd	bd	bd	0.10	0.28	3.22	bd	0.19	bd	0.49	0.01	0.01
	max	bd	bd	bd	bd	bd	bd	bd	bd	bd	bd	2.89	0.97	6.60	bd	96	13.1	1.03	0.02	0.01

Colored (colorless to blue, pink) sapphires from Gorgona/Xanthi display significant differences regarding the concentration of some chromophore elements. Colorless to blue sapphires are mostly characterized by significant variations of their Ti content which reflects their zoned coloration. Bluish areas display high Ti values, which are up to 6 times higher compared to colorless/white areas and reach up to 6462 ppm. Iron content is quite fixed in both colorless and blue areas, reaching up to 1339 ppm, but is slightly higher in the pink varieties (up to 1782 ppm). Pink sapphires are also characterized by much less Ti (up to 810 ppm) and significantly higher Cr (up to 1082 ppm) concentrations, compared to the colorless/blue areas where Cr values are generally less than 298 ppm. Vanadium and Ga content remains relatively fixed, regardless of the color, with maximum values of up to 227 and 121 ppm, respectively. Mg values are generally higher in the colorless to blue grains, with values up to 536 ppm, while in the pink grains is generally below 65 ppm. Si and Ca display values up to 369 and 217 ppm, respectively, while other elements, are generally close to or below the detection limit, with the exception of Sn and Ta, which reach up to 213 and 4 ppm, respectively, in the colorless to blue grains.

Iron is the most abundant trace element in analyzed sapphires form Ikaria island, with values reaching up to 7324 ppm. Ti displays high values as well, up to 4508 ppm. Cr, Mg and V values are relatively constant with maximum values reaching up to 313, 54 and 164 ppm, respectively. Gallium shows elevated concentrations, up to 114 ppm, resulting in very high values of the Ga/Mg critical ratio. Si and Ca display wide variations between 353–6492 ppm and 312–4158 ppm, with the highest values correlating and thus indicating submicroscopic silicate inclusions (possibly margarite). Other trace elements display generally low concentrations except for Sn, whose values reach up to 23 ppm.

Corundums from Naxos island are mostly characterized by high Fe concentrations, which reach up to 6678 ppm. Higher Fe values are related to areas with more intense blue coloration. Ti is considerably lower, reaching values up to 966 ppm, with the exception of one colorless to blue corundum, where higher values were detected, up to 3222 ppm. Chromium varies significantly, between 1 and 851 ppm. The higher Cr values characterize corundum grains with pink and purple hues. Mg and Ga values range between 2–208 ppm and 42-184 ppm, respectively. The higher Ga concentrations are remarked in the metabauxite-hosted sapphires, that accordingly display the highest Ga/Mg ratio values compared to any other Greek corundums (this study). Vanadium content is quite fixed in all samples and varies between 10 and 65 ppm, with the majority of the measured values being in the range of 20–40 ppm. Si and Ca reach values up to 2537 and 2418 ppm, respectively, probably reflecting the presence of submicroscopic inclusions. Sodium is generally low, but in some cases values up to 271 ppm were detected, probably due to submicroscopic inclusions of silicate minerals (plagioclase and/or paragonite). Almost all corundums contain traces of Be, which ranges between 0.10 and 38 ppm. Manganese is generally close to or below detection, but in the case of a corundum from metabauxites reaches up to 110 ppm. Most samples are characterized by high Sn values, which reach up to 1812 ppm, while Nb is also enriched, reaching up to 769 ppm. Finally, Ta reaches up to 549 ppm, while the rest trace elements are mostly below detection (Table 4).

6. Fluid Inclusions

The fluid inclusions (FI) in the studied corundums range in size between <1 μm and 55 μm. Only one type of two-phase vapor-liquid CO_2 (LcarVcar) FI is observed within all samples [79]. (Figure 6a–d,g). They often have an elongated negative crystal shape of the host corundum (Figure 6a,k). They occur as isolated, and based on their similarity in shape and the mode of occurrence, they are considered as primary FI (Figure 6a,c,f,j,k). However, the most significant criterion for identifying primary FI is when they occur along the growth zones of the corundum (Figure 6i), based on the criteria suggested by Bodnar [80].

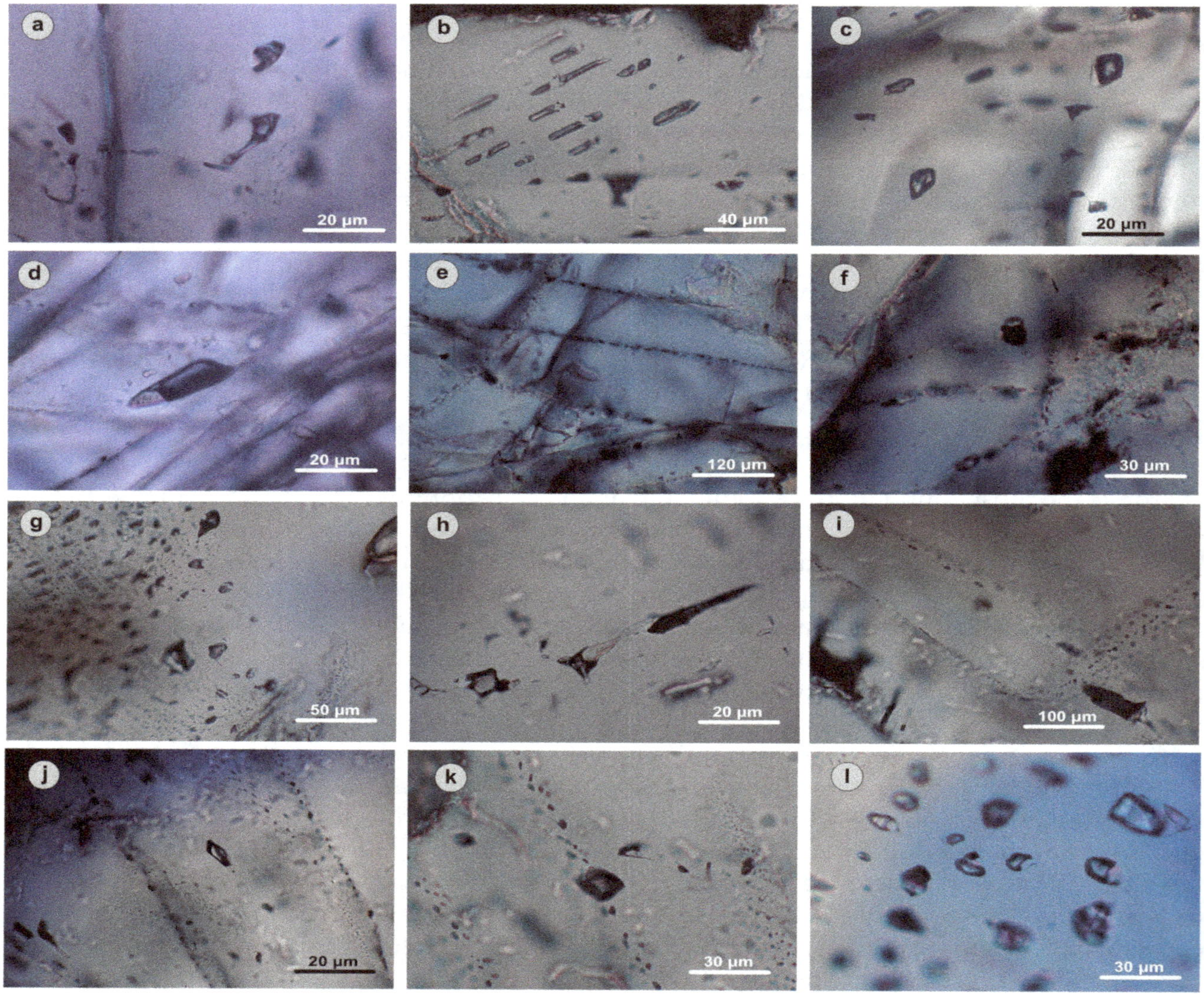

Figure 6. Photomicrographs of FI in corundums from the Rhodope massif of Greece, Paranesti and Gorgona (**a**–**f**) and from Naxos island the Attic-Cycladic massif (**g**–**l**). (**a**) Isolated primary two-phase vapor-liquid inclusions composed of $CO_2 \pm (CH_4$ and/or $N_2)$. The fluid inclusion at the upper part has a negative crystal shape, and the inclusion underneath shows necking-down deformation (Drama, thin-section Dr2). (**b**) Pseudosecondary two-phase carbonic FI along a fracture healed during crystal growth (Xanthi, thin-section Go5a). (**c**) Primary two-phase carbonic FI (Xanthi, thin-section Go5a). (**d**) Two-phase carbonic inclusion which shows evidence of post-entrapment modification (brittle crack and halo of neoformed inclusions) (Xanthi, thin-section Go5b). (**e**) Tiny FI along fracture plane trails, but due to their small size no phases can be identified (Xanthi, thin-section Go1b). (**f**) Isolated two-phase carbonic inclusion (Xanthi, thin-section Go1b); (**g**) Oriented cluster of pseudosecondary carbonic FI along fractures healed during crystal growth (thin-section, Nx1a). (**h**) Isolated primary two-phase carbonic inclusion with necking-down deformation (thin-section, Nx1a). (**i**) FI along growth zones of the corundum (thin-section, Nx1a). (**j**,**k**) Primary two-phase carbonic FI (thin-section, Nx2a). (**l**) Pseudosecondary two-phase carbonic FI along a fracture healed during crystal growth (thin-section, Nx2b).

Oriented clusters of $L_{car}-V_{car}$ FI along planes are very common in the studied samples. They are related to micro-cracks and subsequent fractures healed during crystal growth, and therefore are considered as pseudosecondary (Figure 6b,g,l). Numerous tiny FI (length <2 μm) along trails are usually observed, but due to their small size, the nature of the present phase(s) cannot be identified (Figure 6e). They are considered as secondary FI. The majority of the FI appears stretched and necked or empty due to leaking phenomena (Figure 6a,d,h). Microthermometric results were based on primary and pseudosecondary inclusions, without necking-down and post-entrapment modification evidence. However, many measurements were unsuccessful due to the black color of the carbonic fluid in the inclusions and the difficulty to observe phase changes [79].

Table 5 presents the microthermometric data of all studied FI from Drama, Xanthi, Ikaria and Naxos. The melting temperatures of CO_2 ($TmCO_2$) range between −57.3 and −56.6 °C, at or slightly lower than the triple point of CO_2 (−56.6 °C). This indicates that the fluid is dominated by CO_2 with very small quantities of CH_4 and/or N_2. Temperatures of Tm were not obtainable in the sample Dr1a from Paranesti (Drama). Clathrate nucleation was not observed in any measured fluid inclusion, demonstrating that liquid water (H_2O) was not incorporated in the whole process of the corundums formation. All FI homogenized to the liquid carbonic phase (LCO_2) at temperatures ($ThCO_2$) varying from 27.3 to 31.0 °C. This Th is close to the critical temperature of pure CO_2 and corresponds to densities of the source fluids from 0.46 to 0.67 g/cm^3. Figure 7 shows the histograms of the $ThCO_2$ of the FI from the four different studied corundum occurrences in Greece (Paranesti, Gorgona, Naxos, Ikaria).

Table 5. Microthermometric data of studied primary and pseudosecondary fluid inclusions from corundum crystals in Greece.

Sample	Locality	*n*	Tm CO_2 (°C)	Th CO_2 (°C)	Densities (g/cm^3)
Dr1a	Paranesti,	7	−57.2 to −56.7	28.7–31.1	0.46–0.64
Dr2	Drama	7	−57.1 to −56.6	27.4–30.4	0.57–0.67
Go1a		4	−57.0 to −56.8	29.2–30.2	0.58–0.62
Go1b	Gorgona,	3	-56.7 to -56.6	27.8–30.5	0.57–0.66
Go5a	Xanthi	6	−57.2 to −56.8	28.2–30.9	0.53–0.65
Go5b		6	−57.2 to −56.7	27.4–29.2	0.62–0.67
Ik1a	Ikaria	3	−56.9 to −56.6	30.1–30.8	0.54–0.59
Ik1b		6	−57.0 to −56.6	27.4–28.3	0.65–0.67
Nx1a		8	−57.2 to −56.6	27.3–31.0	0.51–0.67
Nx1b		10	−57.3 to −56.7	28.1–30.5	0.57–0.65
Nx2a	Naxos	4	−57.2 to −56.6	27.9–29.3	0.62–0.66
Nx2b		6	−57.1 to −56.8	28.9–31.0	0.51–0.63
Nx4b		4	−57.2 to −56.9	27.5–30.4	0.57–0.67

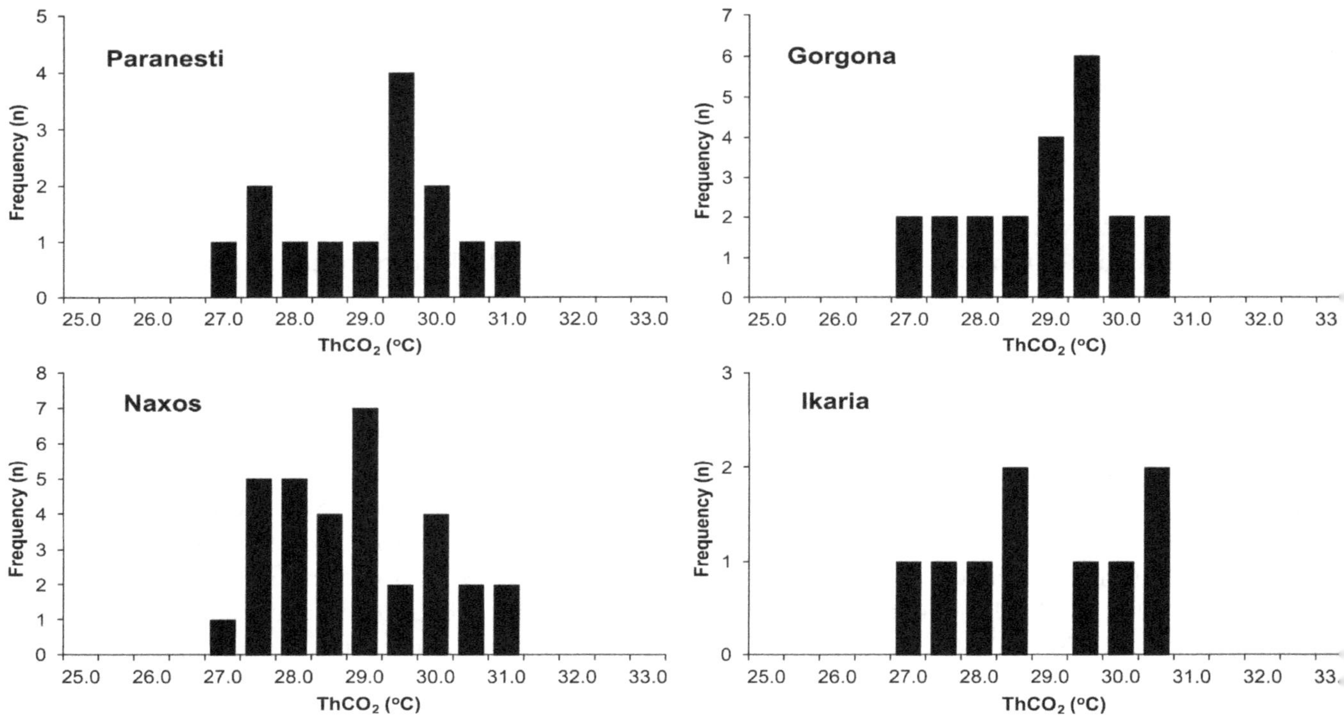

Figure 7. Histograms of homogenization temperatures of CO_2 ($ThCO_2$) to the liquid phase (LCO_2) in the fluid inclusion from the Rhodope massif (Paranetsi, Gorgona) and the Attic-Cycladic massif (Naxos, Ikaria) from Greece.

7. Oxygen Isotope Data

Oxygen isotope compositions (Table 6) confirm their geological typology, i.e., with, respectively $\delta^{18}O$ of 4.9 ± 0.2‰ for sapphire in plumasite, 20.5‰ for sapphire in marble and 1‰ for ruby in mafics (Figure 8). The desilicated blue sapphire-bearing pegmatite from Naxos have similar oxygen isotope values to those from desilicated pegmatites in mafic host rocks from elsewhere [2,7,14,81,82]. The O-isotope composition of sapphire is buffered by the $\delta^{18}O$ value of the mafic host rock. The ruby in pargasite schists from Paranesti has a very low $\delta^{18}O$-value of 1‰ that can be interpreted in different ways: (i) inherited pre-metamophic reactions between sea-water and hot basic/ultrabasic rocks before subduction and metamorphism; (ii) syn-metamorphism depletion in ^{18}O related to hydration during amphibolite facies metamorphic conditions; (iii) post-amphibolite facies metamorphism with recrystallization under the effect of metasomatism of the metamorphosed mafic/ultramafic rocks with a high depletion in ^{18}O; and (iv) metamorphic/metasomatic conditions involving deeply penetrating meteoric waters along major crustal structures, see Wang et al. [83]. At the moment, the absence of more O-isotope data on this type of ruby precludes any of these possible hypotheses. The two $\delta^{18}O$-values of ruby in marble, between 20.5‰ and 22.1‰, are in agreement with the range of values found for this type of ruby worldwide [2,14].

Table 6. Oxygen isotope values of corundum from Greece. $\delta^{18}O$ corundum (‰, V-SMOW) (after Wang et al. [83]).

Sample	$\delta^{18}O$	Description
Nx2a	4.8	Blue to colorless sapphire in plumasite
Nx3a	5.1	Blue to colorless sapphire in plumasite
Dr1b	1.0	Red ruby in pargasite-schist
Go5a	22.1	Pink-purple sapphire in marble
Go5b	20.5	Pink sapphire in marble
Ik1a	22.4	Blue sapphire in metabauxite

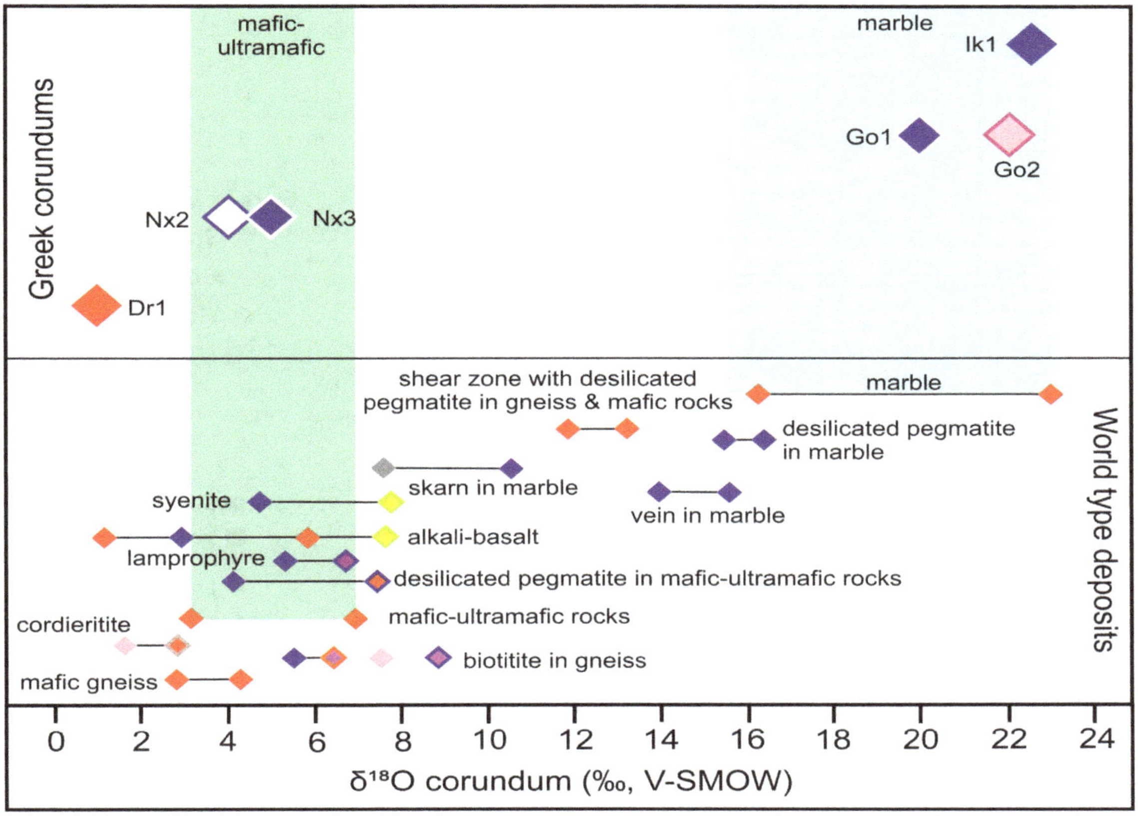

Figure 8. Oxygen isotopic composition of Greek corundum (after Wang et al. [83]) compared with the oxygen isotopic ranges from corundum deposits worldwide from the data of Giuliani et al. [2,7,14,81,82]. Color in diamonds represents the color of the studied corundums. Colorless sapphires are represented by white diamonds.

. Discussion

.1. *Trace Elements Fingerpring: Metamorphic versus Magmatic Origin of Greek Corundum*

Chromophore and genetic indicator elements (Fe, Cr, Ti, V, Ga and Mg) are commonly used to listinguish corundum from different primary sources using elemental diagrams [16,84–88]. In the Cr + V)/Ga versus Fe/Ti diagram (Figure 9a, [84,85]), displaying the fields for metamorphic and nagmatic corundums, the majority of the samples plot in the field of metamorphic corundum, xhibiting a large variation of Fe/Ti ratios. Some sapphires from Naxos island that plot in the nagmatic field have very low Cr/Ga ratios. Rubies from Paranesti/Drama show high Cr/Ga and e/Ti ratios, followed by the pink and purple sapphires from Naxos island. Both colorless to blue nd pink sapphires from Gorgona/Xanthi and blue sapphires from Ikaria island are characterized by elatively fixed Cr/Ga, but display variations with respect to their Fe/Ti ratios.

In the Fe – Cr × 10 – Ga × 100 discrimination diagram (Figure 9b; after Sutherland et al. [84]), the ubies from Paranesti are extremely rich in Cr_2O_3 and display a linear trend inside the metamorphic orundum field. Sapphires from Gorgona/Xanthi, Naxos and Ikaria islands are scattered in both the netamorphic and magmatic corundum fields. Colored sapphires from Gorgona/Xanthi are scattered n both the metamorphic and magmatic fields and display two trends, along the Ga–Fe and Ga–Cr lines, vith the pink varieties plotting closer to the Cr_2O_3 edge, indicating a direct relationship between color nd Cr_2O_3 content. Colorless to blue sapphires from Naxos island mostly plot along the Fe–Ga line, in he magmatic corundum field, while the pink and purple varieties display a linear trend towards the Cr_2O_3 edge, inside the metamorphic field, which is subparallel to the trend of pink sapphires from Gorgona/Xanthi. Blue sapphires from Ikaria island clearly plot in the metamorphic field.

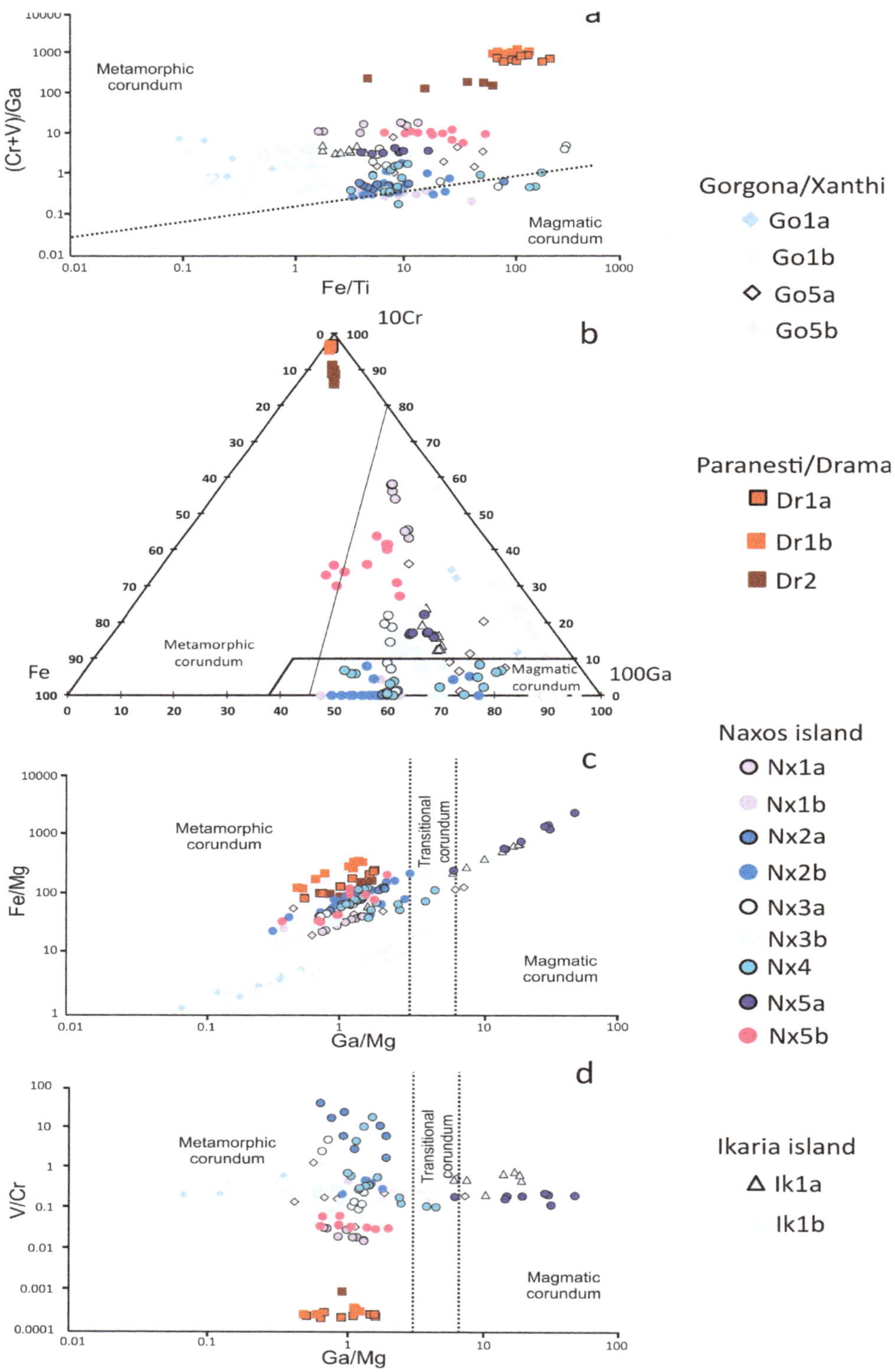

Figure 9. Greek corundum LA-ICP-MS analyses plotted on a (**a**) (Cr + V)/Ga versus Fe/Ti discrimination diagram separating the fields for magmatic and metamorphic corundums (after Sutherland et al. [84] and Harris et al. [85]); (**b**) trace element Fe − Cr × 10 − Ga × 100 discrimination diagram after Sutherland et al. [84]; (**c**) Fe/Mg versus Ga/Mg diagram discrimination separating the fields for magmatic, transitional and metamorphic corundums (after Peucat et al. [16], Sutherland et al. [84]; (**d**) V/Cr versus Ga/Mg discrimination diagram separating the fields for magmatic, transitional and metamorphic corundums (after Sutherland et al. [86]).

In the Fe/Mg versus Ga/Mg plot (Figure 9c), most samples plot in the metamorphic corundum field, except for sapphires from metabauxites of Naxos and Ikaria islands and two analyses of pink sapphires fom Gorgona/Xanthi. A few blue sapphires from Naxos island and pink sapphires from Gorgona/Xanthi plot in the in-between area of transitional corundum. Two distinct trends can be remarked in the metamorphic corundum field: Sapphires from Gorgona/Xanthi form a linear trend, with the pink varieties exhibiting higher both Fe and Ga concentrations. The rest of the analyzed corundums plot in a relatively small area, parallel to the previous trend, but are also characterized by higher Fe values, the highest of which characterize the rubies from Paranesti/Drama.

The V/Cr versus Ga/Mg diagram is useful for deciphering the genetic environments of corundum (Figure 9d). The majority of the analyzed samples plot in the metamorphic corundum field, with the exception of a few analyses that fit in the transitional field and the metabauxite-hosted sapphires from Ikaria and Naxos islands that have a magmatic signature (Figure 9d). In this diagram, rubies from Paranesti/Drama plot close to the Ga/Mg axis, as they are characterized by very low V content. On the other hand, colored sapphires from the other localities plot in two vertical trends. The first one, is characterized by fixed Ga/Mg content and refers to pink, purple and the majority of the colorless to blue sapphires from Naxos island, and expresses a decrease in the Cr content from the purple to the pink and finally the blue varieties and subsequent increase of their V content. The second trend is characterized by stable V/Cr ratio, but shows a significant variation regarding its Ga/Mg content. Gallium concentration increases by three orders from the blue sapphires of Gorgona/Xanthi towards the pink varieties of the same locality and finally to the metabauxite-hosted sapphires of Ikaria and Naxos islands which show the highest Ga values.

In the ($FeO–Cr_2O_3–MgO–V_2O_3$) versus ($FeO + TiO_2 + Ga_2O_3$) diagram after Giuliani et al. [87,88], which further differentiates the metamorphic environments (Figure 10), all rubies from Paranesti/Drama clearly fall in the field of rubies with mafic/ultramafic origin. Pink sapphires from Gorgona/Xanthi, plot in an overlapping area between the marble-hosted rubies and the metasomatic corundum fields, while the rest of sapphires from this locality, which are colorless to blue in color, plot in a linear trend along the y axis of the diagram, indicating variable contents of Fe, Ti and Ga. The rest of the studied samples (Naxos and Ikaria sapphires) plot mostly in the field of metasomatic corundum, with some analyses plotting in the borders with the field of syenite-related sapphires. In general, the trends observed in Figures 9 and 10 reflect variations in chromophore element concentrations, even in samples from the same locality. This could be attributed to changes in chemistry and physicochemical conditions of the corundum-forming environment, perhaps due to metasomatic processes as described by Harris et al. [85].

8.2. Comparison with Corundum Deposits around the World

Various plots, based on trace element fingerprints, have been proposed as a useful tool for distinguishing the origin and genesis of gem corundum deposits (Figures 11 and 12; [4,16,89–93]).

In the Mg $\times$ 100 – Fe – Ti $\times$ 10 diagram (Figure 11a), rubies from Paranesti/Drama plot both in the magmatic and metamorphic fields, and mostly along the Fe–Mg line. Analyses that plot in the magmatic field are also of metamorphic origin and indicate the variation of the Fe/Mg ratio, thus suggesting a limitation of this chemical diagram. Sapphires from Gorgona/Xanthi and the majority of the Naxos island's sapphires plot in the metamorphic field. The blue and pink sapphires from Gorgona/Xanthi, as well as some blue sapphires from Naxos island, plot along the Mg–Ti line of low-Fe metamorphic sapphires (e.g., Ilakaka/Madagascar and Ratnapura Balangoda/Sri Lanka), subparallel to the so-called Kashmir trend. Other colored sapphires from Naxos plot subparallel to the Kashmir trend of metasomatic blue sapphires, but they are characterized by elevated contents of Fe in the purple- and Mg in the pink-colored varieties, and are scattered through the Bo Phloi/Thailand atypical blue sapphires. Metabauxite-hosted sapphires from Naxos and Ikaria islands display a Ti-rich magmatic trend and plot preferably along the Fe–Ti line and follow the Fe–Ti Pailin (Cambodia) of magmatic blue sapphires.

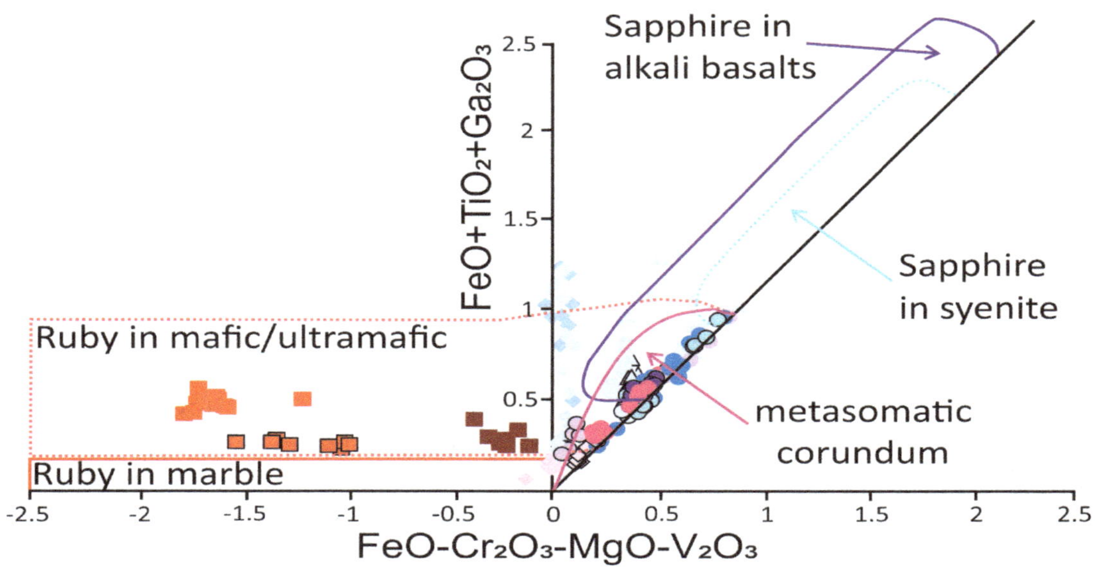

Figure 10. Greek corundum LA-ICP-MS analyses plotted on a $FeO–Cr_2O_3–MgO–V_2O_3$ versus $FeO + TiO_2 + Ga_2O_3$ discrimination diagram after Giuliani et al. [87,88]. Symbols as in Figure 9.

In the Fe × 0.1 − (Cr + V) × 10 − Ti plot, after Peucat et al. [16], rubies from Paranesti/Drama, a well as pink and purple sapphires plot close to the Cr + V edge, as they are enriched in Cr (Figure 11b Moreover, pink sapphires from Gorgona/Xanthi, along with metabauxite-hosted sapphires from Ikari and Naxos islands, plot close to the same edge, but due to their relative enrichment in V, rather tha Cr. The blue sapphires from Gorgona/Xanthi, along with some blue sapphires from Naxos, plote clos to the Ti edge, in a linear array parallel to the (Cr + V)–Ti line. Sapphires from Naxos are scarce i the diagram, mainly reflecting their variable content of Ti, with the weak-colored areas plotting clos to or along the Fe–(Cr + V) line. The majority of the Naxos sapphires form a linear array, paralle to the Mogok marble-hosted trend, but they differ in respect to their lower Fe and higher Cr + V contents. Some Naxos sapphires plot close to the fields for Colombia and Umba blue sapphires hoste in desilicated pegmatites.

In the Fe versus Ga/Mg diagram (Figure 12a; after Peucat et al. [16], Zwaan et al. [89]), the majorit of the analyzed corundums are scattered in the field of metamorphic sapphires, and specifically i the plumasitic sub-field. Some sapphires from Naxos island, with blue and pink color plot in th field of alluvial sapphires from Yogo Culch Montana, while blue sapphires from Gorgona/Xanth are characterized by very low Ga/Mg ratios and thus plot slightly outside the metamorphic fiel Metabauxite-hosted sapphires from Naxos and Ikaria islands plot along the Main Asian Field (MAF of sapphires related to alkali basalt. They form a linear array indicating variations in their Ga/M ratio and their Fe content is slightly higher compared to sapphires from Pailin (Cambodia) and Ilme (Russia). Finally, a few analyses of pink sapphires from Gorgona/Xanthi are plotted in the MAF fiel and are characterized by slightly higher both Fe and Ga/Mg values, compared to the rest pink an colorless to blue sapphires of this location.

The Cr_2O_3 versus Fe_2O_3 diagram (Figure 12b), adapted from Schwarz et al. [90], shows differen types of African deposits: marble-type from Mong Hsu and Mogok (Myanmar), desilicated pegmatit from Mangari (Kenya), amphibolitic-type from Chimwadzulu (Malawi), Songea and Winza (Tanzania and basaltic-type from Thai (Thailand-Cambodia border region). The majority of the Paranesti/Dram rubies plot outside the fields of rubies from elsewhere, because they contain significantly more Cr_2O_3 Some samples, though, poorer in Cr_2O_3, fit well into the Winza (Tanzania) ruby field. Sapphire sample plot mostly along the Fe_2O_3 axis, with the majority of the compositions being comparable with colore sapphires from Winza.

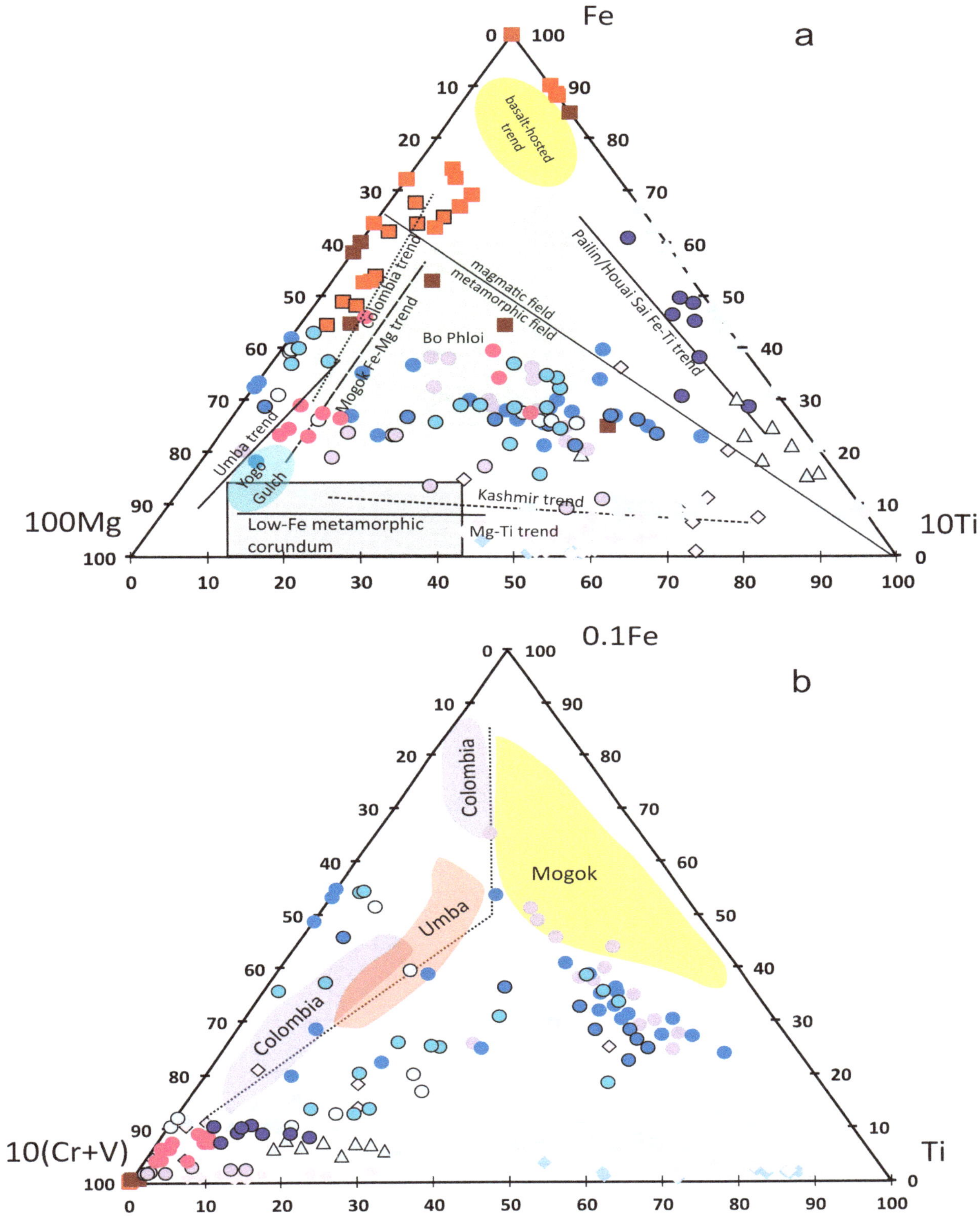

Figure 11. Greek corundum LA-ICP-MS analyses plotted on a (**a**) Mg × 100 – Fe – Ti × 10 discrimination diagram. Also shown are the metamorphic Fe–Mg and Fe–Ti trends of Mogok and low-Fe sapphires of Pailin, Ilakaka and Ratnapura Balangoda, the Umba and Kashmir trends of metasomatic blue sapphires, and the Bo Phloi scattering of atypical blue sapphires (modified after Peucat et al. [16]); (**b**) Fe × 0.1 – (Cr + V) × 10 – Ti diagram for Colombia, Mogok and Umba blue sapphires along with the plots from the Greek corundums, modified after Peucat et al. [16]. Symbols as in Figure 9.

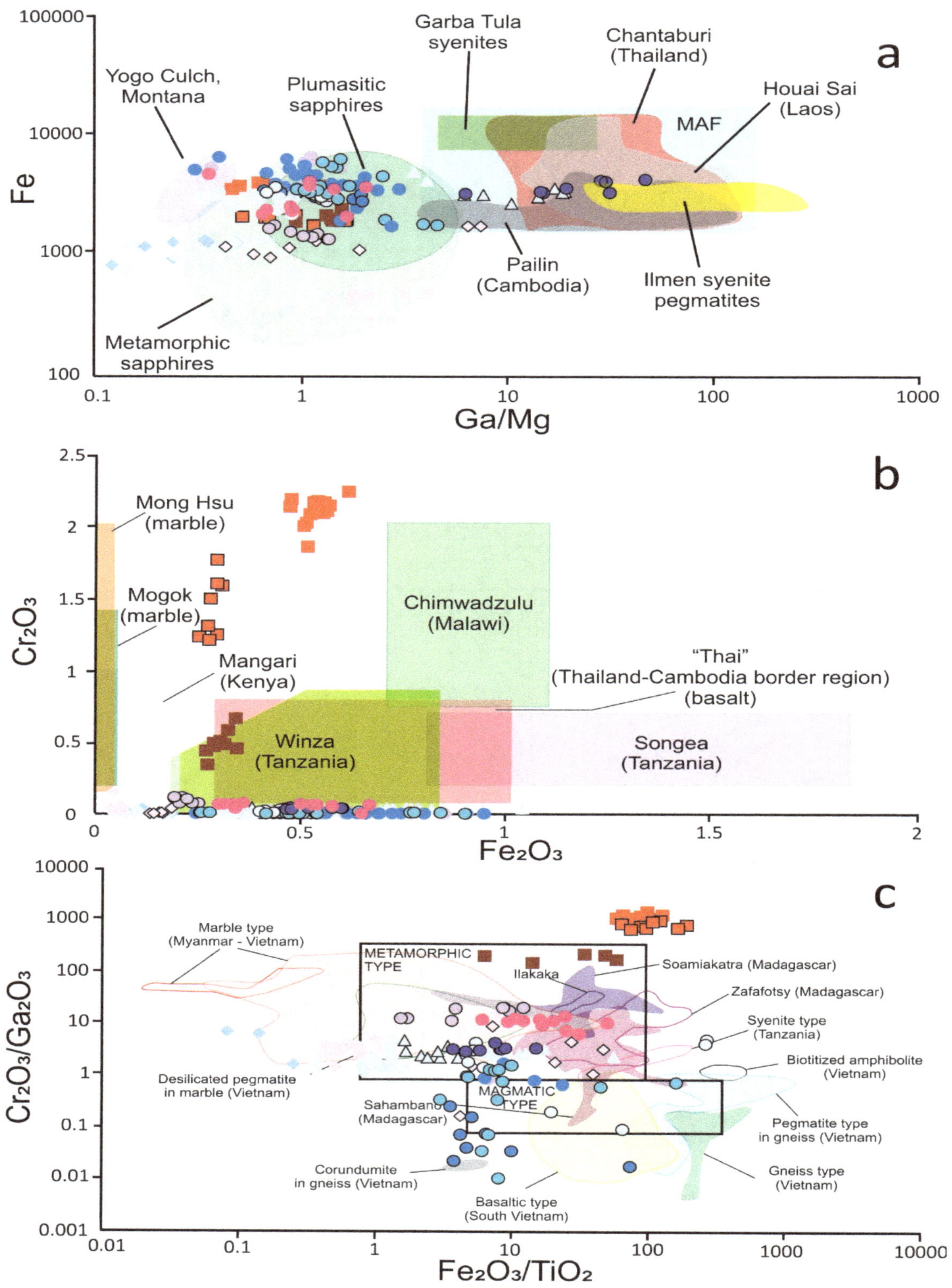

Figure 12. Greek corundum LA-ICP-MS analyses plotted on a (**a**) Fe versus Ga/Mg discrimination diagram (after Peucat et al. [16], Zwaan et al. [89]); (**b**) Cr_2O_3 versus Fe_2O_3 diagram demonstrating the fields of different African deposits (modified after Schwarz et al. [90]); (**c**) Cr_2O_3/Ga_2O_3 versus Fe_2O_3/TiO_2 plot demonstrating composition of diverse corundum deposits (modified after Rakotondrazafy et al. [91], Sutherland et al. [4], Pham Van et al. [92], Simonet et al. [93]. Symbols as in Figure 9.

The Cr_2O_3/Ga_2O_3 versus Fe_2O_3/TiO_2 diagram (Figure 12c; modified from Rakotondrazafy et al. [91]), displays the geochemical fingerprint of corundums from different types. The majority of analyzed corundums plot in the field of metamorphic origin. It is clear that rubies from Paranesti/Drama show higher Cr_2O_3 values, and thus cannot plot in fields of rubies from elsewhere.

Some rubies from Paranesti also fall close to the Soamiakatra rubies, which are hosted in clinopyroxenite enclaves of alkali basalts. Pink and purple sapphires from Naxos island display variation of their Fe_2O_3/TiO_2 ratio. The pink sapphires plot in a common field of the Soamiakatra and Sahambano sapphires (Madagascar), while the purple varieties are scattered in a sector overlapping both the marble- and the desilicated pegmatite in marble-types of Vietnam.

Colorless to blue sapphires from Naxos plumasites are scattered along the lower limit of the metamorphic field, indicating significant variation of their Fe_2O_3/TiO_2 values, while the Cr_2O_3/Ga_2O_3 values stay relatively fixed. Pink-colored sapphires from Gorgona/Xanthi plot partly inside the marble-hosted corundum from Vietnam, while the blue varieties plot in the extension of this trend, slightly outside the field, indicating minor Fe_2O_3 content. Metabauxite-hosted corundum from Naxos island plot slightly outside the magmatic corundum field, and their ratios are not comparable with any other corundum of the plot, except a few analyses, which plot close to the Vietnamese field of gneiss-hosted corundumite. Ikaria metabauxite-hosted samples plot well into the metamorphic corundum field and show a wide range of Fe_2O_3/TiO_2 values, covering the fields between marble-hosted and desilicated pegmatite in marbles (Vietnam), Sahambano and Zafafotsy (Magadascar) corundum deposits.

3.3. Fluid Characterization

The fluid inclusions study in the corundum (rubies and sapphires) from the four Greek occurrences revealed the presence of CO_2-dominated fluids with very small quantities of CH_4 and/or N_2, and relatively low densities, varying between 0.46 and 0.67 g/cm^3 [79]. Primary and pseudosecondary water-free carbonic fluid inclusions represent the main fluid, which was incorporated during the crystallization of rubies and sapphires. The absence of any significant change in the fluid composition and density of the primary and pseudosecondary inclusions possibly suggests that (i) the fluid was homogeneous and related to the same source; and (ii) the host rocks were not affected by the circulation of any external fluids. The CO_2-rich fluids are likely of metamorphic origin and probably derived from devolatilization of marble in most cases (Naxos and Ikaria).

Most of the corundum in different geological environments worldwide generally contains pure or nearly pure CO_2-bearing fluid inclusions. Previous studies have shown that in metamorphic complexes, CO_2-bearing fluids were incorporated in the corundum formation from granulite facies rocks in Sri Lanka [94], in the marble hosted ruby deposits from Luc Yen in North Vietnam [95] and in pegmatoids from the Nestos Shear Zone in Greece [57]. Corundum occurrences with pure or almost pure CO_2-bearing fluids, without any water included, were documented in sapphires from pegmatites in the Kerala district of India [96] and in a corundum bearing skarn from granulites in Southeast Madagascar [97]. In other corundum occurrences, such as in the Kashmir blue sapphires and in the Thailand sapphires, CO_2 is an important component of the source fluids [98,99]. The occurrence of high-density CO_2-rich fluid inclusions in granulite facies rocks shows that large amounts of CO_2 infiltrate the lower crust during the peak of metamorphism [100–102]. However, Hollister et al. [103] have shown that low-density CO_2 fluid inclusions must have been trapped after the peak of metamorphism. Two main sources for input of CO_2 fluids in the lower crust have been suggested [101,104,105]: from the mantle, or from the metamorphism of previously dehydrated crust. High-Al and low-Si protoliths in a high regional metamorphic grade can produce a pure supercritical CO_2 fluid and lead to the formation of corundum. A mantle-derived CO_2 has been suggested for Naxos by Schuiling and Kreulen [106].

In the present study, similar fluids containing almost pure CO_2 have been documented. Water wa not identified in the fluid inclusions, neither by optical microscopy nor by any phase transition during microthermometry. This excludes the possibility of fluid immiscibility for the corundum formation and implies the presence of a primary water-free (or very poor) CO_2 dominated fluid. However, the presence of minor amounts of water in the paleofluid in all studied corundums cannot be excluded due to the presence of hydrous mineral inclusions in the corundum. Already, Buick and Holland [107 have argued that the "primary" fluid inclusions in the metamorphic complex of Naxos have been compositionally modified by selective leakage of H_2O during uplift, and Krenn et al. [57] reported on recrystallization-induced leakage resulting in minor admixture of H_2O from former hydrous inclusion in corundum from pegmatites along the Nestos suture zone, Xanthi area.

It is very likely that the studied primary low-density fluid inclusions (d = 0.46–0.67 g/cm^3) were entrapped after the peak of metamorphism, suggesting that corundum formation took place during retrogression. The pseudosecondary, also low-density, carbonic fluid inclusions were entrapped in trails during the corundums formation process and are related to the evolution of the metamorphic events at a retrograde metamorphic regime of cooling and uplift and the subsequent exhumation also after the peak of metamorphism. These variations in the density of the fluid inclusions can be interpreted as the result of pressure variation associated with successive localized microfracturing a process which was suggested in the case of Luc Yen rubies of North Vietnam [95]. However, a continuous entrapment of the carbonic fluids during growth of corundum with pressure decreasing is not excluded.

8.4. Conditions of Greek Corundum Formation

The Xanthi and Paranesti corundums belong to metamorphic s.s. deposits, according to the classification of Giuliani et al. [2,7] and Simonet et al. [3], and more specifically, to those related to meta-limestones and mafic granulites, respectively. Both occurrences formed during the retrograde metamorphic path of high-temperature/medium-pressure metamorphism of platform carbonates and amphibolites during the Cenozoic collision that resulted in the Nestos Suture Zone. Wang et al. [30 concluded that Paranesti rubies were formed within an ultramafic precursor, most probably an aluminous clinopyroxenite, during amphibolite facies metamorphism. The estimated P–T conditions for their formation are 4 kbar < P < 7 kbar and 580 °C < T < 750 °C, and with subsequent retrogression

The Paranesti (and Gorgona/Xanthi) corundum-bearing assemblages followed a nearly isotherma decompression within the amphibolite facies and then a further evolution towards the greenschis facies, along the path that was proposed by Krenn et al. [57] for the Nestos suture zone (Figure 13) This path records a transition from the kyanite to the sillimanite stability field during retrogression as observed by former kyanite surrounded by fibrolitic sillimanite ([57,108] and this study). In the studied samples from Paranesti, corundum surrounds kyanite, and was probably formed after kyanite in the stability field of sillimanite during the retrogression. Mineralogical observations of the corundums from Paranesti ([30] and this study) indicate that the margarite occurs as reaction rims around ruby grains, suggesting that it was formed subsequently due to retrogression and according to the following reaction:

$$\text{Margarite} = \text{Anorthite} + \text{Corundum} + H_2O \quad (1)$$

In contrast to Paranesti, margarite in the Gorgona/Xanthi corundum-bearing marbles seems to be a prograde mineral ([22,109] and this study], indicating that corundum along with anorthite could have been formed by the breakdown of margarite (see Reaction (1)). In the studied samples, margarite is either in contact or is separated from corundum by a kaolinite rim, which probably represents earlier anorthite. As described by Storre and Nitsch [110] and Chatterjee [111,112], margarite breakdown to anorthite and corundum takes place at temperatures of 610, 625 and 650 °C for pressures of 6, 7 and bar, respectively. On the other hand, and under the assumption that margarite was formed together vith corundum, a more complex reaction:

$$8\ \text{Diaspore} + \text{Pyrophyllite} + 2\ \text{Calcite} = 2\ \text{Margarite} + \text{Corundum} + 2\ CO_2 + 3\ H_2O \quad (2)$$

s described by Okrusch et al. [113] and Haas and Holdaway [114] may explain this assemblage. This eaction requires the metastable persistence of diaspore + pyrophyllite, which should have otherwise eacted to form Al-silicate about 40–60 °C below the lower stability limit of corundum [113]. The ssemblage corundum and chlorite within the Gorgona marbles can be formed according to the ollowing reaction, as experimentally demonstrated by Seifert [115]:

$$\text{Mg-chlorite} + 6\ \text{Corundum} + 2\ \text{Spinel} = \text{sapphirine} + 4\ H_2O \quad (3)$$

ut the absence of sapphirine in the studied samples indicates that P–T did not exceed 6 kbar or temperatures between 620 and 720 °C [115]. The formation of the Xanthi corundum under issociation of Mg-spinel + calcite into corundum + dolomite during the retrograde metamorphism of pinel-bearing dolomites according to the reaction:

$$\text{Spinel} + \text{Calcite} = \text{Corundum} + \text{Dolomite} + CO_2 \quad (4)$$

vas not observed, but cannot be ruled out. This reaction path has been described by Buick and Iolland [116] in spinel-bearing dolomites from the leucocratic core of Naxos, where spinel crystals re separated from the calcite of the host rock by corundum + dolomite mantles, and they are often urrounded by Mg-chlorite + pargasite. The above authors suggested synchronous formation of Ag-chlorite + pargasite in place of spinel + calcite. Decomposition of spinel into corundum during he retrograde metamorphism in marbles as described by Reaction (4) was described as the major orundum-forming mechanism for the rubies at Jegdalek (Afghanistan), Hunza (Pakistan) and Luc 'en (Vietnam) [117].

In summary, conditions of corundum formation in the Gorgona marbles as suggested by iati [22,109] and this study, are in accordance to those estimated by Wang et al. [30] for the Paranesti ubies. As an alternative hypothesis, corundum in the Gorgona marbles could have formed at ower pressures of about 3–4 kbar, during late stages of shear deformation from CO_2-rich fluids as roposed by Krenn et al. [57] for corundum–anorthite assemblages postdating kyanite in desilicated egmatoid veins within gneisses and amphibolites of the Nestos suture zone, about 4 km south of the Gorgona locality.

In the absence of geochronological data for corundum-bearing assemblages within the Nestos uture zone, it remains difficult to state if they all represent the same metamorphic event, or can be ttributed to different P–T conditions corresponding to different ages.

The plumasites from Kinidaros, Naxos contain tourmaline and beryl, in addition to corundum, northite and phlogopite, and originate from desilication of leucogranitic pegmatites. For these egmatites, Matthews et al. [118] reported temperature variations from >700 °C to ~400 °C, based on oxygen isotope fractionation among quartz, tourmaline and garnet, and attributed their formation rom anatectic melts during regional high-temperature metamorphism. According to these authors, rystallization of the pegmatitic magmas should initiate under water-undersaturated conditions, but vith crystallization of anhydrous minerals and ascent, the magma should evolve to water-saturated onditions at 630 to 640 °C. Thus, some of the higher temperatures (T = 650 °C) given by the isotope hermometry could represent the crystallization at reduced water activity. In addition, Siebenaller [119] used fluid inclusion measurements in tourmaline, garnet and beryl from leucogranite pegmatites from Naxos island, to estimate P–T formation conditions for the pegmatites of between 5 kbar/600 °C and 2 kbar/450 °C, along the exhumation path of leucogranite (Figure 13). The absence of andalusite from the studied corundum-bearing assemblages constrains the lower limit of corundum formation along the retrograde path at about 3 kbar and 550 °C.

Finally, metabauxite occurrences on southern Naxos (Kavalaris Hill) correspond to the thermal dissociation of diaspore and the formation of corundum (α-AlOOH $\leftrightarrow$ Al_2O_3 + H_2O, Haas [120]), as described by Feenstra and Wunder [73]. The first corundum occurrence in Naxos metabauxites has been recorded as corundum-in isograd (T~420–450 °C and P~6 kbar, [73]). Similarly, the metabauxites from Ikaria were formed at T in the range 450–550 °C and P ~5–6kbar [75,77]. According to Iliopoulos [75] the latter correspond to corundum-chloritoid-bearing metabauxites from Naxos island (zones II and III of Feenstra [21]; Figure 13).

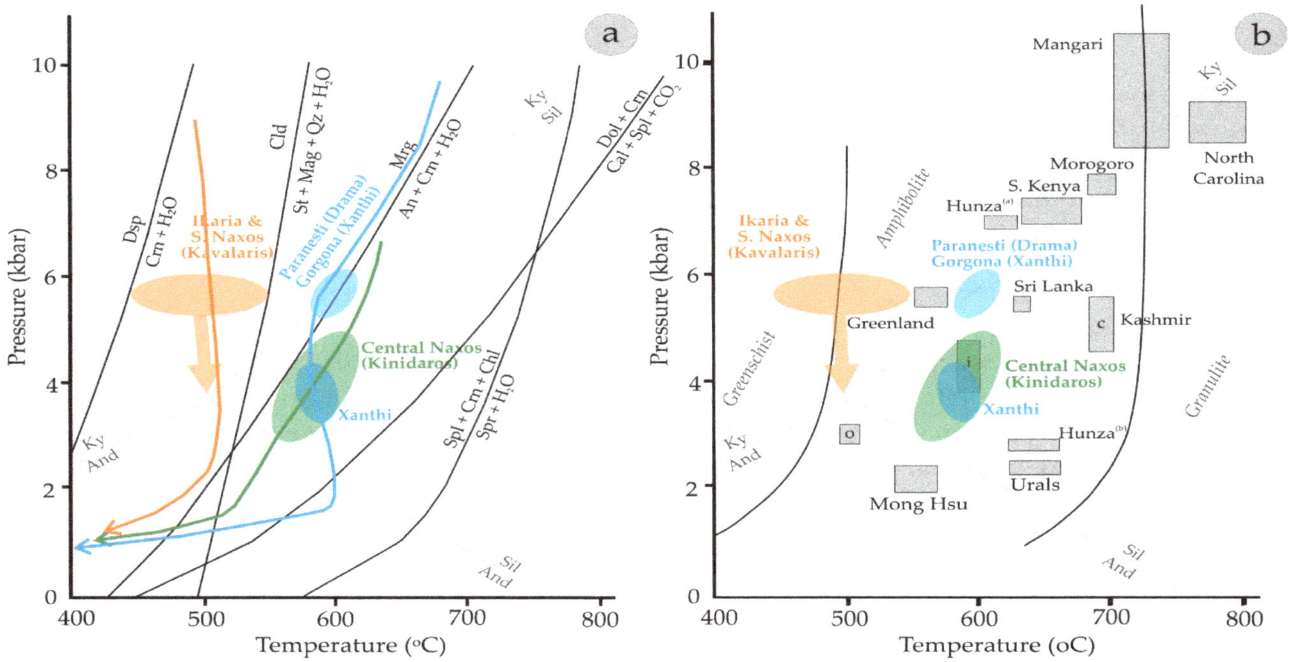

Figure 13. (**a**) P–T diagram showing mineral equilibria related to the formation of corundum in Greece (modified from Garnier et al. [117]). Upper blue area: Paranesti ruby and Gorgona sapphires ([30] and this study); Lower blue area: Alternative conditions for the Gorgona sapphires (for explanation see text); Orange area: Ikaria and southern Naxos metabauxites. The thick orange arrow indicates possible path for formation of vein-type sapphires from both localities; Green area: Central Naxos plumasites. Blue arrow: P–T–t path of high-grade rocks from the Nestos suture zone (from Krenn et al. [57]); Green arrow: P–T–t path of the tourmaline-garnet-beryl-bearing leucogranite from Naxos island (from Siebenaller [119]); Orange arrow: P–T–t path of the lower limit of metamorphic zone III (which hosts corundum-chloritoid-bearing metabauxites) from Naxos island (from Duchêne et al. [64]). The equilibrium curve "chloritoid + O_2 = staurolite + magnetite + quartz", which marks the disappearance of chloritoid, defines an upper limit for the studied chloritoid-bearing metabauxites at southern Naxos (Kavalaris) and Ikaria islands is from Feenstra [21]. Abbreviations: An = anorthite; And = andalusite; Cc = calcite; Cld = chloritoid; Clin = clinochlore; Co = corundum; Do = dolomite; Dsp = diaspore; Ky = kyanite; Ma = margarite; Mag = magnetite; Qz = quartz; Sill = sillimanite; Spr = sapphirine; Sp = spinel; St = staurolite; (**b**) Comparison between the hypothetical P–T conditions for the formation of corundum in Greece and in metamorphic deposits around the world (modified from Giuliani et al. [7], Simonet et al. [3]). Boxes indicate P–T fields of known deposits: North Carolina, Rubies from mafic granulites [121]; Morogoro, corundum-bearing anatexites [122]; Mangari, Southern Kenya, Metasomatic rubies [123–125]; Hunza, rubies in marbles [113]; Sri Lanka, sapphires from granulites [94]; Greenland, metasomatic rubies [126]; Kashmir metasomatic sapphires with three P–T boxes corresponding to the evolution of the fluids in the sapphire crystals from the center (c), to intermediate (i) and outer (o) zones [98]; Urals, rubies in marbles [127], and Mong Hsu, rubies in marbles [128].

However, the studied sapphires from both localities correspond to open space-filling material in extensional fissures and indicate a second, late stage of corundum formation. We interpret this corundum formation to be of metasomatic origin from a low-temperature CO_2-bearing metamorphic fluid, as already proposed for similar sapphire + margarite + tourmaline-filled veins from Naxos island [116]. This is in agreement with the findings of Tropper and Manning [129], who consider corundum-filled hydrofractures in the Naxos metabauxites to be products of retrograde cooling and decompression, consistent with kinematically late textures. This hypothesis is also supported by the fact that trace element fingerprints of Naxos and Ikaria metabauxite-hosted sapphire suggest magmatic affinities (this study). A hydrothermal origin of diaspore has been reported for the gem-quality diaspore crystals (var. zultanite) occurring in fissures of metabauxites in the İlbir Mountains, SW Turkey [130]. According to the above authors, the diaspore mineralization is caused by hydrothermal remobilization of primary bauxite components into crosscutting structures. Figure 13 displays an evolution (thick orange arrow) towards lower P–T conditions, along the P–T–t retrograde path of the metamorphic zone III (hosting corundum-chloritoid-bearing metabauxites) from Naxos island (according to Duchêne et al. [53]). Since the tectonometamorphic evolution of Ikaria is quite similar to that of Naxos [55,74], this path could probably reflect the formation of the vein-type sapphire assemblages in both Naxos and Ikaria islands.

8.5. *Greece: A New Gem Corundum Province?*

Greek corundum occurrences display a wide color variation, ranging from deep red, pink, purple, and blue to colorless, with crystal sizes of up to 5 cm. Among the studied occurrences, some corundums from Gorgona/Xanthi and the plumasite-hosted sapphires from Naxos display significant transparency and homogeneity of color and should be further examined for their suitability as potential cut gemstones. The rest of the studied corundums, especially the vivid-colored varieties, could be considered suitable for their use as cabochons. Although gem-quality corundums are considered to be absent in emery deposits [7], the Naxos and especially Ikaria blue sapphires are of gem (cabochon) quality, and are atypical for other metabauxite-hosted corundums. Future exploration is required in order to establish the potential for economic exploitation of the corundum-bearing areas in Greece, pointing towards a new gem corundum province in the world.

We highlight here the enrichment in Be of blue-colorless sapphires from Naxos plumasites, a feature which has only been reported from Ilakaka, Madagaskar, Sar-e-Sang, Afganistan and Weldborough, Tasmania, and was attributed to nano-inclusons of unidentified Be-rich phases [131]. The Naxos plumasitic sapphires, as well as those from the Xanthi marbles and Ikaria metabauxites, display very high values in Nb, Ta, W and Sn, much higher than those reported from the above occurrences. A further study of the concentration of these elements and especially of Be in the Greek corundums could be a useful tool for determining their origin [131].

9. Conclusions

Gem corundum in Greece covers in a variety of geological environments located within the Rhodope (Xanthi and Drama areas) and Attico-Cycladic (Naxos and Ikaria islands) tectono-metamorphic units. Pargasite-schist hosted a ruby deposit in the Paranesti/Drama area and marble hosted pink to blue sapphires in the Xanthi area, occurring along the UHP-HP Nestos suture zone. Plumasite-hosted sapphires from Naxos island display a wide color variation (blue to colorless and pink). Deep blue colored sapphires from Naxos and Ikaria islands are hosted within extensional fissures in metabauxite lenses within marbles. Various mineral inclusions in corundums are in equilibrium and/or postdate corundum crystallization, and reflect the surrounding mineralogical assemblages. Included in corundums are: spinel and pargasite (Paranesti), spinel, zircon (Xanthi), margarite, zircon, apatite, diaspore, phlogopite and chlorite (Naxos) and chloritoid, ilmenite, hematite, ulvospinel, rutile and zircon (Ikaria).

The chromophores of the studied corundums show a wide range in concentration and a uniqu trace element chemistry with variable critical ratios (Fe/Mg, Ga/Mg, Ga/Cr and Fe/Ti). Be, Nb, Sr Ta and W are anomalously enriched in the plumasite-related sapphires from Naxos.

Based on the geological setting of formation and trace element fingerprints the Paranesti an Xanthi corundum occurrences can be classified as metamorphic s.s hosted mafics/ultramafics an marbles, respectively. Those from central Naxos are of metasomatic origin and are related to desilicate pegmatites crosscutting ultramafic rocks. Finally, blue sapphires from southern Naxos and Ikaria hosted in metabauxites, display an atypical magmatic signature indicating a metasomatic origin b fluid-rock interaction. The study of fluid inclusions in corundum showed that they are CO_2-dominan with low density with very small quantities of CH_4 and/or N_2, and water-free. CO_2-rich fluids ar probably of metamorphic origin and derived from devolatilization of carbonate formations. Gree corundums are characterized by a wide color variation, homogeneity of the color hues and transparenc and could be considered as potential gemstones.

Author Contributions: P.V. and C.M. collected the studied samples. P.V. assisted by C.M., I.G., K.Z., S.M., S.K K.W., S.H., St.K., J.B., F.Z., V.K., M.T. and G.L. obtained and evaluated the mineralogical data. G.G. and A. obtained and evaluated the oxygen isotopic data. V.M. and A.T. conducted the fluid inclusion measurements. P.V C.M., G.G., V.M., A.T. and S.K. wrote the manuscript.

Acknowledgments: The authors would like to thank Evangelos Michailidis for his kind help with the SEM i the University of Athens. Two anonymous reviewers and the Editor are especially thanked for their constructiv comments that greatly improved the manuscript.

References

1. Garnier, V.; Ohnenstetter, D.; Giuliani, G.; Schwarz, D. Saphirs et rubis. Classification des gisements d corindon. *Règne Minéral* **2004**, *55*, 4–47.
2. Giuliani, G.; Ohnenstetter, D.; Garnier, V.; Fallick, A.E.; Rakotondrazafy, M.; Schwarz, D. The geology an genesis of gem corundum deposits. In *Geology of Gem Deposits*; Mineralogical Association of Canada Shor Course Series; Groat, L., Ed.; Mineralogical Association of Canada: Quebec City, QC, Canada, 2007; Volum 37, pp. 23–78.
3. Simonet, C.; Fritsch, E.; Lasnier, B. A classification of gem corundum deposits aimed towards gem exploratio *Ore Geol. Rev.* **2008**, *34*, 127–133. [CrossRef]
4. Sutherland, F.L.; Schwarz, D.; Jobbins, E.A.; Coenraads, R.R.; Webb, G. Distinctive gem corundum fron discrete basalt fields: A comparative study of Barrington, Australia, and west Pailin, Cambodia gem field *J. Gemmol.* **1998**, *26*, 65–85. [CrossRef]
5. Garnier, V.; Giuliani, G.; Ohnenstetter, D.; Schwarz, D.; Kausar, A.B. Les gisements de rubis associés au marbres de l'Asie centrale et du Sud-est. *Règne Minéral* **2006**, *67*, 17–48.
6. Garnier, V.; Maluski, H.; Giuliani, G.; Ohnenstetter, D.; Schwarz, D. Ar–Ar and U–Pb ages of marble-hoste ruby deposits from central and southeast Asia. *Canad. J. Earth Sci.* **2006**, *43*, 509–532. [CrossRef]
7. Giuliani, G.; Ohnenstetter, D.; Fallick, A.E.; Groat, L.; Fagan, A.J. The geology and genesis of gem corundun deposits. In *Geology of Gem Deposits*, 2nd ed.; Mineralogical Association of Canada Short Course Serie Groat, L.A., Ed.; Mineralogical Association of Canada: Quebec City, QC, Canada, 2014; Volume 4 pp. 113–134. ISBN 9780921294375.
8. Graham, I.T.; Khin, Z.; Cook, N.J. The genesis of gem deposits. *Ore Geol. Rev.* **2008**, *34*, 1–2. [CrossRef]
9. Graham, I.; Sutherland, L.; Zaw, K.; Nechaev, V.; Khanchuk, A. Advances in our understanding of the gen corundum deposits of the West Pacific continental margins intraplate basaltic fields. *Ore Geol. Rev.* **2008**, *3* 200–215. [CrossRef]
10. Sorokina, E.S.; Karampelas, S.; Nishanbaev, T.R.; Nikandrov, S.N.; Semiannikov, B.S. Sapphire megacrysts i syenite pegmatites from the Ilmen Mountains, South Urals, Russia: New mineralogical data. *Can. Mine* **2017**, *55*, 823–843. [CrossRef]

1. Muhlmeister, S.; Fritsch, E.; Shigley, E.J.; Devouard, B.; Laurs, M.B. Separating natural and synthetic rubies on the basis of trace-element chemistry. *Gems Gemol.* **1998**, *34*, 80–101. [CrossRef]
2. Sutherland, F.L.; Schwarz, D. Origin of gem corundums from basaltic fields. *Aust. Gemmol.* **2001**, *21*, 30–33.
3. Saminpanya, S. Mineralogy and Origin of Gem Corundum Associated with Basalt in Thailand. Ph.D. Thesis, University of Manchester, Manchester, UK, 2000.
4. Giuliani, G.; Fallick, A.E.; Garnier, V.; France-Lanord, C.; Ohnenstetter, D.; Schwarz, D. Oxygen isotope composition as a tracer for the origins of rubies and sapphires. *Geology* **2005**, *33*, 249. [CrossRef]
5. Abduriyim, A.B.; Kitwaki, H. Determination of the origin of blue sapphire using laser ablation inductively coupled plasma mass spectrometry (LA-ICP-MS). *J. Gemmol.* **2006**, *30*, 23–26. [CrossRef]
6. Peucat, J.J.; Ruffault, P.; Fritsch, E.; Bouhnik-Le Coz, M.; Simonet, C.; Lasnier, B. Ga/Mg ratio as a new geochemical tool to differentiate magmatic from metamorphic blue sapphires. *Lithos* **2007**, *98*, 261–274. [CrossRef]
7. Sutherland, F.L.; Duroc-Danner, J.M.; Meffre, S. Age and origin of gem corundum and zircon megacrysts from the Mercaderes–Rio Mayo area, South-west Colombia, South America. *Ore Geol. Rev.* **2008**, *34*, 155–168. [CrossRef]
8. Papastamatiou, I. The Emery of Naxos. *Geol. Geoph. Res.* **1951**, *1*, 37–68.
9. Andronopoulos, V.; Tsoutrelis, C. Investigation of a Red Corundum Occurence in Xanthi, Greece. Unpublished Report; Greek Geol. Survey, Athens, Greece. 1964; 8p.
0. Ktenas, C. La géologie de l' île de Nikaria (rédigée des restes de l'auteur par G. Marinos). *Geol. Geoph. Res.* **1969**, *13*, 57–85.
1. Feenstra, A. Metamorphism of Bauxites on Naxos, Greece. Ph.D. Thesis, Instituut voor Aardwetenschappen, Rijksuniversiteit te Utrecht, Utrecht, The Netherlands, 1985.
2. Liati, A. Corundum-and zoisite-bearing marbles in the Rhodope Zone, Xanthi area (N. Greece): Estimation of the fluid phase composition. *Mineral. Petrol.* **1988**, *38*, 53–60. [CrossRef]
3. Zannas, I. Corundum occurrence at Stirigma-Gorgona area, northern Xanthi. Petrogenesis and color variations. *Miner. Wealth* **1995**, *97*, 9–18.
4. Iliopoulos, I.; Katagas, C. Corundum bearing metabauxites from Ikaria Island (Greece): Mineralogy and geochemistry. In Proceedings of the 10th International Congress of the Geological Society of Greece, Thessaloniki, Greece, 15–17 April 2004; pp. 423–424.
5. Voudouris, P. The minerals of Eastern Macedonia and Western Thrace: Geological framework and environment of formation. *Bull. Geol. Soc. Greece* **2005**, *37*, 62–77.
6. Voudouris, P.; Melfos, V.; Katerinopoulos, A. Precious stones in Greece: Mineralogy and geological environment of formation. Understanding the genesis of ore deposits to meet the demand of the 21st century. In Proceedings of the 12th Quadrennial IAGOD Symposium, Moscow, Russia, 21–24 August 2006.
7. Voudouris, P.; Graham, I.; Melfos, V.; Zaw, K.; Lin, S.; Giuliani, G.; Fallick, A.; Ionescu, M. Gem corundum deposits of Greece: Diversity, chemistry and origins. In Proceedings of the 13th Quadrennial IAGOD Symposium, Adelaide, Australia, 6–9 April 2010; Volume 69, pp. 429–430.
8. Voudouris, P.; Katerinopoulos, A.; Magganas, A. Mineralogical geotopes in Greece: Preservation and promotion of museum specimens of minerals and gemstones. Sofia Initiative "Mineral diversity preservation". In Proceedings of the IX International Symposium on Mineral Diversity Research and Reservation, Sofia, Bulgaria, 16–18 October 2017; pp. 149–159.
9. Graham, I.; Voudouris, P.; Melfos, V.; Zaw, K.; Meffre, S.; Sutherland, F.; Giuliani, G.; Fallick, A. Gem corundum deposits of Greece: A spectrum of compositions and origins. In Proceedings of the 34th IGC Conference, Brisbane, Australia, 5–10 August 2012.
0. Wang, K.K.; Graham, I.T.; Lay, A.; Harris, S.J.; Cohen, D.R.; Voudouris, P.; Belousova, E.; Giuliani, G.; Fallick, A.E.; Greig, A. The Origin of A New Pargasite-Schist Hosted Ruby Deposit From Paranesti, Northern Greece. *Can. Mineral.* **2017**, *55*, 535–560. [CrossRef]
1. Voudouris, P. Hydrothermal corundum, topaz, diaspore and alunite supergroup minerals in the advanced argillic alteration lithocap of the Kassiteres-Sapes porphyry-epithermal system, western Thrace, Greece. *Neues Jahrbuch für Mineralogie Abhandlungen* **2014**, *191*, 117–136. [CrossRef]
2. Pouchou, J.L.; Pichoir, F. Quantitative analysis of homogeneous or stratified microvolumes applying the model "PAP". In *Electron Probe Quantitation*; Heinrich, K.F.J., Newbury, D.E., Eds.; Plenum Press: New York, NY, USA, 1991; pp. 31–75.

33. Griffin, W.L.; Powell, W.J.; Pearson, N.J.; O'Reilly, S.Y. GLITTER: Data reduction software for laser ablatio ICP-MS. In *Laser Ablation ICP-MS in the Earth Sciences: Current Practices and Outstanding Issues*; Mineralogica Association of Canada Short Course Series; Sylvester, P., Ed.; Mineralogical Association of Canada: Quebe City, QC, Canada, 2008; Volume 40, pp. 308–311.
34. Van Achterbergh, E.; Ryan, C.G.; Jackson, S.E.; Griffin, W.L. Data reduction software for LA-ICP-M Appendix. In *Laser Ablation-ICPMass Spectrometry in the Earth Sciences: Principles and Application* Mineralogical Association of Canada Short Course Series; Sylvester, P.J., Ed.; Mineralogical Association c Canada: St John's, NL, Canada, 2001; Volume 29, pp. 239–243.
35. Driesner, T.; Heinrich, C.A. The system H_2O-NaCl. I. Correlations for molar volume, enthalpy, and isobari heat capacity from 0 to 1000 degrees C, 1 to 5000 bar, and 0 to 1 X-NaCl. *Geochim. Cosmochim. Acta* **2007**, *7* 4880–4901. [CrossRef]
36. Sharp, Z.D. In situ laser microprobe techniques for stable isotope analysis. *Chem. Geol.* **1992**, *101*, 3–1 [CrossRef]
37. Jolivet, L.; Brun, J.P. Cenozoic geodynamic evolution of the Aegean region. *Int. J. Earth. Sci.* **2010**, *99*, 109–13 [CrossRef]
38. Ring, U.; Glodny, J.; Thomson, S. The Hellenic subduction system: High-pressure metamorphism exhumation, normal faulting and large-scale extension. *Earth Plan. Sci.* **2010**, *38*, 45–76. [CrossRef]
39. Jolivet, L.; Faccenna, C.; Huet, B.; Labrousse, L.; Le Pourhiet, L.; Lacombe, O.; Lecomte, E.; Burov, E Denèle, Y.; Brun, J.-P.; et al. Aegean tectonics: Strain localization, slab tearing and trench retreat. *Tectonophysic* **2013**, *597*, 1–33. [CrossRef]
40. Fytikas, M.; Innocenti, F.; Manneti, P.; Mazzuoli, R.; Peccerillo, A.; Villari, L. Tertiary to Quaternary evolutio of volcanism in the Aegean region. In *The Geological Evolution of the Eastern Mediterranean*; Dixon, J.E Robertson, A.H.F., Eds.; Geological Society Special Publications: Oxford, UK, 1984; Volume 17, pp. 687–69
41. Kydonakis, K.; Brun, J.-P.; Sokoutis, D. North Aegean core complexes, the gravity spreading of a thrus wedge. *J. Geophys. Res. Solid Earth* **2015**, *120*, 595–616. [CrossRef]
42. Voudouris, P. *Gemstones and Mineral-Megacrysts in Greece*; University New South Wales: Sydne Australia, 2010.
43. Burg, J.-P. Rhodope: From Mesozoic convergence to Cenozoic extension. Review of petro-structural data i the geochronological frame. *J. Virtual Explor.* **2012**, *42*, 1–44. [CrossRef]
44. Brun, J.-P.; Sokoutis, D. Kinematics of the Southern Rhodope Core Complex (North Greece). *Int. J. Earth Sc* **2007**, *96*, 1079–1099. [CrossRef]
45. Turpaud, P.; Reischmann, T. Characterization of igneous terranes by zircon dating: Implications for UH occurrences and suture identification in the Central Rhodope, Northern Greece. *Int. J. Earth Sci.* **2010**, *9* 567–591. [CrossRef]
46. Mposkos, E.; Kostopoulos, D. Diamond, former coesite and supersilicic garnet in metasedimentary rock from the Greek Rhodope: A new ultrahigh-pressure metamorphic province established. *Earth Plan. Sci. Let* **2001**, *192*, 497–506. [CrossRef]
47. Liati, A. Identification of repeated Alpine (ultra) highpressure metamorphic events by U-Pb SHRIM geochronology and REE geochemistry of zircon: The Rhodope zone of Northern Greece. *Contrib. Minera Petrol.* **2005**, *150*, 608–630. [CrossRef]
48. Gautier, P.; Bosse, V.; Cherneva, Z.; Didier, A.; Gerdjikov, I. Polycyclic alpine orogeny in the Rhodop metamorphic complex: The record in migmatites from the Nestos shear zone (N. Greece). *Bull. Geol. So France* **2017**, *188*, 36. [CrossRef]
49. Bonev, N.; Burg, J.-P.; Ivanov, Z. Mesozoic–Tertiary structural evolution of an extensional gneiss dome—Th Kesebir–Kardamos dome, eastern Rhodope (Bulgaria–Greece). *Int. J. Earth Sci.* **2006**, *95*, 318–340. [CrossRef
50. Menant, A.; Jolivet, L.; Vrielynck, B. Kinematic reconstructions and magmatic evolution illuminating crusta and mantle dynamics of the eastern Mediterranean region since the late Cretaceous. *Tectonophysics* **2016**, *67* 103–140. [CrossRef]
51. Wüthrich, E. Low Temperature Thermochronology of the Northern Aegean Rhodope Massif. Unpublishe Ph.D. Thesis, ETH Zürich, Zürich, Switzerland, 2009.
52. Melfos, V.; Voudouris, P. Cenozoic metallogeny of Greece and potential for precious, critical and rare metal exploration. *Ore Geol. Rev.* **2017**, *59*, 1030–1057. [CrossRef]

53. Duchêne, S.; Aïssa, R.; Vanderhaeghe, O. Pressure-temperature-time evolution of metamorphic rocks from Naxos (Cyclades, Greece): Constraints from thermobarometry and Rb/Sr dating. *Geodin. Acta* **2006**, *19*, 301–321. [CrossRef]
54. Ottens, B.; Voudouris, P. *Griechenland: Mineralien-Fundorte-Lagerstätten*; Christian Weise Verlag: Munchen, Germany, 2018; 480p.
55. Beaudoin, A.; Augier, R.; Laurent, V.; Jolivet, L.; Lahfid, A.; Bosse, V.; Arbaret, L.; Rabillard, A.; Menant, A. The Ikaria hightemperature Metamorphic Core Complex (Cyclades, Greece): Geometry, kinematics and thermal structure. *J. Geodesy* **2015**, *92*, 18–41.
56. Papanikolaou, D.; Panagopoulos, G. On the structural style of Southern Rhodope (Greece). *Geol. Balc.* **1981**, *11*, 13–22.
57. Krenn, K.; Bauer, C.; Proyer, A.; Mposkos, E.; Hoinkes, G. Fluid entrapment and reequilibration during subduction and exhumation: A case study from the high-grade Nestos shear zone, Central Rhodope, Greece. *Lithos* **2008**, *104*, 33–53. [CrossRef]
58. Nagel, T.J.; Schmidt, S.; Janák, M.; Froitzheim, N.; Jahn-Awe, S.; Georgiev, N. The exposed base of a collapsing wedge: The Nestos Shear Zone (Rhodope Metamorphic Province, Greece): The Nestos shear zone. *Tectonics* **2011**, *30*. [CrossRef]
59. Bonneau, M. Correlation of the Hellenic nappes in the south-east Aegean and their tectonic reconstruction. *Geol. Soc. Lond. Spec. Publ.* **1984**, *17*, 517–527. [CrossRef]
60. Scheffer, C.; Vanderhaeghe, O.; Lanari, P.; Tarantola, A.; Ponthus, L.; Photiades, A.; France, L. Syn- to post-orogenic exhumation of metamorphic nappes: Structure and thermobarometry of the western Attic-Cycladic metamorphic complex (Lavrion, Greece). *J. Geodyn.* **2016**, *96*, 174–193. [CrossRef]
61. Grasemann, B.; Schneider, D.A.; Stöckli, D.F.; Iglseder, C. Miocene bivergent crustal extension in the Aegean: Evidence from the western Cyclades (Greece). *Lithosphere* **2012**, *4*, 23–39. [CrossRef]
62. Kruckenberg, S.C.; Vanderhaeghe, O.; Ferré, E.C.; Teyssier, C.; Whitney, D.L. Flow of partially molten crust and the internal dynamics of a migmatite dome, Naxos, Greece. *Tectonics* **2011**, *30*. [CrossRef]
63. Laurent, V.; Huet, B.; Labrousse, L.; Jolivet, L.; Monie, P.; Augier, R. Extraneous argon in high-pressure metamorphic rocks: Distribution, origin and transport in the Cycladic Blueschist Unit (Greece). *Lithos* **2017**, *272*, 315–335. [CrossRef]
64. Altherr, R.; Kreuzer, H.; Wendt, I.; Lenz, H.; Wagner, G.A.; Keller, J.; Harre, W.; Höhndorf, A. A late Oligocene/early Miocene high temperature belt in the Attic-Cycladic crystalline complex (SE Pelagonian, Greece). *Geol. Jahrb.* **1982**, *23*, 97–164.
65. Lister, G.S.; Banga, G.; Feenstra, A. Metamorphic core complexes of cordilleran type in the Cyclades, Aegean Sea, Greece. *Geology* **1984**, *12*, 221–225. [CrossRef]
66. Gautier, P.; Brun, J.P. Crustal-scale geometry and kinematics of late-orogenic extension in the central Aegean (Cyclades and Ewia Island). *Tectonophysics* **1994**, *38*, 399–424. [CrossRef]
67. Jansen, J.B.H.; Schuiling, R.D. Metamorphism on Naxos; petrology and geothermal gradients. *Am. J. Sci.* **1976**, *276*, 1225–1253. [CrossRef]
68. Wybrans, J.R.; McDougall, I. Metamorphic evolution of the Attic Cycladic Metamorphic Belt on Naxos (Cyclades, Greece) utilizing $^{40}Ar/^{39}Ar$ age spectrum measurements. *J. Metam. Geol.* **1988**, *6*, 571–594. [CrossRef]
69. Buick, I.S.; Holland, T.J.B. The P-T path associated with crustal extension, Naxos, Cyclades, Greece. In *Evolution on Metamorphic Belts*; Daly, J.S., Cliff, R.A., Yardley, B.W.D., Eds.; Geological Society Special Publications: Lomdon, UK, 1989; Volume 43, pp. 365–369.
70. Baker, J.; Matthews, A. The stable isotopic evolution of a metamorphic complex, Naxos, Greece. *Contrib. Mineral. Petrol.* **1995**, *120*, 391–403. [CrossRef]
71. Katzir, Y.; Valley, J.W.; Matthews, A.; Spicuzza, M.J. Tracking fluid flow during deep crustal anatexis: Metasomatism of peridotites (Naxos, Greece). *Contrib. Mineral. Petrol.* **2002**, *142*, 700–713. [CrossRef]
72. Andriessen, P.A.M.; Hebedaa, E.H.; Simonb, O.J.; Verschurea, R.H. Tourmaline K/Ar ages compared to other radiometric dating systems in Alpine anatectic leucosomes and metamorphic rocks (Cyclades and southern Spain). *Chem. Geol.* **1991**, *91*, 33–48. [CrossRef]
73. Feenstra, A.; Wunder, B. Dehydration of diasporite to corundite in nature and experiment. *Geology* **2002**, *30*, 119–122. [CrossRef]

74. Laurent, V.; Beaudoin, A.; Jolivet, L.; Arbaret, L.; Augier, R.; Rabillard, A.; Menant, A. Interrelations between extensional shear zones and synkinematic intrusions: The example of Ikaria Island (NE Cyclades, Greece *Tectonophysics* **2015**, *651–652*, 152–171. [CrossRef]
75. Iliopoulos, I. Petrogenesis of Metamorphic Rocks from Ikaria Island. Ph.D. Thesis, University of Patras Patra, Greece, 2005.
76. Ring, U. The Geology of Ikaria Island: The Messaria extensional shear zone, granites and the exotic Ikari nappe. *J. Virtual Explor.* **2007**, *27*, 3.
77. Liati, A.; Skarpelis, N. The metabauxites of Ikaria island, Eastern Aegean, Greece. In Proceedings of the 5th International Symposium on eastern Mediterranean Geology, Thessaloniki, Greece, 14–20 April 2004 pp. 1427–1430.
78. Henry, D.J.; Dutrow, B.L. Compositional zoning and element partitioning in nickeloan tourmaline from metamorphosed karstbauxite from Samos, Greece. *Am. Mineral.* **2001**, *86*, 1130–1142. [CrossRef]
79. Karantoni, V.D. Gemological Study of Corundum, Variety of Ruby and Sapphire from Greece. Unpublished Master's Thesis, School of Geology, Aristotle University of Thessaloniki, Thessaloniki, Greece, 2018.
80. Bodnar, R.J. Introduction to fluid inclusions. In *Fluid Inclusions: Analysis and Interpretation*; Mineralogical Association of Canada Short Course Series; Samson, I.M., Anderson, A.J., Marshall, D.D., Eds.; Mineralogical Association of Canada: Quebec City, QC, Canada, 2003; Volume 32, pp. 1–8.
81. Giuliani, G.; Fallick, A.; Rakotondrazafy, M.; Ohnenstetter, D.; Andriamamonjy, A.; Ralantoarison, T Rakotosamizanany, S.; Razanatseheno, M.; Offant, Y.; Garnier, V.; et al. Oxygen isotope systematics of gem corundum deposits in Madagascar: Relevance for their geological origin. *Mineral. Depos.* **2007**, *42*, 251–270 [CrossRef]
82. Giuliani, G.; Fallick, A.E.; Ohnenstetter, D.; Pegre, G. Oxygen isotope composition of sapphire from the French Massif Central: Implications for the origin of gem corundum in basalt fields. *Mineral. Depos.* **2009**, *44* 221–231. [CrossRef]
83. Wang, K.K.; Graham, I.T.; Martin, L.; Voudouris, P.; Giuliani, G.; Lay, A.; Harris, S.; Fallick, A.E. Fingerprinting Paranesti rubies through in-situ oxygen isotopes. *Minerals*. (under review).
84. Sutherland, F.L.; Zaw, K.; Meffre, S.; Giuliani, G.; Fallick, A.E.; Graham, I.T.; Webb, G.B. Gem corundum megacrysts from East Australia basalt fields: Trace elements, O isotopes and origins. *Aust. J. Earth Sci.* **2009** *56*, 1003–1020.
85. Harris, S.J.; Graham, I.T.; Lay, A.; Powell, W.; Belousova, E.; Zappetini, E. Trace element geochemistry and metasomatic origin of alluvial sapphires from the Orosmayo region, Jujuy Province, Northwest Argentina *Canad. Mineral.* **2017**, *55*, 595–617. [CrossRef]
86. Sutherland, F.; Zaw, K.; Meffre, S.; Yui, T.-F.; Thu, K. Advances in Trace Element "Fingerprinting" of Gem Corundum, Ruby and Sapphire, Mogok Area, Myanmar. *Minerals* **2015**, *5*, 61–79. [CrossRef]
87. Giuliani, G.; Lasnier, B.; Ohnenstetter, D.; Fallick, A.E.; Pégère, G. Les gisements de corindon de France *Le Règne Minéral* **2010**, *93*, 5–22.
88. Giuliani, G.; Caumon, G.; Rakotozamisanny, S.; Ohnenstetter, D.; Rakotondrazafy, M. Classification chimique des corindons par analyse factorielle discriminante: Application à la typologie des gisements de rubis et saphirs. *Rev. Gemmol.* **2014**, *188*, 14–22.
89. Zwaan, J.H.; Buter, E.; Mertz-Kraus, R.; Kane, R.E. Alluvial sapphires from Montana: Inclusions geochemistry and indications of a metasomatic origin. *Gems Gemmol.* **2015**, *51*, 370–391.
90. Schwarz, D.; Pardieu, V.; Saul, J.M.; Schmetzer, K.; Laus, B.M.; Giuliani, G.; Klemm, L.; Malsy, A.; Erel, E. Hauzenberger, C.; et al. Rubies and Sapphires from Winza, Central Tanzania. *Gems Gemmol.* **2008**, *44* 322–347. [CrossRef]
91. Rakotondrazafy, A.F.M.; Giuliani, G.; Ohnenstetter, D.; Fallick, A.E.; Rakotosamizanany, S. Andriamamonjy, A.; Ralantoarison, T.; Razanatseheno, M.; Offant, Y.; Garnier, V.; et al. Gem corundum deposits of Madagascar: A review. *Ore Geol. Rev.* **2008**, *34*, 134–154. [CrossRef]
92. Pham Van, L.; Hoáng Quang, V.; Garnier, V.; Giuliani, G.; Ohnenstetter, D.; Lhomme, T.; Schwarz, D. Fallick, A.E.; Dubessy, J.; Phan Trong, T. Gem corundum deposits in Vietnam. *J. Gemmol.* **2004**, *29*, 129–147
93. Simonet, C.; Paquette, J.L.; Pin, C.; Lasnier, B.; Fritsch, E. The Dusi (Garba Tula) sapphire deposit, Central Kenya—A unique Pan-African corundum-bearing monzonite. *J. Afr. Earth Sci.* **2004**, *38*, 401–410. [CrossRef
94. De Maesschalck, A.A.; Oen, I.S. Fluid and mineral inclusions in corundum from gem gravels in Sri Lanka *Mineral. Mag.* **1989**, *53*, 539–545. [CrossRef]

5. Giuliani, G.; Dubessy, J.; Banks, D.; Quang, V.H.; Lhomme, T.; Pironon, J.; Garnier, V.; Trong Trinhd, P.; Van Longf, P.; Ohnenstetter, D.; et al. CO_2-H_2S-COS-S_8-AlO(OH)-bearing fluid inclusions in ruby from marble-hosted deposits in Luc Yen area, North Vietnam. *Chem. Geol.* **2003**, *194*, 167–185. [CrossRef]
6. Menon, R.D.; Santosh, M.; Yoshida, M. Gemstone mineralization in southern Kerala, India. *J. Geol. Soc. India* **1994**, *44*, 241–252.
7. Rakotondrazafy, A.F.M.; Moine, B.; Cuney, M. Mode of formation of hibonite ($CaAl_{12}O_{19}$) within the U-Th skarns from the granulites of S-E Madagascar. *Contrib. Mineral. Petrol.* **1996**, *123*, 190–201. [CrossRef]
8. Peretti, A.; Mullis, J.; Kündig, R. Die Kaschmir-Saphire und ihr geologisches Erinnerungsvermögen. *Neue Zürcher Zeitung* **1990**, *187*, 59.
9. Srithai, B.; Rankin, A.H. Fluid inclusion characteristics of sapphires from Thailand. In *Mineral Deposits: Processes to Processing. Proceedings of the 5th Biennial SGA Meeting and the 10th Quadrennial IAGOD Symposium, London, UK, 22–25 August 1999*; Stanley, C.J., Ed.; Balkema: Rotterdam, The Netherlands, 1999; pp. 107–110.
00. Touret, J.L.R. Le facies granulite en Norvege meridionale. II. Les inclusions fluids. *Lithos* **1971**, *4*, 423–436. [CrossRef]
01. Newton, R.C.; Smith, J.V.; Windley, B.F. Carbonic metamorphism, granulites and crustal growth. *Nature* **1980**, *288*, 45–50. [CrossRef]
02. Morrison, J.; Valley, J.W. Post-granulite facies fluid infiltration in the Adirondack Mountains. *Geology* **1988**, *16*, 513–516. [CrossRef]
03. Hollister, L.S.; Burruss, R.C.; Henry, D.L.; Hendel, E.M. Physical conditions during uplift of metamorphic terrains, as recorded by fluid inclusions. *Bull. Minéral.* **1979**, *102*, 555–561.
04. Lamb, W.M.; Valley, J.W. Metamorphism of reduced granulites in low-CO_2 vapor-free environment. *Nature* **1984**, *312*, 56–58. [CrossRef]
05. Newton, R.C. Fluids of granulite facies metamorphism. In *Fluid—Rock Interactions during Metamorphism*; Walther, J.V., Wood, B.J., Eds.; Springer: New York, NY, USA, 1986; Volume 5, pp. 36–59.
06. Schuiling, R.D.; Kreulen, R. Are thermal domes heated by CO_2-rich fluids from the mantle? *Earth Plan. Sci. Lett.* **1979**, *43*, 298–302. [CrossRef]
07. Buick, I.S.; Holland, T.J.B. The nature and distribution of fluids during amphibolite facies metamorphism, Naxos (Greece). *J. Metam. Geol.* **1991**, *9*, 301–314. [CrossRef]
08. Mposkos, E.; Krohe, A.; Baziotis, I. Alpine polyphase metamorphism in metapelites from Sidironero complex (Rhodope domain, NE Greece). In Proceedings of the XIX CBGA Congress, Thessaloniki, Greece, 23–26 September 2010; Volume 100, pp. 173–181.
09. Liati, A. Regional Metamorphism and Overprinting Contact Metamorphism of the Rhodope Zone, Near Xanthi (N. Greece): Petrology, Geochemistry, Geochronology. Unpublished Ph.D. Thesis, Technical University of Braunschweig, Braunschweig, Germany, 1986.
10. Storre, B.; Nitsch, K.-H. Zur Stabilität von Margarit im System CaO-Al_2O_3-SiO_2-H_2O. *Contrib. Mineral. Petrol.* **1974**, *43*, 1–24. [CrossRef]
11. Chatterjee, N. Margarite stability and compatibility relations in the system CaO-Al_2O_3-SiO_2-H_2O as a pressure-temperature indicator. *Am. Mineral.* **1976**, *61*, 699–709.
12. Chatterjee, N. The system CaO-Al_2O_3-SiO_2-H_2O: New phase equilibria data, some calculated phase relations, and their petrological applications. *Contrib. Mineral. Petrol.* **1984**, *88*, 1–13. [CrossRef]
13. Okrusch, M.; Bunch, T.E.; Bank, H. Paragenesis and petrogenesis of a corundum bearing marble at Hunza (Kashmir). *Mineral. Depos.* **1976**, *11*, 278–297. [CrossRef]
14. Haas, H.; Holdaway, M.J. Equilibria in the system Al_2O_3-SiO_2-H_2O involving the stability limits of pyrophyllite. *Am. J. Sci.* **1973**, *273*, 449–464. [CrossRef]
15. Seifert, F. Stability of sapphirine: A study of the aluminous part of the system MgO-Al_2O_3-SiO_2-H_2O. *J. Geol.* **1974**, *82*, 173–204. [CrossRef]
16. Markl, G.; Scharlau, A. Grosse Sapphire, Margarit und Tremolit aus den Smirgel-gruben auf Naxos, Griechenland. *Lapis Mineral. Mag.* **2008**, *33*, 26–35.
17. Garnier, V.; Giuliani, G.; Ohnenstetter, D.; Fallick, A.E.; Dubessy, J.; Banks, D.; Hoang Quang, V.; Lhomme, T.; Maluski, H.; Pêcher, A.; et al. Marble-hosted ruby deposits from Central and Southeast Asia: Towards a new genetic model. *Ore Geol. Rev.* **2008**, *34*, 169–191. [CrossRef]
18. Matthews, A.; Putlitz, B.; Hamiel, Y.; Hervig, R.L. Volatile transport during the crystallization of anatectic

melts: Oxygen, boron and hydrogen stable isotope study on the metamorphic complex of Naxos, Greec *Geochim. Cosmochim. Acta* **2003**, *67*, 3145–3163. [CrossRef]
119. Siebenaller, L. Fluid Circulations during Collapse of an Accretionary Prism: Example of the Naxos Islan Metamorphic Core Complex (Cyclades, Greece). Ph.D. Thesis, University of Lorraine, Lorraine, France, 200
120. Haas, H. Diaspore-corundum equilibria determined by epitaxis of diaspore on corundum. *Am. Minera* **1972**, *57*, 1375–1385.
121. Tenthorey, E.A.; Ryan, J.G.; Snow, E.A. Petrogenesis of sapphirine-bearing metatroctolites from the Buc Creek ultramafic body, southern Appalachians. *J. Metam. Geol.* **1996**, *14*, 103–114.
122. Altherr, R.; Okrusch, M.; Bank, H. Corundum- and kyanite-bearing anatexites from the Precambrian c Tanzania. *Lithos* **1982**, *15*, 191–197. [CrossRef]
123. Mercier, A.; Debat, P.; Saul, J.M. Exotic origin of the ruby deposits of the Mangari area in SE Kenya. *Ore Geo Rev.* **1999**, *14*, 83–104. [CrossRef]
124. Key, R.M.; Ochieng, J.O. The growth of rubies in south-east Kenya. *J. Gemmol.* **1991**, *22*, 484–496. [CrossRef
125. Simonet, C. Géologie des Gisements de Saphir et de Rubis. L'exemple de la John Saul Mine, Mangari, Keny Ph.D. Thesis, Université de Nantes, Nantes, France, 2000.
126. Garde, A.; Marker, M. Corundum crystals with blue-red color zoning near Kangerdluarssuk, Sukkertoppe district, West Greenland. *Report Gronlands Geologiske Undersokelse* **1988**, *140*, 46–49.
127. Kissin, A.J. Ruby and sapphire from the Southern Ural Mountains, Russia. *Gems Gemmol.* **1994**, *30*, 243–25 [CrossRef]
128. Peretti, A.; Mullis, J.; Mouawad, F. The role of fluorine in the formation of color zoning in rubies from Mon Hsu, Myanmar (Burma). *J. Gemmol.* **1996**, *25*, 3–19. [CrossRef]
129. Tropper, P.; Manning, C.E. The solubility of corundum in H_2O at high pressure and temperature and it implications for Al mobility in the deep crust and upper mantle. *Chem. Geol.* **2007**, *240*, 54–60. [CrossRef]
130. Hatipoğlu, M.; Türk, N.; Chamberlain, S.; Akgün, M. Gem-quality transparent diaspore (zultanite) in bauxit deposits of the İbir Mountains, Menderes Massif, SW Turkey. *Mineral. Depos.* **2010**, *45*, 201–205. [CrossRef
131. Pardieu, V. Blue sapphires and Beryllium: An unfinished world quest. *InColor* **2013**, *23*, 36–43.

4

Zircon Xenocrysts from Cenozoic Alkaline Basalts of the Ratanakiri Volcanic Province (Cambodia), Southeast Asia—Trace Element Geochemistry, O-Hf Isotopic Composition, U-Pb and (U-Th)/He Geochronology—Revelations into the Underlying Lithospheric Mantle

Paula C. Piilonen [1,*], F. Lin Sutherland [2], Martin Danišík [3], Glenn Poirier [1], John W. Valley [4] and Ralph Rowe [1]

1 Mineralogy Section, Beaty Centre for Species Discovery, Research & Collections, Canadian Museum of Nature, P.O. Box 3443, Station D, Ottawa, Ontario K1P 6P4, Canada; gpoirier@nature.ca (G.P.); rrowe@nature.ca (R.R.)

2 Geoscience, Australian Museum, 1 William Street, Sydney NSW 2010, Australia; linsutherland1@gmail.com

3 John de Laeter Centre, The Institute for Geoscience Research (TIGeR), Curtin University, Perth WA 6845, Australia; m.danisik@curtin.edu.au

4 Wisconsin Secondary Ion Mass Spectrometer Laboratory (WiscSIMS), Department of Geoscience, University of Wisconsin, Madison, Wisconsin 53706, USA; valley@geology.wisc.edu

* Correspondence: ppiilonen@nature.ca

Abstract: Zircon xenocrysts from alkali basalts in Ratanakiri Province, Cambodia represent a unique low-Hf zircon within a 12,000 km long Indo-Pacific megacryst zone. Colorless, yellow, brown, and red crystals ({100}, {101}, subordinate {211}, {1103}), with hopper growth and corrosion features range up to 20 cm in size. Zircon chemistry indicates juvenile, Zr-saturated, mantle-derived alkaline melt (Hf 0.6–0.7 wt %, Y <0.2 wt %, U + Th + REE (Rare-Earth Elements) < 600 ppm, Zr/Hf 66–92, Eu/Eu^*_N ~1, positive Ce/Ce^*_N, HREE (Heavy REE) enrichment). Incompatible element depletion with increasing Yb/Sm_N from core to rim at ~ constant Hf suggests single stage growth. Ti-in-zircon temperatures (~570–740 °C) are lower than predicted by crystal morphology (800–900 °C) and decrease from core to rim (ΔT = 10–50 °C). The $\delta^{18}O$ values (4.88 to 5.01‰ VSMOW (Vienna Standard Mean Ocean Water)) are relatively low for xenocrysts from the zircon Indo-Pacific zone (ZIP). The $^{176}Hf/^{177}Hf$ values (+ εHf 4.5–10.2) give $T_{Depleted\ Mantle}$ model source ages of 260–462 Ma and $T_{Crustal}$ ages of 391–754 ma. The source magmas reflect variably depleted lithospheric mantle with little supracrustal input. Zircon U-Pb (0.88–1.56 Ma) and (U-Th)/He (0.86–1.02 Ma) ages are older than host basalt ages (~0.7 Ma), which suggests limited residence before transport. Zircon genesis suggests Zr-saturated, Al-undersaturated, carbonatitic-influenced, low-degree partial melting (<1%) of peridotitic mantle at ~60 km beneath the Indochina terrane.

Keywords: zircon; xenocryst; alkali basalt; Ratanakiri Volcanic Province; trace elements; O-isotopes and Hf-isotopes; U-Pb; (U-Th)/He

1. Introduction

Southeast (SE) Asia is an important source of gem corundum and zircon [1–3]. Recent and paleo-alluvial gem deposits occur in Thailand, Cambodia, Vietnam, and Laos, as deposits in a belt that extends >12,000 km along the west Pacific continental margins south to Australia and north to eastern China and Russia. This belt is known as the zircon megacryst (xenocryst) Indo-Pacific zone (ZIP, Figure 1) [4–14]. The western Pacific margin is marked by dynamic tectonic settings in which plate motions carry both continental and former sea-floor lithosphere over zones of hot upwelling asthenosphere. All SE Asian deposits are associated with late Cenozoic (<30 Ma) intraplate alkaline basaltic volcanism, which is the result of onset of decompression melting and an extensional tectonic regime in SE Asia following the Himalayan orogeny [15,16]. In SE Asia, intraplate alkali basalt magmas erupted through the late Precambrian to early Cenozoic along roughly northward-trending rifted blocks and fold belt terranes flanking the Indochina cratonic block.

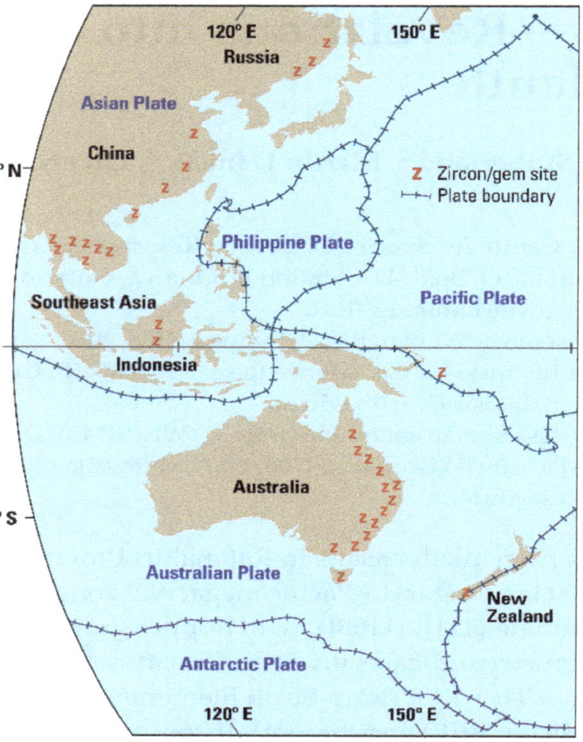

Figure 1. Map showing the 12,000 km long zircon megacryst Indo-Pacific zone (ZIP) along the western Pacific continental margins in Eastern Australasia, Asia, and Russia. The intraplate basalts hosting the megacryst suites are dominantly alkaline in composition and involve deeper mantle generation of magmas that may intersect pre-existing gem-bearing felsic or metamorphic bodies [11]. Consult this reference for zircon locality details.

Gem zircon and corundum xenocrysts, which have been eroded out of the host alkali basalts during the production of lateritic soils are concentrated into economic-grade deposits by secondary processes [17]. Mining for xenocrystic gems has taken place since the early 1400s in Thailand (Chanthaburi-Trat) and the late 1880s in Cambodia (Pailin and Ratanakiri provinces). In some deposits, zircon dominates over corundum, as is the case within the Ratanakiri Volcanic Province (RVP). Mechanized mining continues today in parts of Thailand but in Cambodia, small-scale artisanal mining by Vietnamese, Burmese and local indigenous peoples is common, often utilizing the same primitive techniques used hundreds of years ago.

Zircon is a ubiquitous accessory mineral in a wide range of rock types including mafic and felsic rocks derived from both crustal and mantle sources, lunar rocks, tektites, and metamorphic rocks [18]. It is chemically resistant and can survive weathering and transport processes, which allow it to be concentrated in secondary placer heavy mineral deposits. Zircon is highly refractory and has a low solubility in most melt and fluid compositions, which allows it to survive almost any crustal process including high temperature metamorphism and anatexis. Although the abundance of zircon is low

It has a strong effect on the behavior of many trace elements during crystallization and, thus, is an important accessory mineral in understanding petrogenesis. Furthermore, the low diffusion rates of rare-earth elements (REE), Th, U, Pb, Hf, and O in zircon under most geological conditions conserves both trace element composition zoning and the isotopic signature, which offers a window into its growth, evolution, and recycling events.

Xenocrystic zircon are known from many Cenozoic alkaline basalt provinces world-wide. However, their origins, petrogenesis, and relationship with their host basaltic rocks are still widely debated. Most authors are in agreement with a mantle source for some xenocrystic suites. However, the exact geochemical, isotopic, and mineralogical characteristics of this source are not completely known. Boehnke et al. [19] performed experimental studies on Zr saturation in mafic (basaltic) melts and concluded that such melt compositions require an unrealistically high concentration of Zr (>5000 ppm) to directly crystallize zircon. As such, zircon within ZIP basalts must have crystallized from late-stage, evolved magmas and become entrained in the rising alkali basalt host. Additional experimental studies on Zr saturation in alkaline melts, silicate, or carbonatitic are limited and focused on intermediate to felsic crustal compositions (e.g., [20–22]). More information is needed concerning Zr saturation under upper mantle conditions before inferences can be made about the petrogenesis of zircon xenocrysts. By utilizing trace element geochemistry, isotopic systems that reflect original magmatic conditions (i.e., O and Hf) and comparisons of the crystallization and eruption ages of the xenocrysts within the host basalts. Insights can be gained on the origin of these zircons.

This study presents field and laboratory studies started in 2012 investigating the zircon megacrysts which are being mined from the RVP basaltic gem fields in Northeastern Cambodia.

A recent study on similar zircons [23] has provided valuable new data on one set of RVP zircon megacrysts. However, the present study differs in reporting on a wider range of zircon samples and their color and form variants from several well-defined localities as well as providing further analytical and literature results. Some detailed comparisons of the two data sets are incorporated, which allows for wider discussion and interpretations on the origin of these Cambodian zircons. This study is the first of its kind in Cambodia and provides insight into understanding the composition and the evolutionary history of the subcontinental lithospheric mantle (SCLM) in Northeast Cambodia.

2. General Geology

Southeast Asia including Thailand and Cambodia is an assemblage of distinct terranes amalgamated together to form two dominant crustal blocks known as the Indochina and the Shan-Thai (or Sibumasu) terranes (Figure 2). These two terranes are separated and overprinted by Mesozoic and younger arc systems of the Sukhothai Zone (Fold Belt) and the Sa Kaeo Suture Zone (SKS). Although these two terrains have separate geological histories, they both have their origins at the margin of Gondwana prior to collision during the Mesozoic [24]. The Indochina terrain, which extends from Eastern Thailand, Laos, across Cambodia, and into Southern Vietnam, rifted away from the northern edge of Gondwanaland during the Devonian. This resulted in the opening of Paleotethys [25]. Little is known about the Precambrian tectonic history of Thailand and Cambodia even though gneisses and schists of the Proterozoic Kontum massif in Vietnam extend into Northeast Cambodia near the border with Thailand [26]. Similar high-grade metamorphic rocks have been found in Pailin along the border with Thailand [26,27].

Collision and convergence of the Indochina block with the Asian continent (Shan-Thai terrane in Thailand) started at ~250 Ma and culminated around the Triassic-Jurassic boundary (~210–200 Ma) with the closure of Paleotethys and the end of the Indosinian orogeny [25,26]. The Indochina terrain is composed of continental crust, which has remained intact and stable since the end of the Indosinian orogeny, and is surrounded by younger, post-Jurassic fold belts. Cambodia lies entirely in the Indochina terrain between the Truongson fold belt to the north and the Loei fold belt to the west. Both of these regions contain Carboniferous to Triassic collisional arc-type volcanosedimentary sequences.

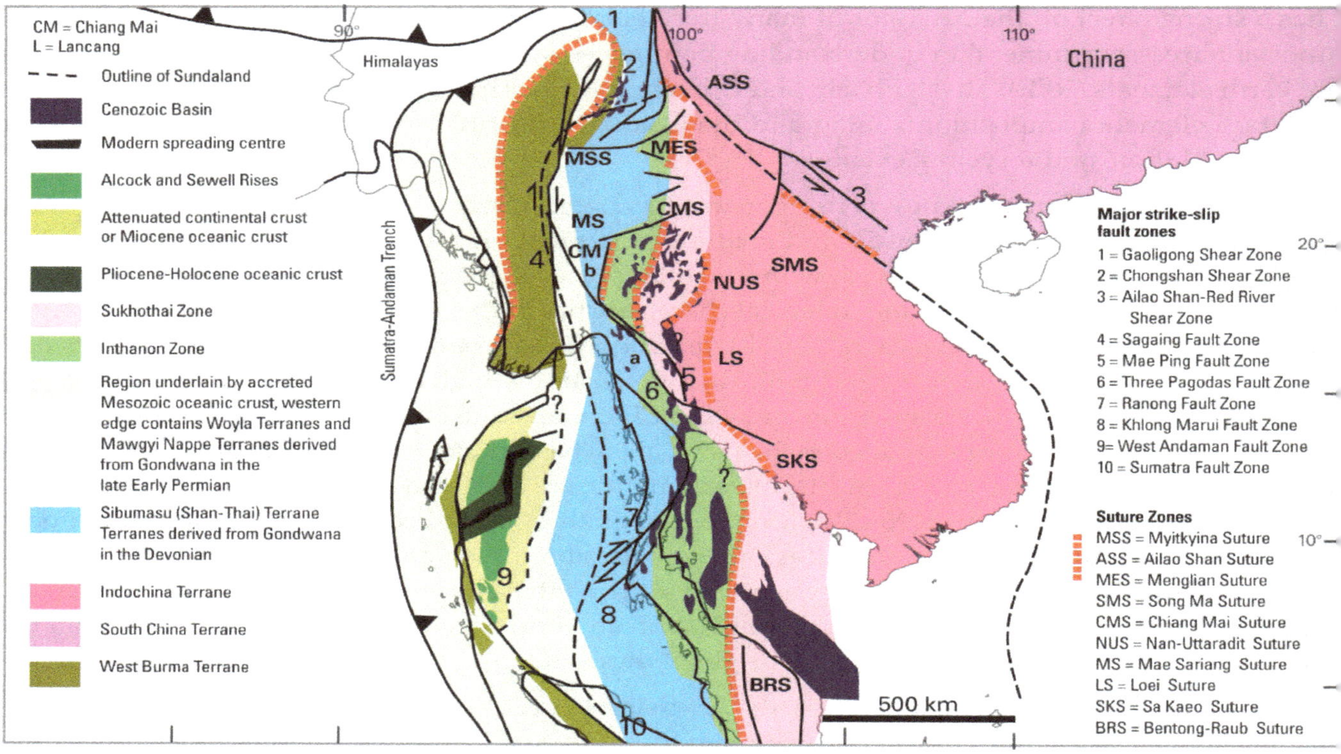

Figure 2. Tectonic map of SE Asia [25]. Cambodia is entirely contained within the Indochina terrain bounded on the northeast by the Song Ma Suture (SMS).

The Cenozoic tectonic evolution of Cambodia and Thailand is related to the collision betwee the Indian and Eurasian plates during the Himalayan Orogeny (55–45 Ma, [28]). Compressiona stresses dominating in the northern part of the collisional zone resulted in the Southeast Chin block and adjacent parts of Southeast Asia (Indosinia) to be rotated and forced to the southeast [29 which results in uplift and transtension and graben structures with predominantly a NW-S strike along with the opening of the Gulf of Thailand, the South China Sea, and the Andama Sea [30]. Uplift, extensional tectonics and rifting throughout SE Asia resulted in lithospheric thinnin which triggered decompressional melting and upwelling of the underlying asthenosphere alon deep-seated faults to form late Cenozoic flood basalts [16,29]. ZIP fields across SE Asia are flanke to the east by extinct marginal spreading basins and offset rifts that formed behind the Pacific islan arc-subduction system [31]. These basins involved thermal rifting and could provide a continuou source of sub-lithospheric melting to promote zircon (and corundum) crystallization during prolonge basaltic events [11,32,33].

Cenozoic basalts occur throughout SE Asia (red regions, Figure 3) ranging in age from 24 Ma t as recently as a 1923 eruption off the Vietnamese coast near the island of Poulo Cécir [34,35]. Know collectively as the SE Asian Volcanic Province [36–38] or the Central-East Asian basaltic province [8 the basalts ascended along rift structures bounded by strike-slip faults within Archean to Paleozoi terrains and were more extensive in Vietnam, Eastern Cambodia, and Southern Laos than elsewhere i the region even though no temporal and spatial correlation has been established [39]. In most areas two separate periods of volcanism are noted: (1) an early phase consisting of SiO_2-rich, Fe-depleted and Ti-depleted quartz and olivine tholeiites and rare trachyandesites, representative of lithospheri sources, and (2) a later, highly alkaline phase consisting of lower SiO_2, high Fe, and Ti olivine tholeiites alkali basalts, basanites, and rare trachybasalts and trachytes, representative of asthenospheric mantl sources [16,40]. Zircon and corundum xenocrysts are associated with late-stage, alkaline phases o volcanism [16,29,41].

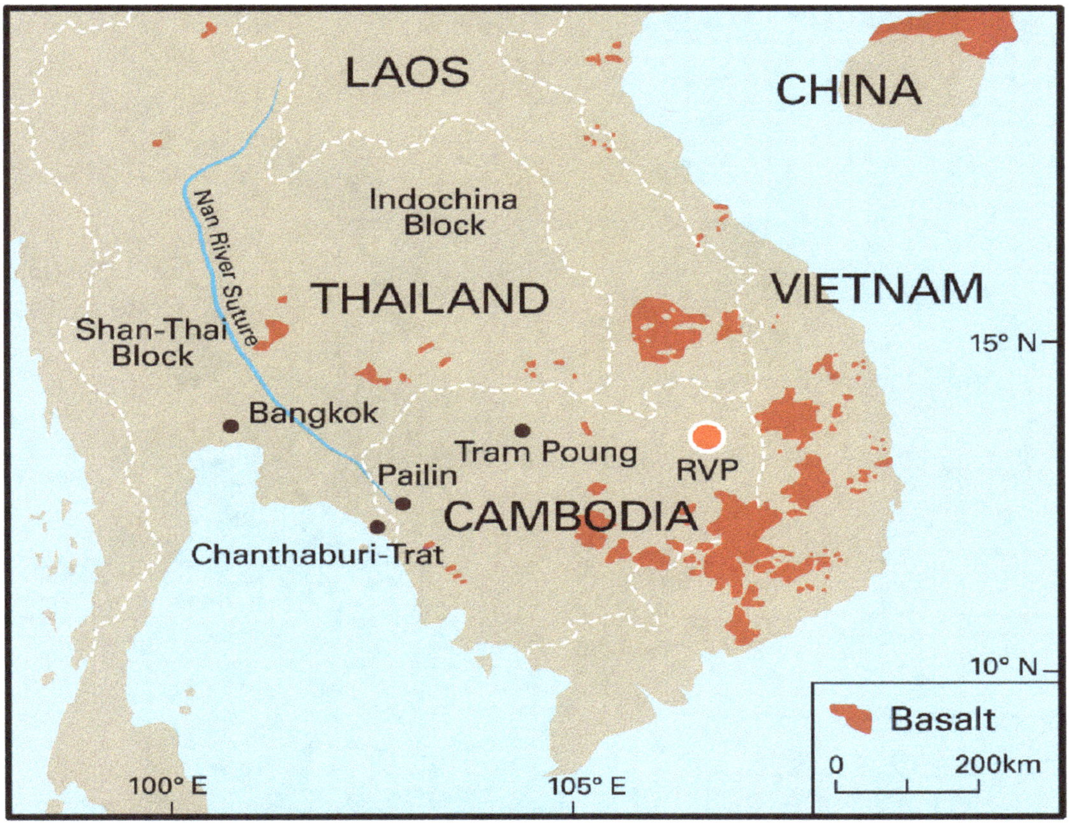

Figure 3. Distribution of Cenozoic alkali basalt provinces (red) in SE Asia (modified after [16]). The location of the Ratanakiri Volcanic Province (RVP) is shown as a red circle.

.1. Ratanakiri Volcanic Province (RVP)

Detailed geological information about Cambodia is difficult to obtain as a result of the destruction f all academic material by the Khmer Rouge. Early work by Lacombe [42,43] concerning the geology, eochemistry, and mineralogy of the RVP basalts is still the most definitive work in existence. Previous :500,000 maps by both France (1970s) and Russia (1990) are out of date. Little work has been completed n the country since that time. Detailed mapping, age dating, and geochemical analyses of the rocks in he RVP is currently underway by the lead author.

Quaternary basaltic rocks cover approximately 10,000 km^2 of the country and overlie Quaternary lluvium and Mesozoic sediments. Small flows and vents occur in the west near Pailin and Poipet nd in the central northern province of Preah Vihear. Two distinctly larger basaltic areas are found in atanakiri province (Figure 4) and comprise the fifth largest basaltic plateau in Indochina: (1) the Bokeo lateau is located between the Sre Pok and the Sesan rivers, herein referred to as the Ratanakiri Volcanic rovince (RVP), and it is host to xenocrystic gem zircon deposits and (2) the smaller, zircon-free Ban hay plateau, northeast of Bokeo and north of the Sesan river near the Vietnamese border [42]. acombe [42] estimated the thickness of the basalt plateaus to 80 m near the center and 40 m closer to he extremities. Initially the flows would have covered an area of around 3200 km^2, but now weathered own to under 2000 km^2 (RVP: 1500 km^2, Ban Chay Plateau: 200 km^2). The initial topographic relief vas quickly filled by the early flows, which allows later flows to spread in all directions and cover a vide area with thin, individual flows (often <2 m thick). Lacombe [42] observed only alkaline basalts in he Ban Lung area (olivine basalts, alkali basalts, basanites, nephelinites, and trachyandesites). He also oted that early flows preceded the last paleomagnetic inversion (0.7 Ma) and the later, more explosive asaltic volcanism occurred during and after the geomagnetic reversal. This falls within the range 17.6–0 Ma) established by Hoang & Flower [40] for related basaltic volcanism in Vietnam and other enozoic volcanism throughout the Northwestern Pacific continental margin.

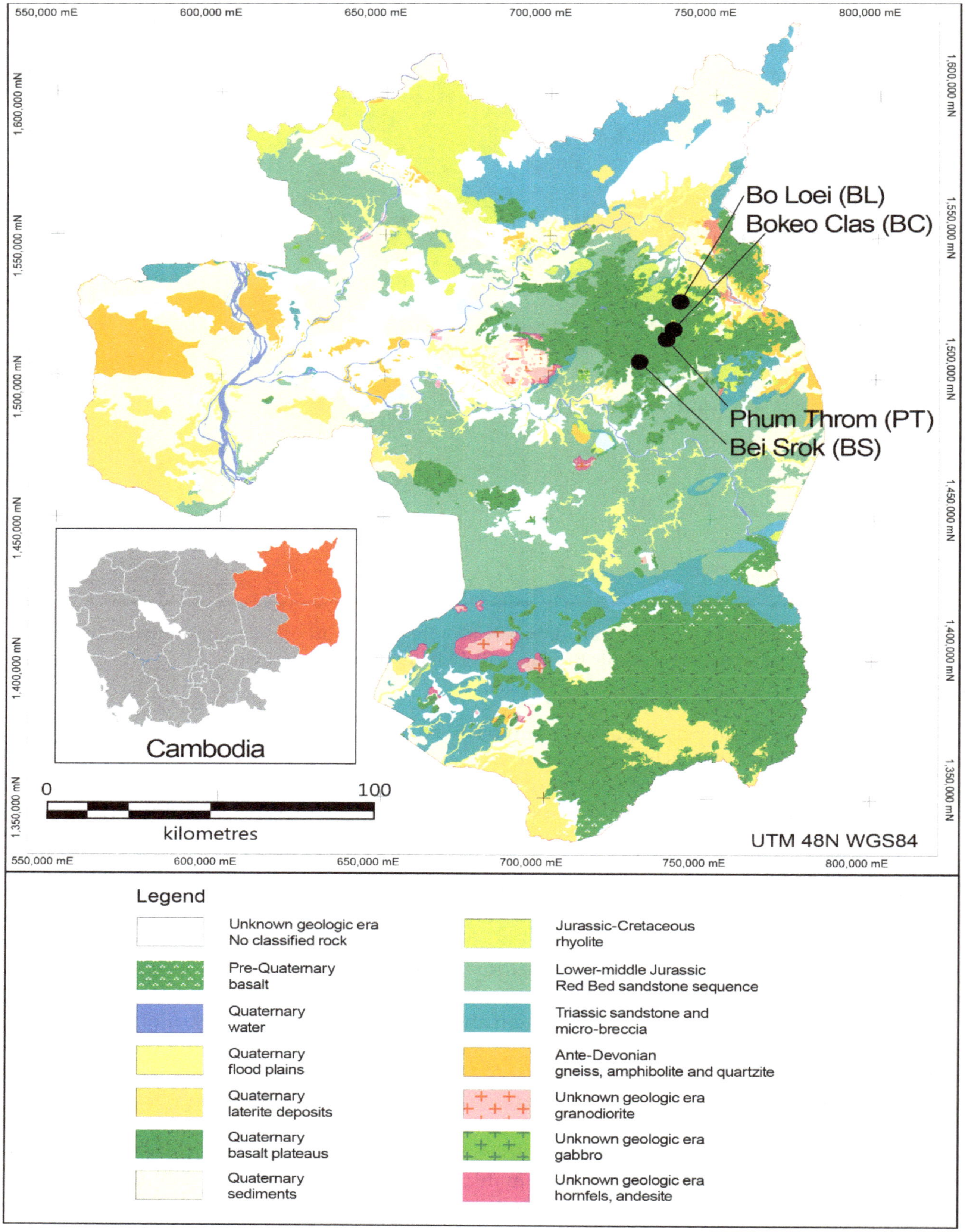

Figure 4. Geological map of Stung Treng, Ratanakiri, and Mondulkiri provinces, Northeast Cambodia. The Ratanakiri Volcanic Province (RVP) and Ban Chay Plateau Quaternary basalts are indicated in green with hatching. Zircon xenocryst localities (black dots) include Bokeo Clas (BC), Phum Throm (PT), Bei Srok (BS), and Bo Loei (BL).

Basaltic volcanism began with fissure flows. A brief, erosional hiatus was followed by increasingly ıore explosive volcanism from both old and new fissures [42]. The final stage was even more explosive nd resulted in craters and deposits of mafic scoria, ignimbrite, welded tuff, tuff, ash flows, and pumice. 'olcanic features include: (1) scoria cones (1–3 km wide × 150 m tall), (2) horseshoe craters, which have een opened as a result of extensive erosion, (3) crater cones, (4) explosive craters, and (5) rift valleys nd volcanotectonic depressions [42,44]. Many of these structures are only recognizable from air photos nd satellite imagery since complete volcanic profiles in the RVP have extensive lateritization and egetative cover.

.2. *Zircon Deposits*

In situ zircon is extremely rare and has only been collected by the senior author in alkali basalt t Phnom Dang, Bokeo (Figure 5), which is an alkali basalt scoria outcrop on the flank of the Phnom)ang scoria cone, and from a tephriphonolite flow 3 km SE of Phnom Dang, Bokeo. Lacombe [43] lso noted the presence of xenocrysts in trachybasalt/alkali basalts and associated pyroclastics at 'hnom Dang, Bo Tum and Bo Loei. At all localities, zircon occurs as single crystals in the alkali basalt vithout other xenocrystic minerals. Gem zircon, which is derived from the alkali basalts, is mined from econdary deposits within a roughly NE–SW trending region ~30 km in strike, 10 km wide, east of he provincial capital of Ban Lung (Figure 4). Unlike deposits in other west Pacific continental margin egions including Western Cambodia and Southeastern Thailand and gem deposits in the RVP ontain dominantly zircon with only trace corundum. Zircon mines in Ratanakiri are small-scale ffairs-gem-quality stones are found in the residual soils and gravels derived from weathered basalts. 'he gemstones are mined by digging vertical pits (10 m deep, 1 m wide) to bedrock and hauling to urface the lateritic soils and basalt gravels in which loose xenocrysts are concentrated. On the surface, ther miners sift through the soil with their hands to extract the zircons. In rare cases, during the vet season, pressured water is used to loosen the lateritic soil and wash out the zircons. Only 20% of ircon are actually valuable as gem stones. Many of the stones are fractured, contain inclusions, or are n unsuitable color. Zircon from Bei Srok (BS) has historically been the most sought after by the gem ndustry due to its dark red color, which, when heat-treated in a reduced environment, changes to a lark, more brilliant blue color (Figure 5). An electron-related or hole-related color center is thought o be the cause of dark red-brown color in the RVP zircon while authors have not yet been able to nequivocally determine the mechanism for the blue coloration upon heat-treating [45]. The highest uality gem zircon is found at the Bei Srok deposit, which is 16 km south of the provincial capital f Ban Lung at the southern end of the gem-bearing region. The Bokeo Clas (BC) and Phum Throm PT) deposits are located 23 km east of Ban Lung near the village of Bokeo and represent the most rolific mining areas. Other less significant occurrences are known throughout the Bokeo plateau vhere locals pan for zircon in the creeks and riverbeds. Rare blue-green corundum has been found in ircon concentrates from Bei Srok.

The Bo Loei (BL) deposit is located 24 km ENE of Ban Lung, 28 km W from the Vietnamese order, and 28 km NNE of Bei Srok. BoLoei marks the northern-most gem occurrence and comprises n alluvial wash containing xenocrysts of zircon, corundum (blue, blue-green, and yellow-orange), nd spinel as well as fragments of alkali basalt, felsic volcanics (rhyolite), and other country rock. 3o Loei is principally a gem zircon deposit with corundum as a by-product (<3% of the total yield, [12]). 'he water table at Bo Loei is higher than that at other Ratanakiri zircon deposits, which allows for anning and concentration of the zircon ore in nearby streams.

(a) (b)

Figure 5. (**a**) Zircon miner at Phum Throm, Ratanakiri, Cambodia. (**b**) Natural and heat-treated faceted RVP zircon (each stone is ~0.75 carats). Photo copyright M. Bainbridge, Canadian Museum of Nature collection.

3. Materials and Methods

3.1. Sample Descriptions

Zircon crystals from the RVP range in size from 2 × 2 × 4 mm to 15 × 20 × 20 cm long (average 7 × 8 × 8 mm). Crystals are subhedral to euhedral, dominantly vitreous to adamantine, transparent grey-yellow, honey yellow, pinkish-orange, light orange-brown, and dark red with minor translucent to opaque milky-grey and yellow crystals. Squat prismatic crystals dominated by {100} and {101} are most typically euhedral. Less common crystal forms include the {211} bipyramid and {110} prism and more often occur in subhedral grains. Prismatic zircon ranges in color from transparent light yellow to dark red to highly-included, translucent milky-yellow and red-brown. Subhedral to anhedral multiform zircon are most often transparent and range in color from light orange-yellow to orange-red to dark red. Sharp, concentric color zoning is evident optically in many crystals, which corresponds to zoning observed in back-scattered electron (BSE) and cathodoluminescence (CL) images. In situ zircon occurs solely as isolated grains within the alkali basalt (Figure 6a,b). The interface between the zircon and the host alkali basalt is characterized by a very thin (1–2 μm) glassy reaction rim with numerous fine fractures, which is indicative of contraction of the glass upon cooling. The presence of such a discrete, quenched glassy interface without evidence for chemical interaction suggests a wide temperature difference between the entrained zircon and the host basaltic magma.

The majority of Ratanakiri zircon display striking resorption and corrosion features along with hopper-type growth patterns (Figure 6c) indicative of disequilibrium with their host magmas. Subhedral to euhedral zircon are often dominated by a short prism capped by a bipyramid on one termination and a hopper-type growth feature on the bottom termination. This hopper-type or corroded habit appears to represent an attachment point with an associated xenocrystic phase or the growth medium. However, multimineralic xenoliths containing zircon have not been found to confirm these possible intergrowths. Late-stage dendritic etching in-filled by baddeleyite is also present on many crystals (Figure 6d), which indicates late-stage interaction with a highly-alkaline corrosive melt or fluid, and remobilization of Zr. Fluid inclusions dominated by H_2O with rare CO_2 bubbles were found by Zeug et al. [45] in a number of RVP zircon xenocrysts. Partially healed fissures surrounded many of the fluid inclusions.

Zircon from Bo Loei is commonly subhedral to anhedral and does not display the etching, dissolution, and corrosion features that are common to other Ratanakiri zircon, which is possibly a result of increased transport in alluvial or fluvial systems.

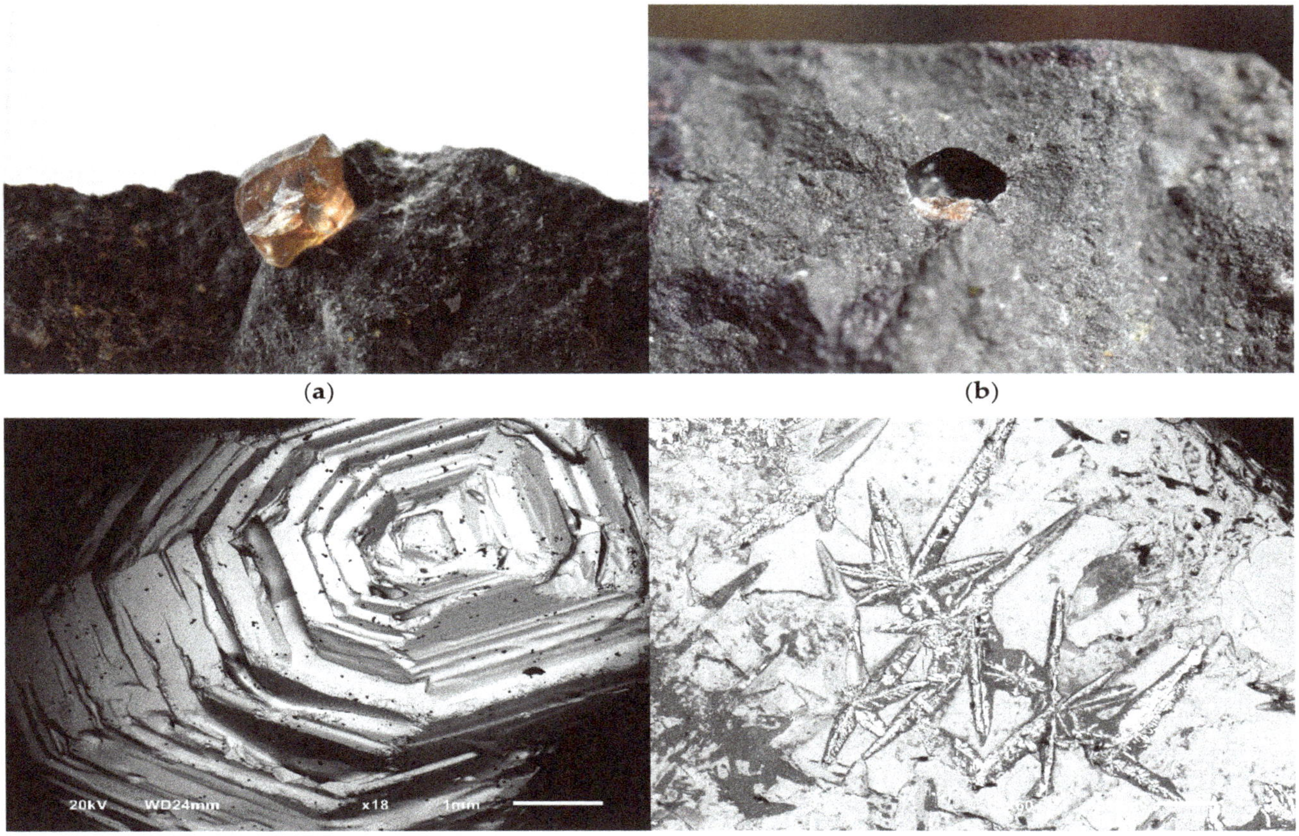

Figure 6. Zircon xenocrysts from the RVP: (**a**) euhedral zircon (7 × 4.5 × 4 mm) in situ from Phnom Dang, Bokeo. (**b**) Glassy reaction rim (1–2 µm wide lining the 5 mm hole) between zircon xenocryst and the host alkali basalt. (**c**) Hopper-type growth feature on the subhedral zircon. (**d**) Late-stage dendritic etching in-filled by baddeleyite on the surface of a zircon xenocryst.

All zircon samples were purchased directly from miners at each locality. Representative zircon grains ranging in size from 4 × 4 × 4 mm to 19 × 11 × 8 mm were hand-picked under a binocular microscope and mounted in epoxy blocks for analysis. The blocks were cut in half, laterally, on a thin section saw to expose the interior center of the grains prior to being polished.

3.2. *Cathodoluminescence Microscopy and Electron Microprobe Analyses*

Cathodoluminescence (CL) and back-scattered electron imaging (BSEI) was done on a JEOL 6610Lv scanning electron microscope (JEOL USA Inc., Peabody, MA, USA) operating at 20 kV with a spot size of 55 µm and a working distance of 20 mm. The SEM was equipped with a monochromatic Gatan miniCL detector (Gatan Inc., Pleasanton, CA, USA). Chemical analyses of the zircon xenocrysts were done with a JEOL Superprobe 8230 (JEOL USA Inc., Peabody, MA, USA) at the University of Ottawa. Operating conditions were as follows: beam diameter of 5 µm, operating voltage 20 kV, and a beam current of 40 nA. A total of six elements were sought and the following standards and X-ray lines were employed: synthetic YIG garnet (Y $L\alpha$), zircon (Zr $L\alpha$, Si $K\alpha$), synthetic UO_2 (U $M\alpha$), synthetic ThO_2 (Th $M\alpha$), and hafnon (Hf $M\alpha$). Count times for Zr, Th, U, and Si were 20 s on peak and 10 s on the background and count times on Hf and Y were 50 s on peak and 25 s on the background. Raw intensities were corrected by using the PAP routine [46]. The Hf concentrations obtained by EMPA (Table 1) were used as an internal standard for trace element determination by laser ablation inductively-coupled plasma mass spectrometry (LA-ICP-MS).

3.3. Trace-Element Analysis

Trace element contents were analyzed by using an Agilent 7700 quadrupole ICP-MS instrumen (Agilent Technologies, Santa Clara, CA, USA) attached to a Photon Machines Excimer 193 nm lase system (Excite, Photon machines Inc., Redmond, WA, USA) at the Macquarie University. The analyse were carried out by using the same laser conditions as for U-Pb dating. Detailed descriptions o analytical and calibration procedures have been given by Belousova et al. [5]. Quantitative results fo 25 elements reported in this paper were obtained through calibration of relative element sensitivitie using the NIST-610 standard glass and the GEMOC GJ-1 and Mud Tank zircon standards [47] as th external calibration standard as well as normalization of each analysis to the EMPA data for Hf as a internal standard. The precision and accuracy of the NIST-610 analyses are 1–2% for REE, Y, Sr, Nb, H Ta, Th, and U at the ppm concentration level, 4% for Ti, 5% for Pb, and 20% for P at ppm concentration

3.4. Hf-Isotope Analysis

Methodology and an analytical condition for the Lu-Hf isotope analysis are provided b Griffin et al. [48]. Hf-isotope analyses were carried out in situ using a New Wave/Merchantek UP-21 laser-ablation microprobe (New Wave Research, Inc., Petach Tikva, Israel), attached to a Nu Plasm multi-collector ICP-MS (Nu Instruments Ltd., Wrexham, UK) at Macquarie University. The analyse were carried out with a beam diameter 55 μm and a 5 Hz repetition rate. Typical ablation times wer 100–120 s and resulted in pits 30–40 μm deep. The ablated sample was transported by He carrier ga from the laser-ablation cell via a mixing chamber to the ICP-MS torch.

Interference of ^{176}Lu on ^{176}Hf is corrected by measuring the intensity of the interference-fre ^{175}Lu isotope and using ^{176}Lu/^{175}Lu = 0.02669 [49] to calculate ^{176}Lu/^{177}Hf. Similarly, the interferenc of ^{176}Yb on ^{176}Hf has been corrected by measuring the interference-free ^{172}Yb isotope and usin ^{176}Yb/^{172}Yb to calculate ^{176}Yb/^{177}Hf. The appropriate value of ^{176}Yb/^{172}Yb was determined b spiking the JMC475 Hf standard with Yb and finding the value of ^{176}Yb/^{172}Yb (0.58669) required t yield the value of ^{176}Hf/^{177}Hf obtained on the pure Hf solution. Detailed discussions regarding th overlap corrections for ^{176}Lu and ^{176}Yb are provided in Pearson et al. [50]. Precision and accurac obtained on the ^{176}Hf/^{177}Hf ratio are illustrated by analyses of standard zircons in Griffin et al. [48] an Pearson et al. [50]. The typical 2SE precision on the ^{176}Hf/^{177}Hf ratios presented here is about 0.00002 which is equivalent to +0.7 eHf unit. The Mud Tank and Temora zircon were used as independen control on reproducibility and instrument stability. Most of the data and the mean value are withi 2 s.d. of the recommended values reported for Mud Tank [^{176}Hf/^{177}Hf = 0.282522 ± 42 (2 s.d.)], [51] and Temora reference material (0.282680 ± 15; [52]).

In order to calculate εHf values, the chondritic values of Bouvier et al. [53] were adopted ^{176}Lu/^{177}Hf (Chondrite Uniform Reservoir (CHUR), today) = 0.0336, ^{176}Hf/^{177}Hf (CHUR today) = 0.282785, and the decay constant for ^{176}Lu of 1.865 x 10^{-11} yr^{-1} ([54]). To calculate mode ages (T_{DM}) based on a depleted-mantle source, we assume that the depleted mantle (DM) reservoi developed from an initially chondritic mantle is complementary to the crust extracted over time T_{DM} ages, which are calculated using the measured ^{176}Lu/^{177}Hf of the zircon, can only give a minimum age for the source material of the magma from which the zircon crystallized. Therefore, we have als calculated a "crustal" model age ($T_{DM}{}^{C}$ in data tables) for each zircon that assumes its parental magm was produced from an average continental crust (^{176}Lu/^{177}Hf = 0.015, Geochemical Earth Referenc Model database, http://www.earthref.org/), which was derived from a depleted mantle.

3.5. Geochronometry

3.5.1. U-Pb Dating by LA-ICP-MS

Zircon U-Pb ages were measured by using an Agilent 7700 quadrupole ICP-MS attached to a Photon Machines Excimer 193 nm laser system. The analyses were carried out with a beam diameter o 50 μm with 5 Hz repetition rate and energy of 8 J/cm^2. The analytical procedures for the U-Pb dating ave been detailed previously [55]. A very fast scanning data acquisition protocol was employed o minimize signal noise. Data acquisition for each analysis was 3 min (1 min background, 2 min ignal). Ablation was carried out in He to improve sample transport efficiency, provide more stable ignals, and give more reproducible U/Pb fractionation. Provided that constant ablation conditions re maintained, accurate correction for U/Pb fractionation can then be achieved.

Sample analyses were bracketed by pairs of analyses of the GEMOC GJ-1 zircon standard [47]. he other well-characterized zircon standard 91500 was analyzed within the run as an independent ontrol on reproducibility and instrument stability (see data tables). U-Pb ages were calculated from he raw signal data by using the online software package GLITTER (version 4.4.4, ARC National ey Centre, Sydney, Australia, www.glitter-gemoc.com; [56]). GLITTER calculates the relevant sotopic ratios for each mass sweep and displays them as time-resolved data. This allows isotopically omogeneous segments of the signal to be selected for integration. GLITTER then corrects the ntegrated ratios for ablation related fractionation and instrumental mass bias by calibration of each elected time segment against the identical time segments for the standard zircon analyses.

The common-Pb correction procedure of Andersen [57] was employed and the analyses presented ere have been corrected assuming recent Pb-loss with a common-Pb composition corresponding to resent-day average orogenic Pb as given by the second-stage growth curve of Stacey and Kramers [58] or $^{238}U/^{204}Pb=9.74$. No correction has been applied to analyses that are concordant within 2σ nalytical error in $^{206}Pb/^{238}U$ and $^{207}Pb/^{235}U$ or which have less than 0.2% common lead.

.5.2. (U-Th)/He Thermochronology

The age of eruption of the zircon host basalt was determined by (U-Th)/He thermochronology. his method, when applied to zircon, has a closure temperature of ~180 °C (for cooling rate of 0 °C/Ma, ~60 μm diameter diffusion domain equivalent sphere radius, [59]). In the simplest cenario (i.e., simple cooling without subsequent reheating), the ages measured by this method can be nterpreted as true 'eruption ages' recording the passage of zircon-bearing magma to the surface and ssociated cooling. It is important to note that the traditionally used zircon U-Pb geochronology has a losure temperature in excess of ~900 °C and records the time of zircon crystallization in the magma hamber as well as provides a maximum limit for the eruption age, but it cannot directly date the ruption age.

The (U-Th)/He dating of zircon was conducted at the University of Waikato (New Zealand) by ollowing the protocols described in Danišík et al. [60,61]. The large zircon megacrysts (mm–cm sized) ould not be fit into the Nb microtubes (i.e., cylinders ~0.9 mm long, with internal diameter of ~0.6 mm) nd were, thus, first abraded (thereby removing the uppermost ~30 microns of the surface) using an ir-abrasion cell with pyrite as abrasion medium. This step circumvented the need of alpha ejection orrection ([62]). Abraded megacrysts were then crushed in a steel mortar. The 60–150 μm fraction vas separated using sieves and cleaned in an ultrasonic bath with ethanol. Clean shards (3–5 per ample) were individually loaded into Nb microtubes, degassed at ~1250 °C under ultra-high vacuum y using a diode laser, and analyzed for ^{4}He by isotope dilution on a Pfeiffer Prisma QMS-200 mass pectrometer. Following He measurements, the zircon shards in Nb microtubes were spiked with ^{35}U and ^{230}Th, and dissolved in hydrofluoric, nitric, and hydrochloric acids. The solutions were nalyzed by isotope dilution for U and Th and by external calibration for Sm on a Perkin-Elmer SCIEX LAN DRC II ICP-MS (PerkinElmer, Inc., Waltham, MA, USA). The total analytical uncertainty (TAU) vas calculated as the quadratic addition on He and weighted uncertainties on U, Th, Sm, and He neasurements and is typically ~2% (1σ). The raw zircon (U-Th)/He ages were not corrected for alpha jection (Ft correction) given the abrasion step described above. Replicates (3 to 5 per sample) with ssociated uncertainties were used to calculate the geometric mean ([63]) and error-weighted standard leviation as representative eruption age for each sample. Replicate analyses of the Fish Canyon uff zircon (n = 18) measured over the period of this study as an internal standard yielded a mean U-Th-Sm)/He age of 28.2 ± 0.6 Ma, which is in excellent agreement with the (U-Th-Sm)/He age of the Fish Canyon Tuff zircon mineral standard dated at 28.3 $\pm$ 1.3 Ma [64]. All replicates in each sampl overlap within 1σ error bars and total analytical errors for individual replicates are significantly belov 5%, which is typical for zircon dating procedures.

3.6. $\delta^{18}O$ Isotopic Analyses

Zircons were analyzed by laser fluorination at the University of Wisconsin, Madiso Zircon concentrates were prepared by standard crushing as well as gravimetric and magneti techniques and soaked in cold HF overnight to remove radiation-damaged domains and contaminatio Bulk samples of ~2 mg were pre-treated in BrF_5 overnight, fluorinated by laser heating with a 32 V CO_2 laser operating at a wavelength of 10.6 μm, and analyzed by a gas-source mass spectromete Data are standardized against multiple analyses of the UWG-2 garnet standard performed on the sam day ($\delta^{18}O$ = 5.80‰VSMOW, [65,66]).

4. Results

4.1. Zircon Crystal Morphology and CL Zonation

Morphological studies were done on intact zircon grains as well as by CL after the crystals ha been cut and polished to expose their centers. The recognition of the dominant forms, {100} prism an {101} bipyramid, as well as less pronounced {110} prisms and {211} pyramids in each crystal allow for discrimination of zircon morphological types by using the Pupin [67] classification. On the basi of empirical observations, Pupin [67] argued for a relationship between zircon morphology and th composition of the medium in which the zircon crystallized. Those crystallizing from a peraluminou melt are dominated by {211} pyramids and those crystallizing from peralkline melts are dominated b {101} pyramids. The alkalinity index [Al/(Na+K)] is, thus, designated as index A. The temperature o the growth medium (index T) is the dominant factor in controlling the relative development of prisn faces with higher temperatures favoring {100} prisms and lower temperatures favoring {110} prisms.

RVP zircon xenocrysts occupy a narrow range of forms on the Pupin [67] typological correlatio diagram and fall into two categories: (1) those dominated by the {100} prism and {101} bipyrami belonging to subgroup J5 and (2) those with additional moderately developed {110} prisms and {21 dipyramids belonging to subgroups S19–20 and S24–25. There is no evidence in CL images to sugges that rapid changes in the magma composition or temperature took place during zircon crystallizatio Both morphological populations suggest crystallization in an alkaline environment at temperatures o 800–900 $\pm$ 50 °C.

All zircon display strong, primary, magmatic oscillatory growth zoning when imaged by C (Figure 7), which is a reflection of the non-equilibrium conditions during crystallization [68]. Oscillator growth areas are very fine with bands on the order of <1–25 μm (Figure 7a,b). A number of grains als display distinctive sector zoning upon which the oscillatory zoning is imposed (Figure 7c–e). Zonatio patterns are generally very sharp and crystallographically-controlled. With the exception of one grai all zircon are non-metamict, unaltered, and do not contain relict cores, which confirms their primar igneous origin. Studies by Witter et al. [69] also found RVP zircon to have negligible radiation damag with time-integrated self-irradiation doses on the order of 10^{16} alpha-decay events/gram (α/g) (not that detectable radiation damage requires a minimum of 0.1–0.2 $\times$ 10^{18} α/g). One grain (BL53- Figure 7f) contains a core that is black in CL and does not display any discernible zoning pattern i either CL, BSEI, or optically. Limited yet distinct magmatic zoning is observed in BSE images of RV zircon. However, when the CL and BSE images can be compared, dark regions in CL correspond t bright areas (highest mean atomic weight) in BSE images. Chemical analyses of the individual zone reveal Th and U to be the most important influence, measurable by EMPA, on the observed zonatio patterns—regions rich in Th + U are darkest in CL (both sector and oscillatory zoning) and brightest i BSE images. Optically, zones enriched in Th + U are darker in color and range from dark orange-brow to dark red.

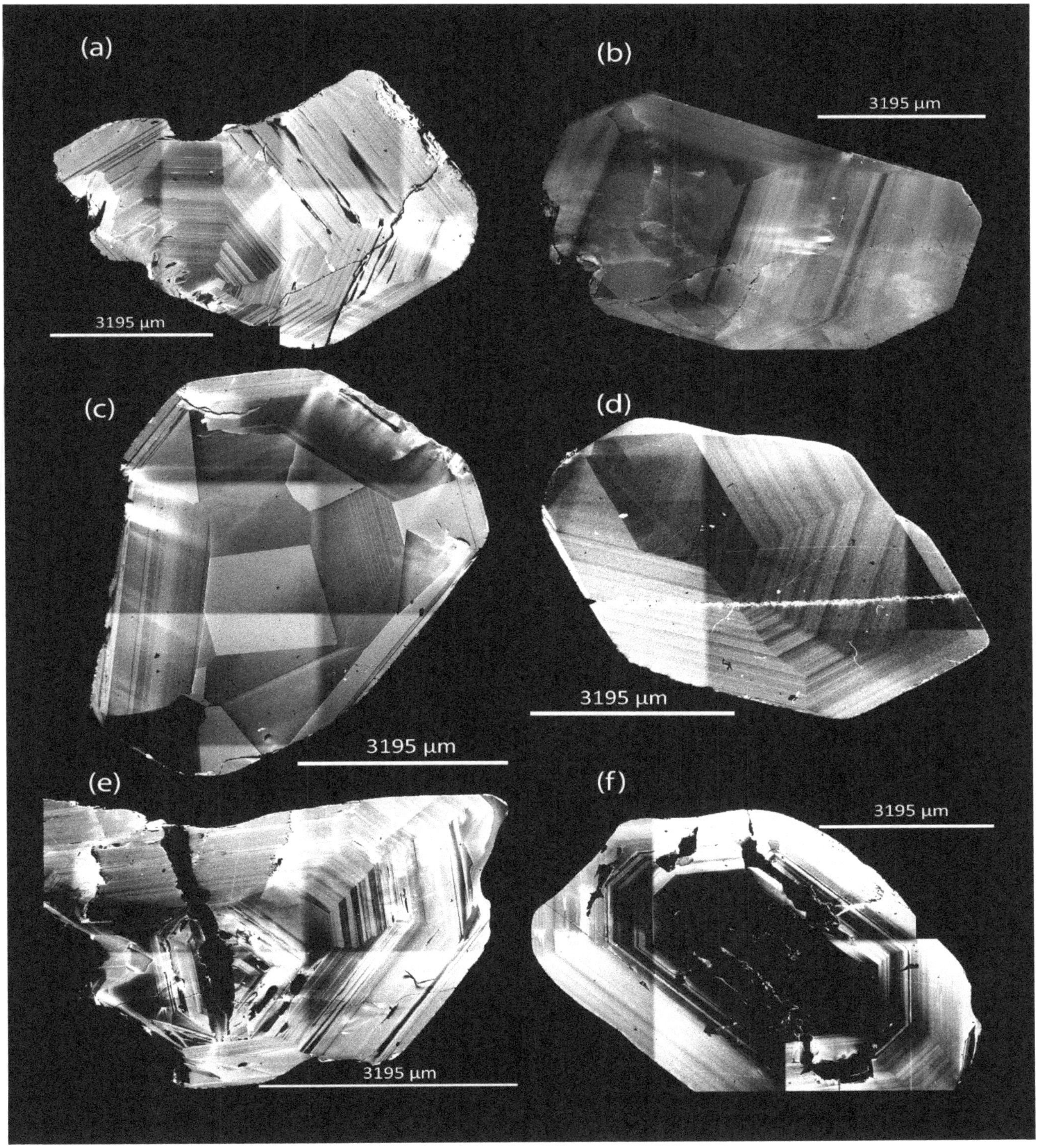

Figure 7. Cathodoluminescence images of RVP zircon xenocrysts showing both oscillatory and sector zonation. Darker zones are enriched in Th, U, and REE: (**a**) concentric zoning in BC43-2, (**b**) concentric zoning in PT49-2, (**c**) sector zoning in PT48-3, (**d**) sector and concentric zoning in BL54-3, (**e**) sector and concentric zoning in BL53-2, and (**f**) dark, CL-inactive core in BL53-1.

These results are in direct contrast to those presented by Cong et al. [23] for six alluvial zircons from an unknown locality in the RVP. Cong et al. [23] indicate that none of the zircon display oscillatory magmatic zoning in CL images and, that in BSE images, the samples are homogeneous. Cathodoluminescence images presented in the Cong et al. [23] study show very weak growth zoning, which is contrary to statements made within the text. In total, more than 50 zircons from the RVP deposits were examined by CL imaging in this study and all showed strong growth zoning. This discrepancy between the two sample sets may simply be the result of incorrect settings during CL imaging by Cong et al. [23]. However, the authors later use these results incorrectly as evidence for thermal homogenization of the zircons in the mantle.

4.2. Zircon Chemistry

Average EMPA data are shown in Table 1 and representative trace element data (ppm) are shown in Table 2. Rim and core analyses are noted when applicable. Figure 8 depicts box plot diagrams for relevant trace elements (a) and ratios (b) in zircon from each locality. Rare-earth element (REE) data were normalized to C-1 chondrite values [70] and plotted on logarithmic multi-element diagrams (La-Lu, Figure 9). Figure 10 depicts the chondrite-normalized trace element spidergram for the RVP zircon (in shaded orange). Data from Cong et al. [23] as well as for zircon from a variety of rock types [5].

Table 1. Average EMPA analyses for RVP zircon xenocrysts.

Sample	BC45	BC46-1	BC46-2	BS47	PT48-1	PT48-2	PT48-3	PT49-1	PT49-2	PT50-1	PT50-2	PT50-3
n *	6	7	7	8	7	4	10	8	7	8	7	7
HfO_2	0.74	0.72	0.80	0.69	0.74	0.77	0.73	0.77	0.80	0.71	0.76	0.79
ZrO_2	66.87	66.04	66.57	66.58	67.04	66.67	67.07	67.50	67.52	67.46	67.39	67.34
Y_2O_3	0.04	0.01	0.03	0.10	0.02	0.03	0.06	0.02	0.03	0.03	0.02	0.03
ThO_2	0.01	0.00	0.01	0.01	0.00	0.01	0.01	0.01	0.01	0.00	0.00	0.01
UO_2	0.01	0.00	0.00	0.02	0.01	0.02	0.01	0.01	0.00	0.01	0.01	0.01
SiO_2	32.34	31.69	32.05	32.62	32.47	32.38	32.49	32.35	32.26	32.32	32.30	32.22
Total	100.01	98.47	99.46	100.01	100.28	99.86	100.37	100.66	100.62	100.53	100.47	100.40
Sample	**BL51-1**	**BL51-2**	**BL52**	**BL53-1**	**BL53-2**	**BL54-1**	**BL54-2**	**BL54-3**	**BL55-1**	**BL55-2**	**BL55-3**	**BL55-4**
n *	**6**	**6**	**8**	**16**	**10**	**10**	**5**	**11**	**5**	**6**	**7**	**8**
HfO_2	0.79	0.72	0.69	0.75	0.73	0.71	0.72	0.75	0.75	0.71	0.71	0.70
ZrO_2	67.66	67.84	68.17	65.79	66.31	66.04	65.80	65.39	66.93	67.20	66.98	67.11
Y_2O_3	0.03	0.04	0.07	0.11	0.03	0.03	0.04	0.17	0.08	0.03	0.06	0.05
ThO_2	0.01	0.00	0.01	0.05	0.01	0.01	0.01	0.07	0.01	0.01	0.01	0.01
UO_2	0.01	0.01	0.01	0.04	0.01	0.01	0.01	0.05	0.03	0.02	0.01	0.01
SiO_2	32.08	32.06	31.99	32.67	32.68	32.38	32.30	32.26	32.28	32.27	32.27	32.35
Total	100.58	100.67	100.94	99.41	99.78	99.19	98.88	98.69	100.07	100.24	100.04	100.23

n * = number of analyses.

Table 2. Average trace element composition, ratios, and sums (ppm) of RVP zircon xenocrysts.

Locality (Code)	Bokeo Clas (BC)			Bei Srok (BS)	Phum Throm (PT)					BoLoei (BL)	
Sample # (n) *	45 (5)	46-1 (4)	46-2 (4)	47 (5)	49-1 (4)	49-2 (4)	50-1 (4)	50-2 (4)	50-3 (3)	53-1 (3)	53-2 (4)
HfO_2 wt %	0.73	0.71	0.80	0.68	0.78	0.79	0.71	0.76	0.80	0.75	0.72
Hf wt %	0.62	0.60	0.67	0.58	0.66	0.67	0.60	0.64	0.68	0.64	0.61
P	58	47	51	91	34	43	51	51	54	53	47
Ti	5.1	4.5	3.7	7.7	3.1	3.1	4.6	3.7	3.8	3.8	4.0
Nb	4.0	1.7	2.3	11.7	1.3	2.0	2.7	2.4	3.2	3.9	2.5
Ta	1.7	0.9	1.3	3.3	0.6	1.1	1.2	1.3	1.6	1.8	1.0
Pb^{204}	1.05	1.16	1.02	1.48	0.97	1.27	1.19	1.18	1.61	1.58	1.28
Pb^{206}	0.05	0.05	0.02	0.13	0.06	0.07	0.03	0.03	0.06	0.18	0.04
Pb^{207}	0.03	0.02	0.02	0.04	0.02	0.02	0.00	0.02	0.03	0.01	0.04
Pb^{208}	0.02	0.06	0.01	0.02	0.02	0.03	0.02	0.01	0.07	0.02	0.01
Th	42	19	13	168	9	17	25	25	30	51	22
U	83	41	39	187	23	43	55	58	61	90	42
Th/U	0.46	0.46	0.33	0.88	0.35	0.35	0.43	0.44	0.38	0.54	0.41
Nb/Ta	2.2	2.0	1.8	3.5	2.3	2.0	2.2	1.8	2.1	2.1	2.4
Yb/Sm_N	71	77	86	25	83	79	59	67	97	52	79
Lu/Gd_N	20	21	24	7	23	22	15	19	28	13	20
Y/Hf	0.00	0.00	0.00	0.00	0.00	0.00	0.00	0.00	0.00	0.00	0.00
Eu/Eu^*_N	1.00	1.01	0.99	0.99	1.04	1.00	1.00	1.03	0.95	1.02	0.99

Table 2. *Cont.*

Locality (Code)	Bokeo Clas (BC)			Bei Srok (BS)	Phum Throm (PT)					BoLoei (BL)	
Sample # (n) *	45 (5)	46-1 (4)	46-2 (4)	47 (5)	49-1 (4)	49-2 (4)	50-1 (4)	50-2 (4)	50-3 (3)	53-1 (3)	53-2 (4)
Y	365.9	150.4	169.9	921.8	119.7	196.1	259.3	245.1	222.7	299.8	206.2
La	bdl	bdl	bdl	bdl	bdl	bdl	bdl	bdl	bdl	bdl	bdl
Ce	1.7	0.7	0.9	5.3	0.6	1.0	1.0	1.1	1.2	2.3	1.2
Pr	0.1	0.0	0.0	0.2	0.0	0.0	0.0	0.0	0.0	0.0	0.0
Nd	0.9	0.3	0.3	3.8	0.3	0.6	0.5	0.7	0.3	0.6	0.3
Sm	1.7	0.6	0.6	6.7	0.6	1.0	1.2	1.3	0.7	1.2	0.8
Eu	1.3	0.5	0.5	4.6	0.4	0.7	0.9	0.9	0.6	1.0	0.6
Gd	8.8	3.5	3.4	29.7	2.8	4.8	6.4	6.1	4.2	7.4	4.8
Tb	2.9	1.1	1.2	8.8	0.9	1.5	2.1	1.9	1.5	2.4	1.6
Dy	37	15	16	103	12	19	27	24	20	31	21
Ho	12	5	6	30	4	7	9	8	7	10	7
Er	55	23	26	120	18	30	39	37	35	43	30
Tm	11	5	5	20	4	6	7	7	7	8	6
Yb	100	44	54	166	36	57	63	70	69	68	52
Lu	18	8	10	25	7	11	11	13	13	12	9
ΣREE	250	107	124	524	86	139	168	172	160	186	135
ΣREE+Y	616	257	294	1446	205	335	427	417	383	486	341
Ce/Ce^*_N	18	11	23	24	14	21	19	13	38	31	37

bdl = below detection limit.

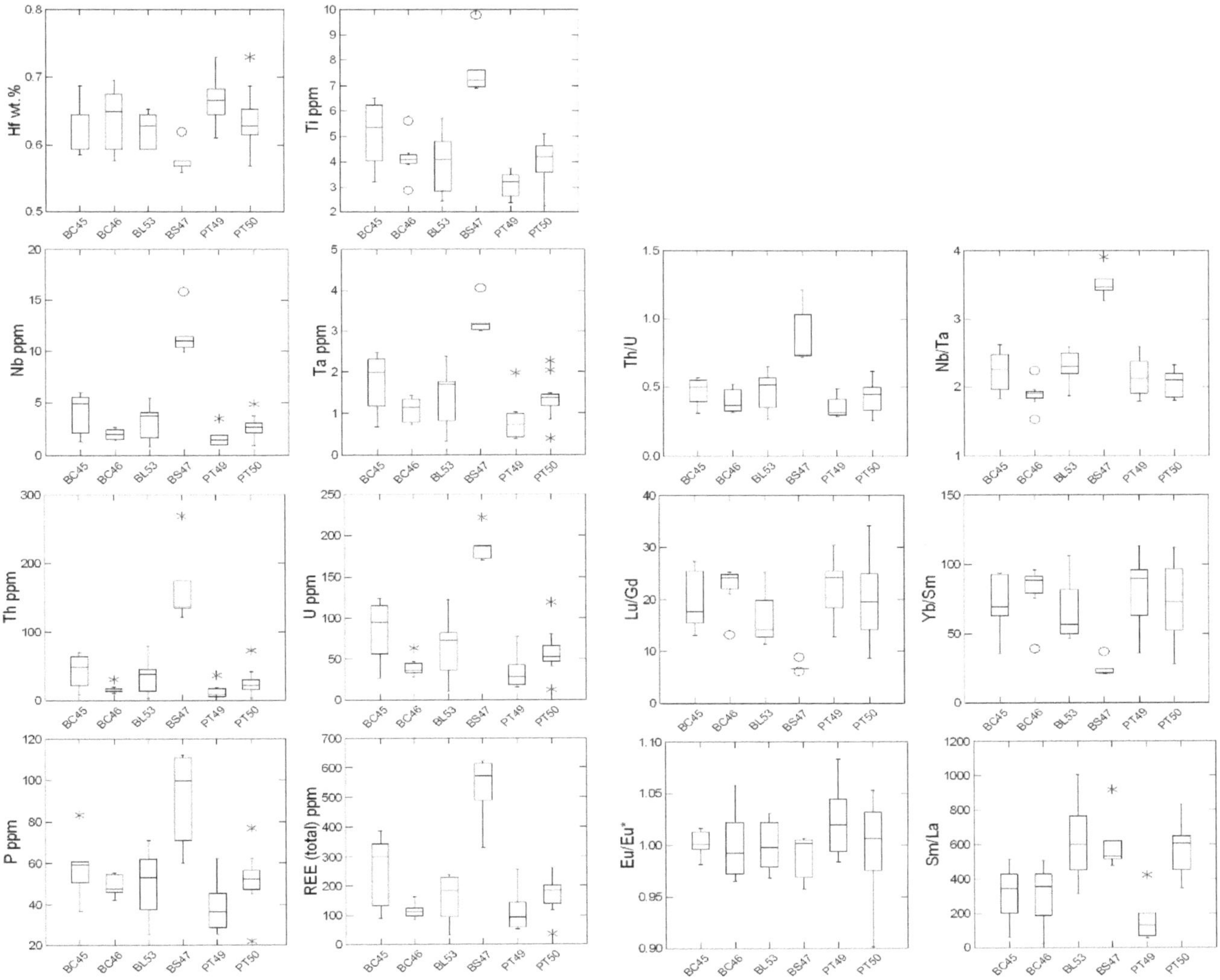

(a) (b)

Figure 8. Boxplot diagrams for relevant trace element (**a**) and trace element ratios (**b**) in RVP zircon xenocrysts. The vertical line and whiskers represent the range in the data values with the horizontal line representing the median of the data set. The box is defined by the first and third quartiles.

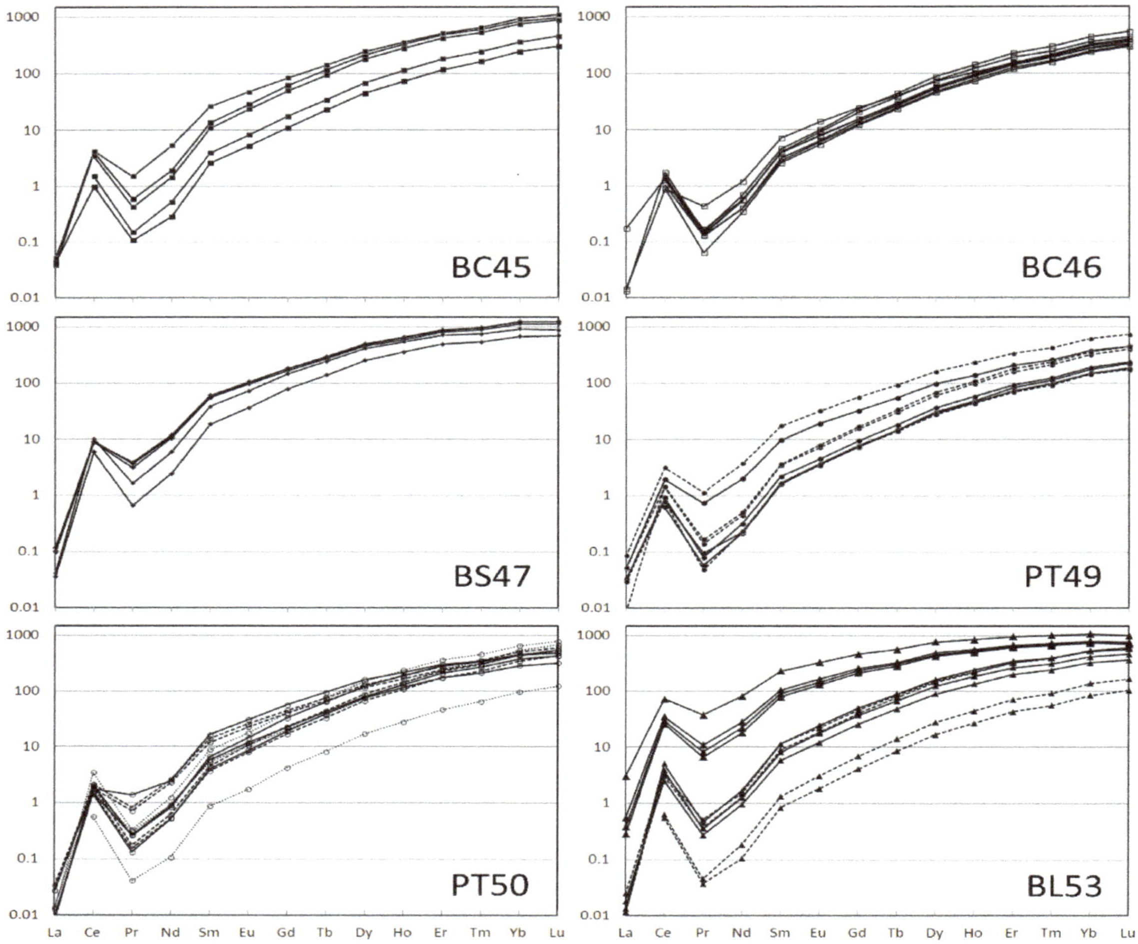

Figure 9. Average chondrite-normalized REE patterns for RVP zircon xenocrysts.

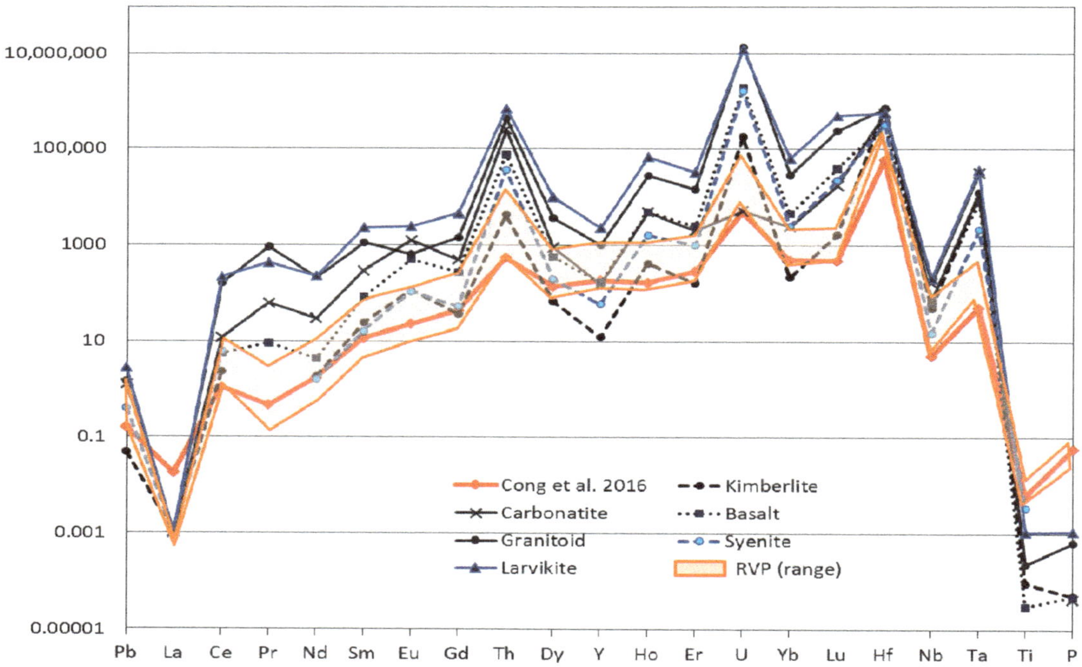

Figure 10. Chondrite-normalized trace element spidergram for zircon from the RVP (orange shaded region) including average data from Cong et al. (red [23]) and zircon from other rock types [5].

Electron microprobe analysis of the zircon xenocrysts gave major element compositional ranges of SiO_2 from 30.93 to 32.87 wt % and ZrO_2 65.06 to 68.36 wt %. Hafnium contents range from 0.66 to 0.86 wt % with an average of 0.74 wt % and do not show significant variation from core to rim. Y_2O_3 contents range up to 0.23 wt % (average: 0.05 wt %). Zircon from Bei Srok has lower HfO_2 than other RVP zircon (range: 0.66 to 0.73 wt %, average: 0.68 wt %). The Zr/Hf ratios (elemental) range from 66 to 88 (average: 79), which is higher than expected for mantle zircon (Zr/Hf = 60–68, Pupin [71]). Zircon from the RVP show wide variations in Ti, Nb, Ta, U, Th, P, Y, and REE contents between both grains and localities. In general, all RVP zircon show heavy REE (HREE) enrichment, have a positive Ce anomaly, and lack a negative Eu anomaly with ΣREE+Y, U, Th, and Ti contents exhibiting a decrease from core to rim in most samples. No compositional difference is observed between morphological types of the xenocrysts. Two distinct compositional populations exist: (1) the main population of RVP zircon from Bokeo Clas, Phum Throm, and BoLoei and (2) zircon from Bei Srok, which have higher overall trace element concentrations than the main RVP population.

Zircon from the main population have trace element concentrations as follows [ppm range (average)]: P 22–88 (average 51), Ti 1.6–6.5 (average 3.7), Nb 0.8–6 (average 2.6), Ta 0.3–3 (average 1.3), Th 2.6–135 (average 28), U 10–157 (average 55), and Y 43–560 (average 227). Average Nb/Ta and Th/U ratios are 2.11 and 0.44, respectively (Figure 5b). The bulk of the zircon are characterized by low ΣREE contents (31–386 ppm, average 154 ppm), which is typical of xenocrysts from alkaline mantle sources [5]. Total REE(+Y) contents decrease from the core to the rim in all samples. The chondrite-normalized REE patterns (Figure 10) show strong light REE (LREE) depletion, which is a pronounced positive Ce anomaly (Ce/Ce^*_N = 11–38, average 27, Eu/Eu^*_N = 1, and heavy REE (HREE) enrichment characterized by a steep, straight slope with Yb/Sm_N ranging from 28–113 (average 75) and Lu/Gd_N = 9–34 (average 20). Normalized Ce and Eu anomalies were calculated as $Ce/Ce^*_N = \frac{Ce}{\sqrt{La \times Pr}}$ and $Eu/Eu^*_N = \frac{Eu}{\sqrt{Sm \times Gd}}$, respectively.

Zircon from Bei Srok (BS47) has higher overall trace element contents as compared to those from other RVP localities [Figure 8, range (average) ppm]: P 60–112 (average 91), Ti 7–10 (average 8), Nb 10–16 (average 12), Ta 3–4 (average 3), Th 123–269 (average 168), U 170–221 (average 187), and Y 574–1076 (average 922). Total REE contents range from 328 to 621 ppm (average 524 ppm) and decrease from core to rim. Bei Srok zircon has significantly higher Nb/Ta (average 3.53) and Th/U (average 0.88) as compared to zircon from BC, PT, and BL (Figure 9b). The chondrite-normalized REE pattern for BS zircon (Figure 9) shows strong LREE depletion with a pronounced positive Ce anomaly (Ce/Ce^*_N = 13–38, average 24), Eu/Eu^*_N = 0.99, and middle REE (MREE)/HREE enrichment characterized by a steep, slightly concave-down curvature with Yb/Sm_N = 21–37 (average 25) and Lu/Gd_N = 6–9 (average 7).

Ti-in-Zircon Thermometer

Watson and Harrison [72] and Watson et al. [73] used high-pressure/high-temperature experimental methods coupled with analyses of natural zircons to determine the exact relationship between Ti content and crystallization temperature in zircon TiO_2. The resultant Ti-in-zircon thermometer (Equation (1)) has the capability to provide temperature with a precision of ± 5 °C depending on the Ti concentration and the analytical method employed [73].

$$T\ (^{\circ}C)_{zircon} = [5080/(6.01 - \log(Ti_{ppm})] - 273 \quad (1)$$

A number of limiting factors must be considered when utilizing the Ti-in-zircon thermometer including the variation of TiO_2 or SiO_2 activity in the melt, the pressure fluctuation, changes in the Ti content of the zircon as a result of solid state diffusion or post-crystallization alteration, and the assumption that the temperature dependence of Ti in zircon is an equilibrium process and obeys Henry's Law [74]. Cherniak and Watson [75] demonstrated that Ti is strongly retained by the zircon structure in both anhydrous (1 atm) and H_2O-CO_2-bearing (1.1–1.2 GPa) systems at temperatures between 1350 and 1550 °C.

Titanium contents in the analyzed zircon are all above the detection limit (0.1 ppm) and range from 2 to 10 ppm (Table 3). All zircon are primary magmatic phases with oscillatory zonation, which is the result of intrinsic non-linear feedback between the parental melt and the growing crystal [68]. Although this zonation suggests local non-equilibrium during growth, which is a result of the continuous process of depletion and replenishment of solute at the crystal-melt boundary, there is no evidence for large-scale disequilibrium due to extrinsic mechanisms in the system. This indicates that Ti contents in the zircon are a primary magmatic feature. In addition, there is no evidence for metamictization or alteration which could be responsible for Ti diffusion. As such, the temperatures obtained by the Ti-in-zircon thermometer are considered representative of the difference in relative crystallization temperatures between zircon from the RVP localities.

Table 3. Ti-in-zircon temperatures for RVP zircon xenocrysts.

Locality	Sample # (n) *	Ti ppm ± 2 SD	T °C ± 2 SD
Bokeo Clas	45 (5)	5.07 ± 2.83	682 ± 46
-	46-1 (4)	4.53 ± 1.45	675 ± 23
-	46-2 (4)	3.75 ± 1.20	661 ± 26
Bei Srok	47 (5)	7.68 ± 2.38	718 ± 24
Phum Throm	49-1 (4)	3.05 ± 1.04	646 ± 24
-	49-2 (4)	3.14 ± 1.07	648 ± 26
-	50-1 (4)	4.64 ± 0.81	677 ± 14
-	50-2 (4)	3.67 ± 0.36	660 ± 7
-	50-3 (3)	3.81 ± 2.87	659 ± 63
BoLoei	53-1 (3)	3.77 ± 1.23	661 ± 26
-	53-2 (4)	4.03 ± 3.52	662 ± 69

* n = number of analyses, SD = standard deviation.

Average Ti-in-zircon temperatures are 663 ± 28 °C (range: 601–739 °C, Table 3). In general, there is a decrease in the calculated temperature from core to rim in all samples with ΔT ranging from 10–55 °C. Wider variation in temperature between core and rim may simply reflect the larger size of the zircon grains themselves, which is a possible reflection of residence time within the parental melt. No difference in calculated temperatures is observed between the various morphological types which is contrary to what is predicted based on Pupin's T versus A diagram [67]. Temperatures based on morphological studies suggest crystallization at 800 to 900 °C, which is significantly higher than that predicted by Ti-in-zircon temperatures. The lower T for Ti-in-zircon results than predicted for natural zircon may result from pressure and Ti^{4+} substitution effects [76]. The relatively low Ti-in-zircon for RVP samples may reflect the likely high-pressure mantle origin of these xenocrysts, which has also been suggested for kimberlite zircons [77].

4.3. Geochronology Results

U-Pb age dates were obtained for 11 zircon grains from the four mining regions with three to nine analyses per grain (Table S1). All samples lack any significant U-Pb age differences between core and rim, which confirms the lack of inherited cores and a single growth event for all the zircon. The young zircon xenocrysts are depleted in ^{207}Pb and subsequently $^{207}Pb/^{235}U$ ages are very poor with large associated errors. As a result, only $^{206}Pb/^{238}U$ is reported in this case (Table 4). Weighted mean $^{206}Pb/^{238}U$ ages range from 0.88 ± 0.22 Ma (PT50) to 1.56 ± 0.21 Ma (PT49) at the 95% confidence level (1σ error, Table 4). Seven U-Pb ages determined by ID-TIMS on RVP zircon during a gemological study [45] gave an average age of 0.92 ± 0.07 Ma (range: 0.83–1.03 Ma), which is in agreement with analyses in this study (Table 4). Analyses of a zircon inclusion in corundum from the BL deposit gave a U-Pb age of 0.855 ± 0.098 Ma, which suggests a temporal, if not a genetic, relationship between the two xenocrystic phases [12]. Similar age ranges are noted for zircon xenocrysts from Eastern

;ondwana margins including Western Cambodia (Table 4). Fission track dating of zircon from Bokeo y Carbonnel et al. [78] yielded ages ranging from 1.10–1.41 Ma, which is also in agreement with the ges from this study.

Table 4. U-Pb, fission track and (U-Th)/He age dates for ZIP xenocrysts.

Locality	U-Pb	Fission Track/(U-Th)/He	Reference
Ratanakiri Volcanic Province (This Study)	-	-	-
Bokeo Clas (BC45)	0.98 ± 0.12	-	This study
Bokeo Clas (BC46)	1.07 ± 0.19	0.91 ± 0.02	-
Bei Srok (BS47)	1.143 ± 0.073	1.02 ± 0.02	-
Phum Throm (PT49)	1.56 ± 0.21	0.93 ± 0.02	-
Phum Throm (PT50)	0.88 ± 0.22	1.02 ± 0.02	-
-	-	0.86 ± 0.02	-
BoLoei (BL53)	0.978 ± 0.054	0.96 ± 0.06	-
-	-	0.96 ± 0.09	-
RVP (unknown locality)	0.92 ± 0.07	-	[45]
-	-	-	-
Pailin, Cambodia	2.74 ± 0.47	2.14 ± 0.02	This study
-	2.27 ± 0.15–2.60 ± 0.20	-	[78]
	1.77 ± 0.15–2.0 ± 0.20		[79]
Chanthaburi-Trat, Thailand	1.19 ± 0.29–2.22 ± 0.22	0.90 ± 0.04–2.13 ± 0.04	This study
-	2.57 ± 0.20	-	[78]
Dak Nong, Vietnam	1.05 ± 0.05	-	[7]
-	7.13 ± 0.88	-	-
Xuan-Loc, Vietnam	0.57–0.70	-	[78]
-	0.2–0.9	-	-
Ban Huai Sai, Laos	1.14 ± 0.07–1.46 ± 0.06	-	[38]
-	2.4 ± 0.7–4.3 ± 1.0	-	[11]
Penglai, Hainan Island, China	4.4 ± 0.1	-	[80]
-	-	4.06 ± 0.35	[81]
Mingxi, China	1.2 ± 0.1	-	[13]
NE Australia	2.61 ± 0.16–2.92 ± 0.16	-	[10]

A total of 37 crystal shards of zircon crystals from the four mining regions were dated by the U-Th)/He method (Table S2). Reproducibility of (U-Th)/He ages (ZHe) from each locality (typically -6 ages) was excellent with all replicates overlapping within analytical uncertainty. Final ZHe ges, calculated for each locality as error-weighted means and standard deviation as uncertainty, ange between 0.86 ± 0.02 and 1.02 ± 0.02 Ma (Table 4). These are slightly younger or overlap with ncertainty via the corresponding U-Pb ages measured by LA-ICP-MS, which represent a minimum ime for zircon crystallization in accordance with the closure temperature concept [82] and also onfirms the accuracy of both dating methods. The age and similarity of ZHe and U-Pb ages points o an extremely fast cooling of the crystals from magmatic temperatures through the zircon helium losure isotherm and suggests that the dated crystals erupted to the surface approximately in the same ime between 1.02 and 0.86 Ma. As with the U-Pb ages, there seems to be no correlation between the ZHe eruption ages and the spatial distribution of dated samples.

.4. *Hf Isotope Signatures*

Zircon from the RVP deposits show a moderate range in the Hf isotope composition Table S3)—the ^{176}Hf/^{177}Hf ratio ranges from 0.282919–0.283065 (average 0.282982 ± 15, n = 20), which corresponds to positive εHf values ranging from 4.77 to 9.92 (average 7.00 ± 0.25). Zircon from Bokeo Clas, Bei Srok, and BoLoei are the most radiogenic with the highest ^{176}Hf/^{177}Hf ratios BC45 = 0.283065, εHf 9.92, BS47 = 0.283039, εHf 9.01, and BL53 = 0.283021, εHf 8.37), which all exhibit within-grain variation greater than their 2SE uncertainties. While variations in ^{176}Hf/^{177}Hf within single grains are observed, there is no consistent relationship between core and rim compositions.

Estimated minimum model ages (T_{DM}) for the source material range from 260 to 401 Ma with crusta model ages ($T_{DM}{}^{C}$) ranging from 391 to 633 Ma.

Zircon from Phum Throm (PT) are the least radiogenic and exhibit less within-grain an between-grain variation (Figure 8). The $^{176}Hf/^{177}Hf$ ratio ranges from 0.282919 to 0.282985 (averag 0.282964 ± 6) with εHf 4.77–7.11 (average 6.34 ± 0.59). Estimated minimum model ages (T_{DM}) for th source material are higher for PT zircon and range from 372 to 462 Ma (average 403 Ma) with crusta model ages ($T_{DM}{}^{C}$) ranging from 605 to 754 Ma (average 654 Ma).

4.5. Oxygen Isotope Ratios

Oxygen isotope ratios provide valuable information about the melt from which the zirco crystallized, which often allows for the discrimination between mantle and crustal sources. Values c $\delta^{18}O$ in RVP zircon are homogeneous with $\delta^{18}O$: 4.93 ± 0.05‰ VSMOW (n = 5, Table 5). These value are consistent with those obtained by SIMS for zircon megacrysts from an unknown RVP locality: δ^{18}C = 4.5–5.5‰ VSMOW, weighted mean = 5.0 ± 0.18‰ VSMOW, n = 210 [23]. The slow rate of diffusion c ^{18}O in non-metamict zircon, coupled with the large crystal size and primary magmatic composition zoning indicate the $\delta^{18}O_{zrn}$ to represent initial magmatic values [83,84]. These values are lov but, within the $\delta^{18}O_{zrn}$ range for igneous zircon in high temperature equilibrium with a mantle sourc ($\delta^{18}O_{zrn}$ = 4.7–5.7‰ VSMOW) and olivine from mantle xenoliths ($\delta^{18}O_{ol}$ = 5.15 ± 0.13‰ [85]). Similarl RVP $\delta^{18}O_{zrn}$ are lower than those observed for other alkali basalt-hosted zircon xenocrysts in SE Asi including in Western Cambodia at Pailin (5.95 ± 0.02‰), Chanthaburi-Trat and Bo Phloi (Thailanc 5.82–6.16‰, Table 5) and Penglai, Hainan Island (South China, $\delta^{18}O_{zrn}$ = 5.31 ± 0.18‰ [80]). The lowe $\delta^{18}O$ values observed in RVP zircon are akin to values observed in zircon from monomineralli anorthoclase xenoliths, which are called anorthoclasites, in alkali basalt at Elie Ness, Scotlan (4.49–5.03‰, average 4.78 ± 0.16‰; [86]) and in zircon from kimberlite [77,87]. The $\delta^{18}O_{zrn}$ values fo RVP zircon suggest crystallization from a homogeneous igneous reservoir in the SCLM underlying th RVP with little to no contamination by supracrustal lithologies.

Table 5. $\delta^{18}O$ data for xenocrystic zircon from the RVP and Thailand.

Sample	mg	μml	μml/mg	$\delta^{18}O$ raw	$\delta^{18}O$ VSMOW
Cambodia-RVP	-	-	-	-	-
Bokeo Clas, BC46	2.85	32.7	11.5	4.93	4.95
Bei Srok, BS47	3.14	34.8	11.1	4.87	4.89
Phum Throm, PT49	2.82	30.7	10.9	4.99	5.01
Phum Throm, PT50	3.40	36.7	10.8	4.86	4.88
Bo Loei, BL55	3.26	35.1	10.8	4.90	4.92
Cambodia-Pailin	2.97	32.1	10.8	5.93	5.95
Thailand	-	-	-	-	-
Khao Ploi Waen, Chanthaburi-Trat	3.80	41.0	10.8	6.14	6.16
Tok Phrom, Chanthaburi-Trat	3.05	32.3	10.6	5.80	5.82
Bo Phloi, Kanchanaburi	3.46	37.1	10.7	5.86	5.88
Standards	-	-	-	-	-
UWG-2, garnet	3.39	45.6	13.5	5.78	-
UWG-2, garnet	2.29	31.4	13.7	5.79	-
UWG-2, garnet	2.16	28.3	13.1	5.80	-
UWG-2, garnet	2.70	36.0	13.3	5.77	-

* n = 4, x = 5.78, ± 0.02‰ 2 std. dev. Note: Thailand analyses performed under the same analytical conditions as RVP zircon.

. Discussion

.1. *Classification and Comparisons of RVP Zircon Xenocrysts*

The extensive substitution of trace elements (up to 50) into the zircon structure is useful in tudies of the compositional variation in igneous zircons as a petrogenetic indicator. Such data can reveal fractionation processes, the nature of source rocks, and provenance of zircons from econdary deposits [5,88–92]. Shnukov et al. [92] and Pupin [93] utilized Hf and Y concentrations to lassify magmatic zircons by focusing on early-stage and late-stage magmatic phases from both mantle and crustal rocks of orogenic and anorogenic origins. Belousova et al. [5] analyzed trace element compositions of zircons from nine different igneous rock types ranging from ultramafic to granitic and syenitic in nature as a way to statistically discriminate between zircon populations and assist in he identification of their source regions. The study suggested that compositional variation in HfO_2, Y, Nb, Ta, Lu, Ce/Ce^*_N, ΣREE, Yb/Sm_N, and Th/U are the most useful petrogenetic indicators across he represented range of rock types and that the provenance or parent rock type of a zircon can be letermined at confidence levels of 75% or more.

Zircon xenocrysts from the RVP typify zircons that crystallize from a juvenile, Zr-saturated, REE+Y-enriched alkaline magma or melt with low HfO_2 (<0.9 wt %), Y_2O_3 (<0.2 wt %), U, Th, ΣREE (<600 ppm), $Eu/Eu^*_N \approx 1$, and steep, and HREE-enriched REE profiles [4,5,94]. RVP zircon xenocrysts exhibit the depletion of ΣREE+Y, U, and Th with increasing Yb/Sm_N from core to rim at almost a constant Hf content. Such zoning favors zircon growth as a single growth stage in an alkaline environment rather than growth involving fractional crystallization [22,94,95]. In depolymerized alkaline melts, the partition coefficients for Hf and Zr between zircon and melt are approximately equal over a wide *T* range and melt compositions [22]. As such, crystallization of zircon will not fractionate Zr and Hf significantly, which results in near constant Zr/Hf in the zircon/melt. An accompanying trend of decreasing REE, Y, U, and Th and increasing Yb/Sm_N suggests that RVP zircon crystallized during a single magmatic event with no influx of additional melt components. The lack of a negative Eu anomaly n these zircons also suggests crystallization in a system where plagioclase did not crystallize or at pressures above the plagioclase stability zone (~0.80 GPa in fertile mantle lherzolite [96]), which lends urther support to a sodium-potassium-rich melt. Increasing Yb/Sm_N from core to rim may indicate he depletion of the melt in LREE as a result of a co-crystallizing phase (i.e., apatite) or selective absorption of HREE during the crystallization period [94]. RVP zircon xenocrysts found as isolated crystals within the host alkali basalt, contain corroded REE-rich apatite, and REE carbonates as rare nclusions in larger crystals. Lacombe [43] also noted a carbonate mineral within the RVP zircon as a urther sink for LREE and a likely growth environment enriched in CO_2.

Utilizing the classification by Pupin [93] based on HfO_2 and Y_2O_3 contents, all RVP zircons plot n field 1c, mantle-derived hawaiite and alkali basalt, which lies within a narrow range of Hf contents and a spread of Y contents (18–2011 ppm). In Shnukov's Y versus Hf diagram, RVP zircons almost entirely fall within the carbonatite field (VII), with rare samples in the alkaline rocks and alkaline metasomatite field (VI) [92]. Such affinities also apply to zircon xenocrysts from nearby Ban Huai Kai, Laos [11], and Pailin, and Cambodia (Piilonen, unpublished data, Figure 11). The statistical classification scheme by Belousova et al. [5] predicts a cabonatitic provenance for almost all RVP zircon xenocrysts in agreement with the Y versus Hf discrimination diagram except for BS47 in which a higher Lu content (>20.7 ppm) suggests an origin from a basaltic parental melt.

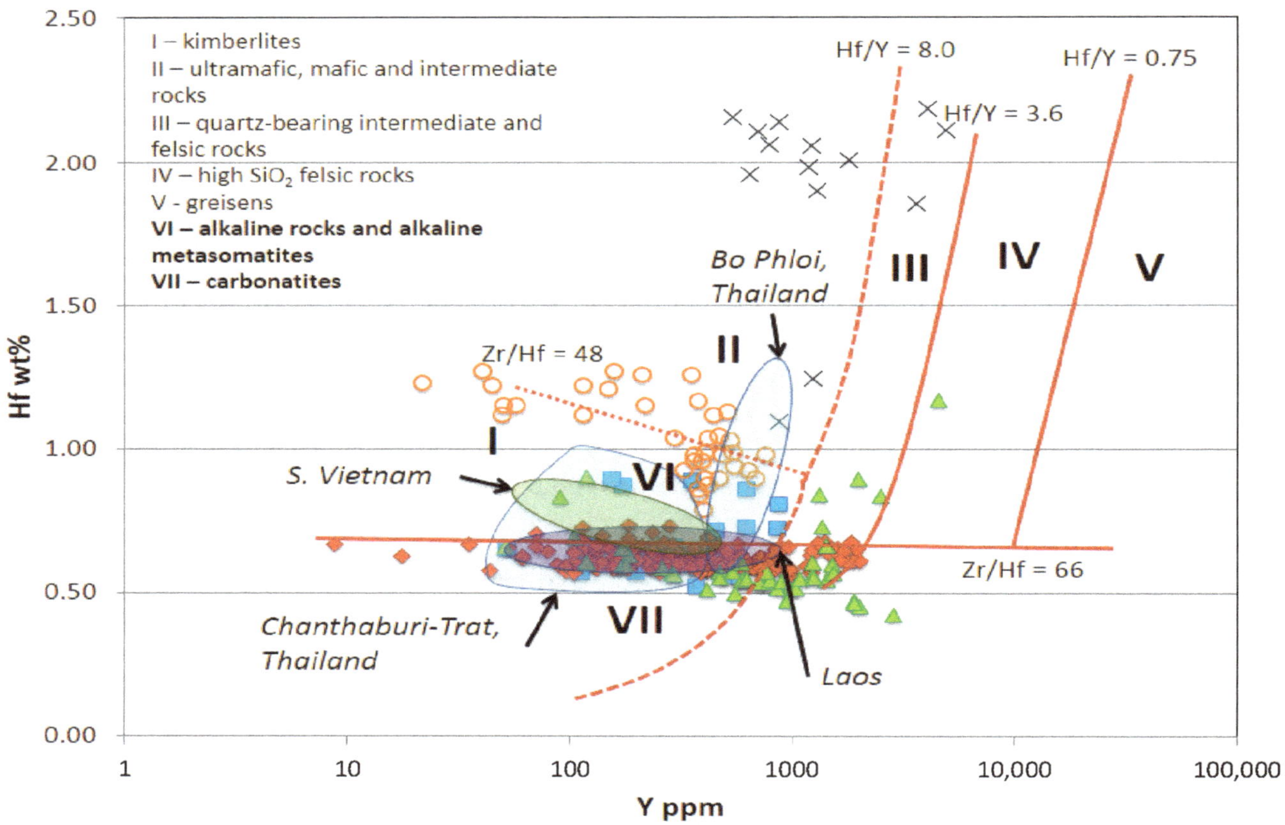

Figure 11. Y versus Hf discrimination diagram for RVP zircon, zircon xenocrysts from ZIP localities, and zircon inclusions in corundum (modified after [92]). Red diamonds: RVP zircon, blue squares: Pailin, Cambodia (Piilonen, unpublished data), blue field: Thailand (Piilonen, unpublished data), green triangles: Australia [11], orange circles: zircon xenocrysts from carbonatites [5], black crosses: zircon inclusions in corundum xenocrysts [10,12,38,97].

Isolated RVP zircon xenocrysts differ from many other zircon xenocryst suites that are commonly spatially associated with corundum xenocrysts either in multimineralic xenoliths or as discrete crystals in heavy mineral separates, e.g., Bo Phloi and Chanthaburi-Trat regions in Thailand (Piilonen, unpublished data, [98,99]), South Vietnam [100], Loch Roag, Scotland [97], and Lava Plains, NE Australia [10,12]. Such zircons associated with corundum, either as xenocrysts or as corundum-hosted inclusions, have consistently higher Hf (>1 wt % and up to 4.5 wt % HfO_2) and often negative Eu anomalies, which suggests evolved, likely peraluminous corundum-bearing sources (Figure 11). Zircon inclusions in blue-green-yellow magmatic corundum xenocrysts (BGY suite) not only show enriched Hf and noticeable negative Eu anomaly but also have enrichments in Y (up to 1 wt % Y_2O_3), U (up to 1 wt % UO_2), Th (up to 2.1 wt % ThO_2), and REE up to 12,060 ppm [11,101–103]. These corundum-hosted zircons are dominated by the {110} prism over the {100} prism unlike the RVP zircon xenocrysts and are a morphology thought to be related to high U, Th, Y, REE, and P contents in the parental melt [104]. Plotted on the Y versus Hf discrimination diagram (Figure 12), the zircon compositions hosted in RVP corundum fall across Type II and III sources (II—ultramafic, mafic and intermediate rocks, III—quartz-bearing intermediate and felsic rocks).

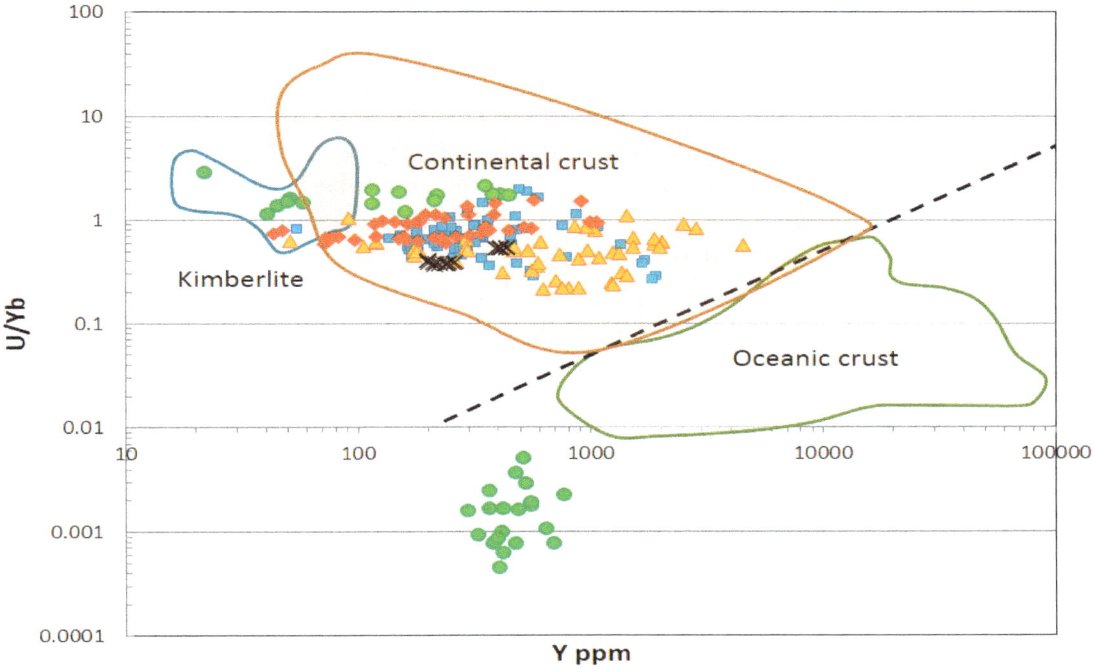

(a)

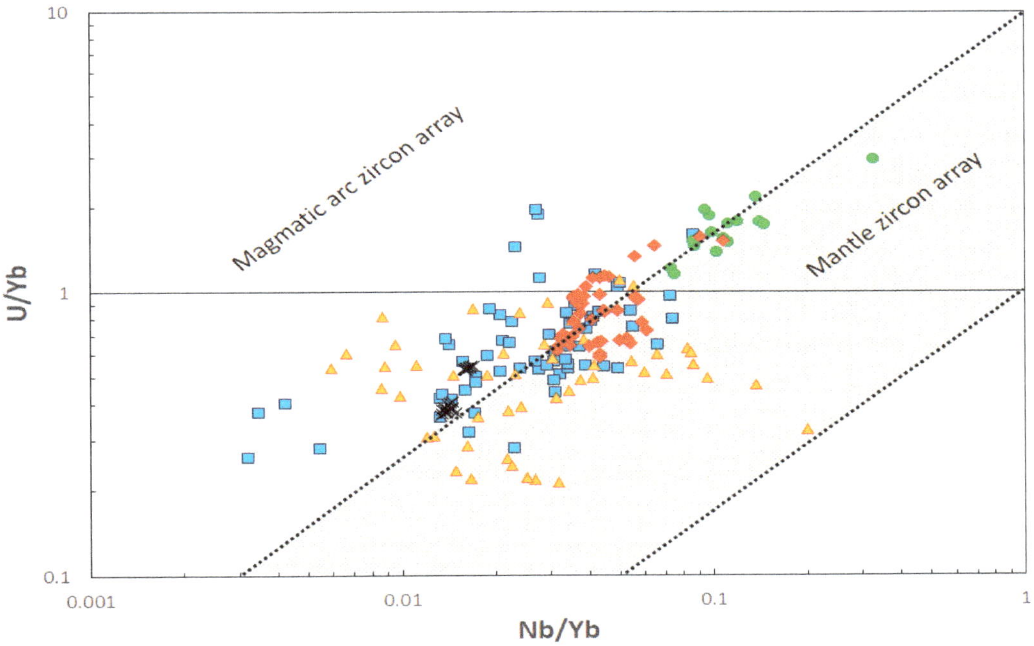

(b)

Figure 12. Y versus U/Yb (a) and Nb/Yb versus U/Yb. (b) Discrimination diagrams for zircon from the RVP (red diamonds, black X's [23]), Thailand (blue squares, Piilonen, unpublished data), other ZIP localities (orange triangles, Australia, Laos, Vietnam [11]), and carbonatites (green circles [5]). Fields for continental and oceanic crust and kimberlite from Grimes et al. [105]. Fields for mantle and magmatic arc arrays from Grimes et al. [106].

.2. *Source Affinities*

All RVP zircon xenocrysts have trace element and O-Hf isotopic compositions indicative of a primary, alkaline magma derived from a variably-depleted lithospheric mantle source with limited supracrustal contamination. Their Hf isotopic compositions lie between those for depleted juvenile mantle and a chondritic reservoir (CHUR, Figure 13). This supports either (1) a juvenile mantle source mixed with older crustal material or (2) recycled material with a longer crustal residence time. Their Y versus Yb/U are more chemically aligned to continental crust rather than recycled oceanic crust values (Figure 11), which is observed for zircon xenocrysts from other mantle-derived rocks including some kimberlites and carbonatites [105]. The variance in the Hf isotope composition between RVP samples and within grains (εHf = 4.77–9.92) may, therefore, reflect localized source variation or unmixed crustal or metasomatic components. Griffin et al. [48] suggested that low and variable εHf in some kimberlitic zircon may be due to crystallization from DM-derived or the ocean island basalt (OIB)-type magmas undergoing assimilation/fractionation reactions with a non-radiogenic lithosphere. The SCLM underneath the RVP is a complex region of multiple rift-related and back-arc-related collisional remnants, the Song Ma suture zone, and back margins of the Himalayan collision zone. As the zircon xenocrysts did not crystallize from the host alkali basalt magma, an additional Mantle-derived melt is needed. A DM-derived melt interacting with SCLM underneath the RVP would encounter various reservoirs with non-radiogenic Hf, widely varying ages, and Lu/Hf ratios will result in variable $^{176}Hf/^{177}Hf$ in the crystallizing zircon.

Similarities of U-Pb (0.88–1.56 Ma) and (U-Th)/He ages (0.86–1.02 Ma) yielded by RVP zircon crystals suggest their minimal mantle residence time (0.18–0.86 Ma) and a rapid ascent to the near-surface temperatures facilitated by the alkali basalts. The U-Pb and (U-Th)/He ages, however, are clearly older than the geomagnetic reversal at 0.7 Ma that predates the later, explosive phase of basaltic volcanism in the area [42]. The U-Pb and (U-Th)/He, thus, suggest an earlier onset of basaltic volcanism in the area starting at least at ~1.02 ± 0.02 Ma. Crystal morphology, compositional zoning and REE patterns indicate zircon growth from a single magma without fractional crystallization or injection of additional melts.

Model Hf ages for RVP zircon give dates significantly older than given by U-Pb and (U-Th)/He methods and correspond to major tectonic events in the SE Asia region. The estimated minimum T_{DM} model ages (333–403 Ma) correspond to the Alleghenian orogeny and suturing of the Indochina and South China blocks (~340 Ma). The older estimated minimum $T_{DM}{}^{C}$ model ages suggest that the source material for RVP zircon crystallization separated from the depleted mantle around 500–650 Ma during a major crust-building collision between East and West Gondwana to form late Neoproterozoic Pannotia (~650 Ma), which then broke-up in the early Paleozoic (~550 Ma).

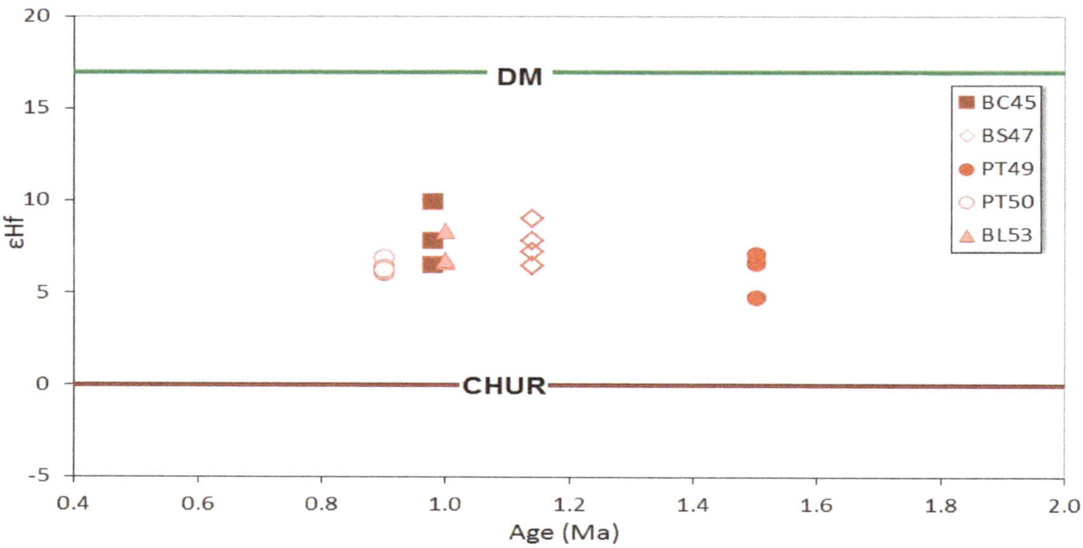

Figure 13. U-Pb Age (Ma) versus εHf for RVP zircon. CHUR: Chondritic uniform reservoir, DM: Depleted mantle.

5.3. Carbonatitic Signatures in Ratanakiri Zircons

Geochemical signatures in RVP zircon suggest that the crystallizing melts had carbonatitic links. Chemical ranges for carbonatite-derived zircons, however, are loosely defined and are much wider than the main carbonatite field (VII) in their Hf-Y discrimination diagram. They also fall within fields (II) ultramafic, mafic, and intermediate rocks as well as (VI) alkaline rocks and alkaline metasomites (Figure 9) [5]. Some zircons analyzed from carbonatitic dikes proved to represent crystallization from a silicate melt instead of a carbonatitic melt and are probably xenocrysts from wall rock contamination [107]. Strict chemical characterization of 'carbonatitic zircons' is clearly complex and partly unreliable [108,109]. Saava et al. [109] surveyed more than 100 carbonatitic zircons and found only 45% could be assigned to a carbonatitic source and that large compositional variation existed between both individual grains and localities. This variation may represent a constantly-evolving carbonatitic melt and its interaction with co-magmatic silicate sources or reflect variable generation in continental carbonatites from the SCLM and, in cases of oceanic island carbonatites, generation in the asthenosphere [110,111].

Carbonatites seem absent in Cambodian surface geology even though they appear in adjacent Northwestern Vietnam where rift-related carbonatites are dated at 28–44 Ma (biotite K-Ar ages) and 30–32 Ma (U-Th-Pb isochron ages) in South Nam Xe [112]. Geochemically, these carbonatites suggest involvement of an enriched mantle component in their genesis. These HSFE-deficient rocks, however, are low in Zr and lack zircon. Cenozoic carbonatite intrusions (12–40 Ma) lie 1600 km to the north within the SW China craton (Mianning-Denchang intrusive belt [113]). For the RVP magmatic sources, with a predisposition to crystallize carbonatitic-type zircon, their exact nature remains uncertain.

The absence of surficial carbonatites does not preclude a carbonatized mantle source [114]. Several types of continental carbonatitic-influenced mantle sources are known to have the potential to generate zircon megacrysts of an RVP type. Asian examples include: (1) stratified mantle deduced from ultramafic xenoliths that show initial silicate and then carbonatitic metasomatism in kamafugites from Western Qinlan, Central China [115], and (2) Pacific slab-induced carbonatitic metasomatism in garnet-bearing lherzolite xenoliths from Cenozoic basalts in NE China [116–118]. The latter is related to alkali basalts at Changle where zircon megacrysts have Hf-Y ranges similar to those in RVP zircons and the carbonatitic-modified mantle is recognized [13]. Elsewhere, upper mantle xenoliths with carbonatitic reactions are recorded in young basalt fields in SE Australia [119]. Vent deposits in this field include zircon megacrysts that show transitional silicate-carbonatitic affinities [120].

The extent to which carbonatitic altered mantle sources have contributed to zircon megacryst suites along Western Pacific continental margins remain uncertain. This uncertainty stems from the patchy spread of carbonatitic fields within the classification and regression trees CART discrimination model and the availability of inclusion and geochemical data furnished from the zircon-related host basalts [11]. A study of mantle xenoliths from young basalts on the Leizhou Peninsula, which is adjacent to the Pengali, Hainan Island mantle zircon xenocryst suites detailed a range of mantle melting and metasomatic effects [13,121]. The study identified trace element diffusion effects within the mantle and showed that HFSE depletions in metasomatized lherzolite do not necessarily require a carbonatitic metasomatizing agent.

5.4. Variation in Zircon-Bearing Xenocryst Assemblages in SE Asia

The enigmatic associations of zircon and corundum xenocrysts in intraplate alkaline basalt provinces continue to raise questions about their petrogenesis. The highly zircon-dominated RVP xenocryst suite, with minor corundum, is rare among the many basaltic gem deposits along West Pacific continental margins; corundum is common elsewhere in Cambodia (Pailin), through much of SE Asia and along Eastern Australia [8]. The lack of corundum in RVP alkali basalts suggests that particular mantle conditions favored RVP zircon genesis.

Some clues to these conditions may be offered by a study of zircon and corundum (sapphire)-bearing albititic dykes exposed within mantle assemblages in the French Pyrenees [122,123]. Some dykes contain only zircon megacrysts, some contain zircon + corundum, and some contain only corundum. Zircon in corundum-free dykes have systematically lower Hf (0.46–0.68 wt % Hf) and lower Y (740–1850 ppm), which is similar to that observed in the RVP zircon while zircon accompanied by corundum has higher values (Hf up to 1.8 wt %, Y up to 13,000 ppm). Geochemically these dykes indicate a mantle origin, which is the result of very low degrees of partial melting (<1%) of a harzburgite source that has undergone metasomatism by a carbonatitic fluid phase prior to crystallization of the mega-crysts. To answer this dichotomy in megacryst generation, Pin et al. [123] attributed zircon formation to more CO_2-rich fluid conditions relative to more hydrous conditions for corundum formation within the feldspathic melts in the dykes during high pressure crystallization of the assemblage on the liquidus.

5.5. Models for Generation of RVP Zircon-Rich, Corundum-Poor Gem Suite

Explanations for the distinct zircon generations (low-Hf isolated zircon versus high-Hf zircon associated with corundum) along with associated megacrysts (anorthoclase) include low degrees of partial melting of the upper mantle coupled with metasomatism by a carbonatitic melt/fluid, the immiscibility of a silico-carbonatitic melt, or the interaction of carbonatitic and syenitic melts within the SCLM [86,100,124]. Both alkali basalt magmas and their mantle source regions are known sources of CO_2 and, in intraplate continental basalts, CO_2 is the main volatile [125].

At pressures ~20 kbar, CO_2 solubility increases dramatically, which promotes low-T melting of peridotite (<1000 °C) as well as significantly lowering of SiO_2 contents in resultant partial melts [114]. Very small degrees of partial melting (<1%) of a peridotite or a harzburgite at upper mantle conditions can produce early liquids strongly enriched in Si, Na, and Al and depleted in Fe, Mg, Ca, and Ti, which is similar in composition to syenite melts [122,123]. If such alkali-rich partial melt further undergoes enrichment by an H_2O-rich or CO_2-rich metasomatizing fluid, the behavior and activity of Si and Al are affected. The addition of H_2O to the metasomatizing fluid increases corundum saturation but does not affect Zr solubility [19]. The degree of partial melting, Zr content, alkalinity and the degree of Al saturation of this silicate melt will be the determining factors in producing low-Hf or high-Hf zircon and corundum. Any carbonatitic melt fraction will reject HFSE (Zr, Hf) and preferentially incorporate LREE elements, which will result in an HFSE-HREE-enriched silicate melt. Foley et al. [114] performed partial melting experiments on peridotite with H_2O and CO_2 in which early melts are carbonate-rich and progress to carbonated silicateelts with further melting. Hafnium along with Nb and Ta show limited compatibility with early carbonate-dominant melts but are enriched in carbonated silicate melts at higher degrees of partial melting. This suggests that low-Hf zircon may crystallize in a carbonate-dominant environment while high-Hf zircon and associated Nb-Ta oxide phases are products of a carbonated silicate melt that has undergone Al-saturation. Saturation of a peralkaline melt in zircon requires orders of magnitude higher Zr content than in peraluminous melts [22,126]. Higher Zr contents in early peralkaline, carbonate-dominated melts may be responsible for the dominance of zircon over corundum in certain deposits such as in the RVP gem field.

The minor high-Hf zircon included in rare RVP corundum xenocrysts distinctly differs in morphology and geochemistry to the low-Hf mantle-derived zircon and suggests a separate limited genesis. The high Hf-zircon resemble examples in Dak Nong, Vietnam sapphires that have crystal O isotope values and contain fluid inclusions suggestive of lower pressure crystallization from a hydrous CO_2/CO_3^{2-} -saturated melt [7,100]. As seen in the RVP zircon relationship, the Dak Nong corundum accompanies lower-Hf zircon xenocrysts, which were assigned to higher-pressure origin from melt enriched in but not saturated in CO_2.

. Conclusions

Xenocrystic zircon from Cenozoic alkali basalts of the Ratanakiri Volcanic Province represent unique suite within the larger zircon megacryst Indo-Pacific zone (ZIP). Crystal morphology, ompositional zoning, and trace element geochemistry indicate single-stage growth within a rimary, mantle-derived alkaline melt at temperatures between 601 and 739 °C. The RVP zircon ave lower $\delta^{18}O$ isotope compositions than observed in xenocrysts from other ZIP suites and a noderate range of Hf isotopic compositions, which suggest crystallization from a juvenile, alkaline thospheric mantle magma source with limited supracrustal contamination. The xenocrysts have een dated by U-Pb (0.88–1.56 Ma) and (U-Th)/He (0.86–1.02 Ma) methods and give ages that are imilar to those obtained for zircon xenocrysts from other alkali basalt provinces in SE Asia. The RVP ircon predate some of the alkali basalts (~0.7 Ma), which indicates very short mantle residence imes before entrainment in the erupting magma and an earlier beginning of basaltic volcanism. 'he SCLM underneath the RVP is a complex region consisting of remnants of multiple rift-arc-related nd back-arc-related collisional margins including the Song Ma suture zone and back margins of he Himalayan collision zone. The overriding conclusion is of crystallization from a metasomatized carbonatitic"-influenced melt from very low partial melting of a peridotite SCLM source at about 0 km beneath the Indochina terrain.

uthor Contributions: Conceptualization, P.C.P. Methodology, P.C.P. and M.D. Formal Investigation, P.C.P., 1.D., G.P., J.W.V., and R.R. Resources, P.C.P. Data Curation, P.C.P. Writing-Original Draft Preparation, P.C.P. Vriting-Review & Editing, P.C.P., F.L.S., and M.D. Project Administration, P.C.P. Funding Acquisition, P.C.P. and 1.D.

cknowledgments: The authors would like to thank Mike & Delayne Weeks, JP Dau, Dennis Oullette, 'raig Tammas Richardson, Adrian Mann, Samath Ma and Viseth Keo of Angkor Gold Corporation and all heir dedicated staff for help with logistics, translation, accommodations, transportation, coffee, access to drill ore and geologic maps, and for making Ban Lung a second home while working in Cambodia. Thanks to all the ircon miners and their families in Ratanakiri for their help in obtaining zircons for study and for taking the time o discuss their operations. Graphical assistance provided by Reni Barlow. This paper is dedicated to Votha Un 1976–2014) who was a fantastic gem guide and friend to all of us who choose to work and travel in Cambodia. 'hanks are extended to two anonymous reviewers whose input helped to improve this manuscript.

eferences

. Shor, R.; Weldon, R. Ruby and sapphire production and distribution: A quarter century of change. *Gems Gemol.* **2009**, *45*, 236–259. [CrossRef]

. Smith, M.H.; Balmer, W.A. Zircon Mining in Cambodia. *Gems Gemol.* **2009**, *45*, 152–154.

. Schwarz, D. Coloured gemstones—Mines and Markets. In Proceedings of the 34th International Gemmological Conference, Vilnius, Lithuania, 25–30 August 2015; pp. 110–112.

. Abduriyim, A.; Sutherland, F.L.; Belousova, E.A. U-Pb age and origin of gem zircon from the New England sapphire fields, New South Wales, Australia. *Aust. J. Earth Sci.* **2012**, *59*, 1067–1081. [CrossRef]

. Belousova, E.A.; Griffin, W.L.; O'Reilly, S.Y.; Fisher, N. Igneous zircon: Trace element composition as an indicator of source rock type. *Contrib. Mineralog. Petrol.* **2002**, *143*, 602–622. [CrossRef]

. Chen, T.; Ai, H.; Yang, M.; Zheng, S.; Liu, Y. Brownish Red Zircon from Muling, China. *Gems Gemol.* **2011**, *47*, 36–41. [CrossRef]

. Garnier, V.; Ohnenstetter, D.; Guliani, G.; Fallick, A.E.; Trong, T.P.; Quang, V.H.; Van, L.P.; Schwarz, D. Basalt petrology, zircon ages and sapphire genesis from Dak Nong, southern Vietnam. *Mineral. Mag.* **2005**, *69*, 21–38. [CrossRef]

. Graham, I.; Sutherland, L.; Zaw, K.; Nechaev, V.; Khanchuk, A. Advances in our understanding of the gem corundum deposits of the West Pacific continental margins intraplate basaltic fields. *Ore Geol. Rev.* **2008**, *34*, 200–215. [CrossRef]

. Qiu, Z.; Wu, F.; Yu, Q.; Xie, L.; Yang, S. Hf isotopes of zircon megacrysts from the Cenozoic basalts in eastern China. *Chin. Sci. Bull.* **2005**, *50*, 2602–2611. [CrossRef]

0. Sutherland, F.L.; Coenraads, R.R.; Abduriyim, A.; Meffre, S.; Hoskin, P.W.O.; Giuliani, G.; Beattie, R.; Wuhrer, R.; Sutherland, G.B. Corundum (sapphire) and zircon relationships, Lava Plains gem fields,

NE Australia: Integrated mineralogy, geochemistry, age determination, genesis and geographical typin *Mineral. Mag.* **2015**, *79*, 545–581. [CrossRef]

11. Sutherland, F.L.; Graham, I.; Yaxley, G.; Armstrong, R.; Giuliani, G.; Hoskin, P.; Nechaev, V Woodhead, J. Major zircon megacryst suites of the Indo-Pacific lithospheric mrgin (ZIP) and their petrogeneti and regional implications. *Mineral. Petrol.* **2016**, *110*, 399–420. [CrossRef]
12. Sutherland, F.L.; Piilonen, P.C.; Zaw, K.; Meffre, S.; Thompson, J. Sapphire within zircon-rich gem deposit Bo Loei, Ratanakiri Province, Cambodia: Trace elements, inclusions, U-Pb dating and genesis. *Aust. Earth Sci.* **2015**, *62*, 761–773. [CrossRef]
13. Yu, Y.; Xu, X.; Chen, X. Genesis of zircon megacrysts in Cenozoic alkali basalts and the heterogeneity o subcontinental lithospheric mntle, eastern China. *Mineral. Petrol.* **2010**, *100*, 75–94. [CrossRef]
14. Zaw, K.; Sutherland, F.L.; Dellapasqua, F.; Ryan, C.G.; Yui, T.-F.; Mernagh, T.P.; Duncan, D. Contrasts in gem corundum characteristics, eastern Australian basaltic fields: Trace elements, fluid/melt inclusions an oxygen isotopes. *Mineral. Mag.* **2006**, *70*, 669–687. [CrossRef]
15. Fan, P.-F. Tectonic patterns and Cenozoic basalts in the western mrgin of the South China Sea. *Geol. Soc Malays. Bull.* **1995**, *37*, 91–97.
16. Fedorov, P.I.; Koloskov, A.V. Cenozoic volcanism of southeast Asia. *Petrology* **2005**, *13*, 352–380.
17. Taylor, J.G.C.; Buravas, S. Gem deposits at Khao Ploi Waen and Bang Ka Cha, Chanthaburi Province In *Geologic Reconnaissance of the mineral Deposits of Thailand*; United States Government Printing Office Washington, WA, USA, 1951; Volume 984, pp. 144–148.
18. Harley, S.L.; Kelly, N.M. Zircon Tiny but Timely. *Elements* **2007**, *3*, 13–18. [CrossRef]
19. Boehnke, P.; Watson, E.B.; Trail, D.; Harrison, T.M.; Schmitt, A.K. Zircon saturation re-revisited. *Chem. Geo* **2013**, *351*, 324–334. [CrossRef]
20. Watson, E.B. Zircon saturation in felsic liquids: Experimental results and applications to trace elemen geochemistry. *Contrib. Mineral. Petrol.* **1979**, *70*, 407–419. [CrossRef]
21. Watson, E.B.; Harrison, T.M. Zircon saturation revisited: Temperature and composition effects in a variety o crustal mgma types. *Earth Planet. Sci. Lett.* **1983**, *64*, 295–304. [CrossRef]
22. Linnen, R.L.; Keppler, H. Melt composition control of Zr/Hf fractionation in mgmatic processe *Geochim. Cosmochimi. Acta* **2002**, *66*, 3293–3301. [CrossRef]
23. Cong, F.; Li, S.-Q.; Lin, F.-C.; Shi, M.-F.; Zhu, H.-P.; Siebel, W.; Chen, F. Origin of Zircon Megacrysts from Cenozoic Basalts in Northeastern Cambodia: Evidence from U-Pb Age, Hf-O Isotopes, and Inclusions. *J. Geo* **2016**, *124*, 221–234. [CrossRef]
24. Ridd, M.F.; Barber, A.J.; Crow, M.J. Introduction to the geology of Thailand. In *Geology of Thailand*; Ridd, M.F Barber, A.J., Crow, M.J., Eds.; The Geological Society: London, UK, 2011; pp. 1–17.
25. Searle, M.P.; Morley, C.K. Tectonic and thermal evolution of Thailand in the regional context of SE Asia. In *Geology of Thailand*; Ridd, M.F., Barber, A.J., Crow, M.J., Eds.; The Geological Society: London, UK, 2011 pp. 539–571.
26. Workman, D.R. Geology of Laos, Cambodia, South Vietnam and the eastern part of Thailand. *Overseas Geo Miner. Resour.* **1977**, *50*, 1–34.
27. Sotham, S. Geology of Cambodia. *CCOP Tech. Bull.* **1997**, *26*, 13–23.
28. Khain, V.E. *Tectonics of the Continents and Oceans*; Nauchny Mir: Moscow, Russia, 2001.
29. Mukasa, S.B.; matthew Fischer, G.; Barr, S.M. The Character of the Subcontinental Mantle in Southeast Asia Evidence From Isotopic and Elemental Compositions of Extension-Related Cenozoic Basalts in Thailand In *Earth Processes: Reading the Isotopic Code*; Basu, A., Hart, S., Eds.; Geophysical Monograph-American Geophysical Union: Washington DC, WA, USA, 1996; Volume 95, pp. 233–252.
30. Le Pichon, X.; Fournier, M.; Jolivet, L. Kinematics, topography, shortening, and extrusion in the India-Eurasia collision. *Tectonics* **1992**, *11*, 1085–1098. [CrossRef]
31. Miller, M.S.; Kennett, B.L.N.; Toy, V.G. Spatial and temporal evolution of the subducting Pacific plate structure along the western Pacific margin. *J. Geophys. Res.* **2006**, *111*, B02401. [CrossRef]
32. Sutherland, F.L. Alkaline rocks and gemstones, Australia: A review and synthesis. *Aust. J. Earth Sci.* **1996**, 4 323–343. [CrossRef]
33. Sutherland, F.L.; Fanning, C.M. Gem-bearing basaltic volcanism, Barrington, New South Wales: Cenozoi

evolution, based on basalt K-Ar ages and zircon fission track and U-Pb isotope dating. *Aust. J. Earth Sci.* **2001**, *48*, 221–237. [CrossRef]

34. Patte, E. Etude de l'ile des Cendres, volcan apparu au large de la côte d'Annam. *Bull. Serv. Géol. l'Indoch.* **1925**, *13*, 1–19.
35. Saurin, E. La néotectonique de l'Indochine. *Rev. Géogr. Phys. Géol. Dyn.* **1967**, *9*, 143–151.
36. Barr, S.M.; MacDonald, A.S. Geochemistry and petrogenesis of Late Cenozoic alkaline basalts of Thailand. *Geol. Soc. Malays. Bull.* **1978**, *10*, 25–52.
37. Barr, S.M.M.; Alan, S. Geochemistry and Geochronology of Late Cenozoic Basalts of Southeast Asia. *GSA Bull.* **1981**, *92*, 1069–1142. [CrossRef]
38. Sutherland, F.L.; Bosshart, G.; Fanning, C.M.; Hoskin, P.W.O.; Coenraads, R.R. Sapphire crystallization, age and origin, Ban Huai Sai, Laos: Age based on zircon inclusions. *J. Asian Earth Sci.* **2002**, *20*, 841–849. [CrossRef]
39. Barr, S.M.J.; Dodie, E. Trace element characteristics of Upper Cenozoic basaltic rocks of Thailand, Kampuchea and Vietnam. *J. Southeast Asian Earth Sci.* **1990**, *4*, 233–242. [CrossRef]
40. Hoang, N.; Flower, M. Petrogenesis of Cenozoic Basalts from Vietnam: Implication for Origins of a 'Diffuse Igneous Province'. *J. Petrol.* **1998**, *39*, 369–395. [CrossRef]
41. Barr, S.M.; Charusiri, P. Volcanic rocks. In *The Geology of Thailand*; Ridd, M.F., Barber, A.J., Crow, M.J., Eds.; The Geological Society: London, UK, 2011; pp. 415–438.
42. Lacombe, P. La mssif basaltique quaternaire à zircons-gemmes de Ratanakiri (Cambodge nord-oriental)—Premiére partie. *Bull. Rech. Géol. min.* **1969**, *3*, 31–91.
43. Lacombe, P. La mssif basaltique quaternaire à zircons-gemmes de Ratanakiri (Cambodge nord-oriental)-Troisiéme partie: Minéralogie et géologie des gisements de zircons. *Bull. Rech. Géol. Min.* **1970**, *4*, 33–79.
44. Whitford-Stark, J.L. A survey of Cenozoic volcanism on mainland Asia. In *Geological Society of America Special Paper*; Geological Society of America: Boulder, CO, USA, 1987; Volume 213.
45. Zeug, M.; Nasdala, L.; Wanthanachaisaeng, B.; Balmer, W.A.; Corfu, F.; Wildner, M. Blue Zircon from Ratanakiri, Cambodia. *J. Gemmol.* **2018**, *36*, 112–132. [CrossRef]
46. Pouchou, J.L.; Pichoir, F. *'PAP'j(rZ) Procedure for Improved Quantitative Microanalysis*; San Francisco Press: San Francisco, CA, USA, 1984.
47. Elhlou, S.; Belousova, E.A.; Griffin, W.L.; Pearson, N.J.; O'Reilly, S.Y. Trace element and isotopic composition of GJ-red zircon standard by laser ablation. *Geochim. Cosmochim. Acta* **2006**, *70*, A158. [CrossRef]
48. Griffin, W.L.; Pearson, N.J.; Belousova, E.; Jackson, S.E.; van Achterbergh, E.; O'Reilly, S.Y.; Shee, S.R. The Hf isotope composition of cratonic mntle: LAM-MC-ICPMS analysis of zircon megacrysts in kimberlites. *Geochim. Cosmochi. Acta* **2000**, *64*, 133–147. [CrossRef]
49. DeBievre, P.; Taylor, P.D.P. Table of the isotopic composition of the elements. *Int. J. Mass Spectrom. Ion Process.* **1993**, *123*, 149–166. [CrossRef]
50. Pearson, N.J.; Griffin, W.L.; O'Reilly, S.Y. Mass fractionation correction in laser ablation-multiple collector ICP-MS: Implications for overlap corrections and precise and accurate in situ isotope ratio measurement. In *Laser-Ablation ICP-MS in the Earth Sciences*; Sylvester, P.J., Ed.; mineralogical Association of Canada: Quebec City, QC, Canada, 2008; Volume 40, pp. 93–116.
51. Griffin, W.L.; Pearson, N.J.; Belousova, E.A.; Saeed, A. Reply to "Comment to short-communication 'Comment: Hf-isotope heterogeneity in zircon 91500' by W.L. Griffin, N.J. Pearson, E.A. Belousova and A. Saeed [Chemical Geology 233 (2006) 358-363]" by F. Corfu. *Chem. Geol.* **2007**, *244*, 354–356. [CrossRef]
52. Kemp, A.I.S.; Wormald, R.J.; Whitehouse, M.J.; Price, R.C. Hf isotopes in zircon reveal contrasting sources and crystallization histories for alkaline to peralkaline granites of Temora, southeastern Australia. *Geology* **2005**, *33*, 797–800. [CrossRef]
53. Bouvier, A.; Vervoort, J.D.; Patchett, P.J. The Lu-Hf and Sm-Nd isotopic composition of CHUR: Constraints from unequilibrated chondrites and implications for the bulk composition of terrestrial planets. *Earth Planet. Sci. Lett.* **2008**, *273*, 48–57. [CrossRef]
54. Scherer, E.; Münker, C.; Mezger, K. Calibration of the Lutetium-Hafnium Clock. *Science* **2001**, *293*, 683–687. [CrossRef] [PubMed]
55. Jackson, S.E.; Pearson, N.J.; Griffin, W.L.; Belousova, E.A. The application of laser ablation

microprobe-inductively coupled plasma-mass spectrometry (LAM-ICPMS) to in situ U-Pb zircon geochronology. *Chem. Geol.* **2004**, *211*, 47–69. [CrossRef]
56. Griffin, W.L.; Powell, W.J.; Pearson, N.J.; O'Reilly, S.Y. GLITTER: Data reduction software for laser ablation ICP-MS. In *Laser Ablation ICP-MS in the Earth Sciences*; Sylvester, P.J., Ed.; Mineralogical Association of Canada: Quebec City, QC, Canada, 2008; pp. 204–207.
57. Andersen, T. Correction of common lead in U-Pb analyses that do not report 204Pb. *Chem. Geol.* **2002**, *192*, 59–79. [CrossRef]
58. Stacey, J.S.T.; Kramers, J.D. Approximation of terrestrial lead isotope evolution by a two-stage model. *Earth Planet. Sci. Lett.* **1975**, *26*, 207–221. [CrossRef]
59. Reiners, P.W.; Spell, T.L.; Nicolescu, S.; Zanetti, K.A. Zircon (U-Th)/He thermochronometry: He diffusion and comparisons with $^{40}Ar/^{39}Ar$ dating. *Geochim Cosmochim. Acta* **2004**, *68*, 1857–1887. [CrossRef]
60. Danišík, M.; Kuhlemann, J.; Dunkl, I.; Evans, N.J.; Székely, B.; Frisch, W. Survival of Ancient Landforms in a Collisional Setting as Revealed by Combined Fission Track and (U-Th)/He Thermochronometry: A Case Study from Corsica (France). *J. Geol.* **2012**, *120*, 155–173. [CrossRef]
61. Danišík, M.; Štěpančíková, P.; Evans, N.J. Constraining long-term denudation and faulting history in intraplate regions by multisystem thermochronology: An example of the Sudetic Marginal Fault (Bohemian Massif, central Europe). *Tectonics* **2012**, *31*. [CrossRef]
62. Farley, K.A.; Wolf, R.A.; Silver, L.T. The effects of long alpha-stopping distances on (U-Th)/He ages. *Geochim Cosmochim. Acta* **1996**, *60*, 4223–4229. [CrossRef]
63. Vermeesch, P. HelioPlot, and the treatment of overdispersed (U-Th-Sm)/He data. *Chem. Geol.* **2010**, *271*, 108–111. [CrossRef]
64. Reiners, P.W. Zircon (U-Th)/He Thermochronometry. *Rev. Mineral. Geochem.* **2005**, *58*, 151–179. [CrossRef]
65. Valley, J.W.; Chiarenzelli, J.R.; McLelland, J.M. Oxygen isotope geochemistry of zircon. *Earth Planet. Sci. Lett.* **1994**, *126*, 187–206. [CrossRef]
66. Valley, J.W.; Kitchen, N.; Kohn, M.J.; Niendorf, C.R.; Spicuzza, M.J. UWG-2, a garnet standard for oxygen isotope ratios: Strategies for high precision and accuracy with laser heating. *Geochim. Cosmochim. Acta* **1995**, *59*, 5223–5231. [CrossRef]
67. Pupin, J.-P. Zircon and granite petrology. *Contrib. Mineral. Petrol.* **1980**, *73*, 207–220. [CrossRef]
68. Shore, M.; Fowler, A.D. Oscillatory zoning in minerals: A common phenomenon. *Can. Mineral.* **1996**, *34*, 1111–1126.
69. Witter, A.; Nasdala, L.; Wanthanachaisaeng, B.; Bunnag, N.; Škoda, R.; Balmer, W.A.; Geister, G. Zeug, M. Mineralogical characterization of gem zircon from Ratanakiri, Cambodia. In Proceedings of the CORALS 2013, Vienna, Austria, 3–6 July 2013.
70. McDonough, W.F.; Sun, S.S. The composition of the Earth. *Chem. Geol.* **1995**, *120*, 223–253. [CrossRef]
71. Pupin, J.P. *Transactions of the Royal Society of Edinburgh, Earth Sciences*; RSE Scotland Foundation: Edinburgh, UK, 2000.
72. Watson, E.B.; Harrison, T.M. Zircon Thermometer Reveals minimum Melting Conditions on Earliest Earth. *Science* **2005**, *308*, 841–844. [CrossRef] [PubMed]
73. Watson, E.B.; Wark, D.A.; Thomas, J.B. Crystallization thermometers for zircon and rutile. *Contrib. Mineral. Petrol.* **2006**, *151*, 413–433. [CrossRef]
74. Fu, B.; Page, F.Z.; Cavosie, A.J.; Fournelle, J.; Kita, N.T.; Lackey, J.S.; Wilde, S.A.; Valley, J.W. Ti-in-zircon thermometry: Applications and limitations. *Contrib. Mineral. Petrol.* **2008**, *156*, 197–215. [CrossRef]
75. Cherniak, D.J.; Watson, E.B. Ti diffusion in zircon. *Chem. Geol.* **2007**, *242*, 470–483. [CrossRef]
76. Crisp, L. The effect of pressure on Ti-in-zircon and Zr-in-rutile. In Proceedings of the Petrology and geochemistry seminars, Research School of Earth Sciences, Canberra, Australia, 29 May 2015.
77. Page, F.Z.; Fu, B.; Kita, N.T.; Fournelle, J.; Spicuzza, M.J.; Schulze, D.J.; Viljoen, F.; Basei, M.A.S.; Valley, J.W. Zircons from kimberlite: New insights from oxygen isotopes, trace elements, and Ti in zircon thermometry. *Geochim. Cosmochim. Acta* **2007**, *71*, 3887–3903. [CrossRef]
78. Carbonnel, J.-P.; Duplaix, S.; Selo, M. La méthode des traces de fission de l'U apliquée á la géochronologie. Datation du magmatisme récente de I'Asie du Sud-Est. *Rev. Géogr. Phys. Geolo. Dyn.* **1972**, *14*, 29–46.
79. Carbonnel, J.-P.; Selo, M.; Poupeau, G. Fission track age of the gem deposit of Pailin (Cambodia) and recent tectonics in the Indochinan province. *Mod. Geol.* **1973** *4*, 61–64.

80. Li, X.-H.; Long, W.-G.; Li, Q.-L.; Liu, Y.; Zheng, Y.-F.; Yang, Y.-H.; Chamberlain, K.R.; Wan, D.-F.; Guo, C.-H.; Wang, X.-C.; et al. Penglai Zircon Megacrysts: A Potential New Working Reference Material for Microbeam Determination of Hf-O Isotopes and U-Pb Age. *Geostand. Geoanal. Res.* **2010**, *34*, 117–134. [CrossRef]
81. Li, Y.; Zheng, D.; Wu, Y.; Wang, Y.; He, H.; Pang, J.; Wang, Y.; Yu, J. A Potential (U-Th)/He Zircon Reference Material from Penglai Zircon Megacrysts. *Geostand. Geoanal. Res.* **2017**, *41*, 359–365. [CrossRef]
82. Dodson, M.H. Closure temperature in cooling geochronological and petrological systems. *Contrib. Mineral. Petrol.* **1973**, *40*, 259–274. [CrossRef]
83. Page, F.Z.; Ushikubo, T.; Kita, N.T.; Riciputi, L.R.; Valley, J.W. High-precision oxygen isotope analysis of picogram samples reveals 2 μm gradients and slow diffusion in zircon. *Am. Mineral.* **2007**, *92*, 1772–1775. [CrossRef]
84. Bowman, J.R.; Moser, D.E.; Valley, J.W.; Wooden, J.L.; Kita, N.T.; Mazdab, F.K. Zircon U-Pb isotope, $\delta^{18}O$ and trace element response to 80 m.y. of high temperature metamorphism in the lower crust: Sluggish diffusion and new records of Archean craton formation. *Am. J. Sci.* **2011**, *311*, 719–772. [CrossRef]
85. Mattey, D.; Lowry, D.; Macpherson, C. Oxygen isotope composition of mntle peridotite. *Earth Planet. Sci. Lett.* **1994**, *128*, 231–241. [CrossRef]
86. Upton, B.G.J.; Hinton, R.W.; Aspen, P.; Finch, A.; Valley, J.W. Megacrysts and Associated Xenoliths: Evidence for Migration of Geochemically Enriched Melts in the Upper Mantle beneath Scotland. *J. Petrol.* **1999**, *40*, 935–956. [CrossRef]
87. Valley, J.W.; Kitchen, N.; Kohn, M.J.; Niendorf, C.R.; Spicuzza, M.J. Oxygen Isotopes in Zircon. In *Reviews in Mineralogy and Geochemistry*; Hanchar, J.M., Hoskin, P.W.O., Eds.; Mineralogical Society of America: Chantilly, VA, USA, 2003; Volume 53, pp. 343–385.
88. Belousova, E.A.; Griffin, W.L.; O'Reilly, S.Y. Zircon Crystal Morphology, Trace Element Signatures and Hf Isotope Composition as a Tool for Petrogenetic Modelling: Examples From Eastern Australian Granitoids. *J. Petrol.* **2006**, *47*, 329–353. [CrossRef]
89. Heaman, L.M.; Bowins, R.; Crocket, J. The chemical composition of igneous zircon suites: Implications for geochemical tracer studies. *Geochim. Cosmochim. Acta* **1990**, *54*, 1597–1607. [CrossRef]
90. Hoskin, P.W.O.; Ireland, T.R. Rare earth element chemistry of zircon and its use as a provenance indicator. *Geology* **2000**, *28*, 627–630. [CrossRef]
91. Seifert, W.; Rhede, D.; Tietz, O. Typology, chemistry and origin of zircon from alkali basalts of SE Saxony (Germany). *N. Jahrb. Mineralo. Abh.* **2008**, *184*, 299–313. [CrossRef]
92. Shnukov, S.E.; Andreev, A.V.; Savenok, S.P. Admixture elements in zircons and apatites: A tool for provenance studies of terrigenous sedimentary rocks. In Proceedings of the European Union of Geosciences (EUG 9), Strasbourg, France, 23–27 March 1997.
93. Pupin, J.-P. Granite genesis related to geodynamics from Hf-Y in zircon. *Earth Environ. Sci. Trans. R. Soc. Edinb.* **2000**, *91*, 245–256. [CrossRef]
94. Visonà, D.; Caironi, V.; Carraro, A.; Dallai, L.; Fioretti, A.M.; Fanning, M. Zircon megacrysts from basalts of the Venetian Volcanic Province (NE Italy): U-Pb ages, oxygen isotopes and REE data. *Lithos* **2007**, *94*, 168–180. [CrossRef]
95. Hoskin, P.W.O.; Schaltegger, U. The Composition of Zircon and Igneous and Metamorphic Petrogenesis. *Rev. Mineral. Geochem.* **2003**, *53*, 27–62. [CrossRef]
96. Borghini, G.; Fumagalli, P.; Rampone, E. The Stability of Plagioclase in the Upper Mantle: Subsolidus Experiments on Fertile and Depleted Lherzolite. *J. Petrol.* **2010**, *51*, 229–254. [CrossRef]
97. Hinton, R.W.; Upton, B.G.J. The chemistry of zircon: Variations within and between large crystals from syenite and alkali basalt xenoliths. *Geochim. Cosmochim. Acta* **1991**, *55*, 3287–3302. [CrossRef]
98. Coenraads, R.R.; Vichit, P.; Sutherland, F.L. An unusual sapphire-zircon-magnetite xenolith from the Chanthaburi Gem Province, Thailand. *Mineral. Mag.* **1995**, *59*, 465–479. [CrossRef]
99. Khamloet, P.; Pisutha-Arnond, V.; Sutthirat, C. Mineral inclusions in sapphire from the basalt-related deposit in Bo Phloi, Kanchanaburi, western Thailand: Indication of their genesis. *Russ. Geol. Geophys.* **2014**, *55*, 1087–1102. [CrossRef]
100. Izokh, A.E.; Smirnov, S.Z.; Egorova, V.V.; Tuan Anh, T.; Kovyazin, S.V.; Phuong, N.T.; Kalinina, V.V. The conditions of formation of sapphire and zircon in the areas of alkali-basaltoid volcanism in Central Vietnam. *Russ. Geol. Geophys.* **2010**, *51*, 719–733. [CrossRef]

101. Guo, J.; O'Reilly, S.Y.; Griffin, W.L. Corundum from basaltic terrains: A mineral inclusion approach to th enigma. *Contrib. Mineral. Petrol.* **1996**, *122*, 368–386. [CrossRef]
102. Guo, J.; O'Reilly, S.Y.; Griffin, W.L. Zircon inclusions in corundum megacrysts: I. Trace element geochemistr and clues to the origin of corundum megacrysts in alkali basalts. *Geochim. Cosmochim. Acta* **1996**, *6* 2347–2363. [CrossRef]
103. Sutherland, F.L.; Hoskin, P.W.O.; Fanning, C.M.; Coenraads, R.R. Models of corundum origin from alka basaltic terrains: A reappraisal. *Contrib. Mineral. Petrol.* **1998**, *133*, 356–372. [CrossRef]
104. Benisek, A.; Finger, F. Factors controlling the development of prism faces in granite zircons: A microprob study. *Contrib. Mineral. Petrol.* **1993**, *114*, 441–451. [CrossRef]
105. Grimes, C.B.; John, B.E.; Kelemen, P.B.; Mazdab, F.K.; Wooden, J.L.; Cheadle, M.J.; Hanghøj, K.; Schwartz, J. Trace element chemistry of zircons from oceanic crust: A method for distinguishing detrital zirco provenance. *Geology* **2007**, *35*, 643–646. [CrossRef]
106. Grimes, C.B.; Wooden, J.L.; Cheadle, M.J.; John, B.E. "Fingerprinting" tectono-magmatic provenance usin trace elements in igneous zircon. *Contrib. Mineral. Petrol.* **2015**, *170*, 92–117. [CrossRef]
107. Liu, Y.; Williams, I.S.; Chen, J.; Wan, Y.; Sun, W. The significance of Paleoproterozoic zircon in carbonatit dikes associated with the Bayan Obo REE-Nb-Fe deposit. *Am. J. Sci.* **2008**, *308*, 379–397. [CrossRef]
108. Campbell, L.S.; Compston, W.; Sircombe, K.N.; Wilkinson, C.C. Zircon from the East Orebody of the Baya Obo Fe-Nb-REE deposit, China, and SHRIMP ages for carbonatite-related mgmatism and REE mineralizatio events. *Contrib. Mineral. Petrol.* **2014**, *168*, 1041. [CrossRef]
109. Saava, E.V.; Belyatsky, B.V.; Antonov, A.B. Carbonatitic Zircon-Myth or Reality: Mineralogical-Geochemica Analyses. Available online: http://alkaline09.narod.ru/abstracts/Savva_Belyatsky.htm (accessed o 26 October 2018).
110. Bell, K.; Simonetti, A. Source of parental melts to carbonatites—Critical isotopic constraints. *Mineral. Petro* **2010**, *98*, 77–89. [CrossRef]
111. Chakmouradian, A. The geochemistry of continental carbonatites revisited: Two mjor types of continenta carbonatites and their trace-element signatures. In Proceedings of the European Geosciences Union Genera Assembly, Vienna, Austria, 19–24 April 2009; p. 10806.
112. Thi, T.N.; Wada, H.; Ishikawa, T.; Shimano, T. Geochemistry and petrogenesis of carbonatites from Sout Nam Xe, Lai Chau area, northwest Vietnam. *Mineral. Petrol.* **2014**, *108*, 371–390.
113. Hou, Z.; Liu, Y.; Tian, S.; Yang, Z.; Xie, Y. Formation of carbonatite-related giant rare-earth-element deposit by the recycling of mrine sediments. *Sci. Rep.* **2015**, *5*, 1–10. [CrossRef] [PubMed]
114. Foley, S.F.; Yaxley, G.M.; Rosenthal, A.; Buhre, S.; Kiseeva, E.S.; Rapp, R.P.; Jacob, D.E. The composition o near-solidus melts of peridotite in the presence of CO_2 and H_2O between 40 and 60 kbar. *Lithos* **2009**, *11* 274–283. [CrossRef]
115. Su, B.-X.; Zhang, H.-F.; Sakyi, P.A.; Ying, J.-F.; Tang, Y.-J.; Yang, Y.-H.; Qin, K.-Z.; Xiao, Y.; Zhao, X.-M Compositionally stratified lithosphere and carbonatite metasomatism recorded in mntle xenoliths from th Western Qinling (Central China). *Lithos* **2010**, *116*, 111–128. [CrossRef]
116. Deng, L.L.; Liu, Y.; Gao, S. Pacific slab-induced carbonatite mntle metasomatism in the eastern Nort China Craton. In Proceedings of the American Geophysical Union Fall Meeting, San Francisco, CA, USA 15–19 December 2014.
117. Zeng, G.; Chen, L.-H.; Xu, X.-S.; Jiang, S.-Y.; Hofmann, A.W. Carbonated mntle sources for Cenozoi intra-plate alkaline basalts in Shandong, North China. *Chem. Geol.* **2010**, *273*, 35–45. [CrossRef]
118. Deng, L.; Liu, Y.; Zong, K.; Zhu, L.; Xu, R.; Hu, Z.; Gao, S. Trace element and Sr isotop records of multi-episode carbonatite metasomatism on the eastern mrgin of the North China Crato *Geochem. Geophys. Geosyst.* **2017**, *18*, 220–237. [CrossRef]
119. Yaxley, G.M.; Green, D.H.; Kamenetsky, V. Carbonatite Metasomatism in the Southeastern Australia Lithosphere. *J. Petrol.* **1998**, *39*, 1917–1930. [CrossRef]
120. Sutherland, F.L.; Graham, I.T.; Hollis, J.D.; Meffre, S.; Zwingmann, H.; Jourdan, F.; Pogson, R.E. Multipl felsic events within post-10 Ma volcanism, Southeast Australia: Inputs in appraising proposed magmati models. *Aust. J. Earth Sci.* **2014**, *61*, 241–267. [CrossRef]
121. Yu, J.-H.; O'Reilly, S.Y.; Zhang, M.; Griffin, W.L.; Xu, X. Roles of Melting and Metasomatism in the Formatio of the Lithospheric Mantle beneath the Leizhou Peninsula, South China. *J. Petrol.* **2006**, *47*, 355–38 [CrossRef]

22. Monchoux, P.; Fontan, F.; De Parseval, P.; Martin, R.F.; Wang, C.R. Igneous albitite dikes in orogenic lherzolites, western Pyréneés, France: A possible source for corundum and alkali feldspar xenocrysts in basaltic terraines. I. mineralogical associations. *Can. Mineral.* **2006**, *44*, 817–842. [CrossRef]
23. Pin, C.; Monchoux, P.; Paquette, J.L.; Azambre, B.; Wang, C.R.; Martin, R.F. Igneous albititic dikes in orogenic lherzolites, western Pyréneés, France: A possible source for corundum and alkali feldspar xenocrysts in basaltic terraines. II. Geochemical and petrogenetic considerations. *Can. Mineral.* **2006**, *44*, 843–856. [CrossRef]
24. Long, A.M.; Menzies, M.A.; Thirlwall, M.F.; Upton, B.G.J.; Aspen, P. Carbonatite-mantle interaction: A possible origin for megacryst/xenolith suites in Scotland. In Proceedings of the 5th International Kimberlite Conference, Araxá, Brazil, 18 June–4 July 1991; pp. 467–477.
25. Bakumenko, I.T.; Tomilenko, A.A.; Bazarova, T.Y.; Yarmolyuk, V.V. The conditions of formation of volcanics in the Late Mesozoic–Cenozoic West Trans-Baikalian volcanic area (from results of study of melt and fluid inclusions in minerals). *Geokhimiya* **1999**, *12*, 1352–1356.
26. Gervasoni, F.; Klemme, S.; Rocha-Júnior, E.R.V.; Berndt, J. Zircon saturation in silicate melts: A new and improved model for aluminous and alkaline melts. *Contrib. Mineral. Petrol.* **2016**, *171*, 21. [CrossRef]

5

Formation Conditions and $^{40}Ar/^{39}Ar$ Age of the Gem-Bearing Boqueirão Granitic Pegmatite, Parelhas, Rio Grande do Norte, Brazil

Sabina Strmić Palinkaš [1,*], Ladislav Palinkaš [2], Franz Neubauer [3], Ricardo Scholz [4], Sibila Borojević Šoštarić [5] and Vladimir Bermanec [2]

1 Department of Geosciences, Faculty of Sciences and Technology, UiT-The Arctic University of Norway in Tromsø, N-9037 Tromsø, Norway
2 Department of Geology, Faculty of Science, University of Zagreb, Horvatovac 95, HR-10000 Zagreb, Croatia; lpalinkas@geol.pmf.hr (L.P.); vladimir.bermanec@public.carnet.hr (V.B.)
3 Department of Geography and Geology, Paris-Lodron-University of Salzburg, Hellbrunner Str. 34, A-5020 Salzburg, Austria; franz.neubauer@sbg.ac.at
4 Departamento de Geologia, Escola de Minas, Universidade Federal de Ouro Preto, Ouro Preto MG-31400-000, Brazil; r_scholz_br@yahoo.com
5 Faculty of Mining, Geology and Petroleum Engineering, University of Zagreb, Pierottijeva 6, HR-10000 Zagreb, Croatia; sibila.borojevic-sostaric@rgn.hr
* Correspondence: Sabina.s.palinkas@uit.no

Abstract: The Boqueirão granitic pegmatite, alias Alto da Cabeça pegmatite, is situated in Borborema Pegmatitic Province (BPP) in Northeast Brazil. This pegmatitic province hosts globally importan reserves of tantalum and beryllium, as well as significant quantities of gemstones, including aquamarine morganite, and the high-quality turquoise-blue "Paraíba Elbaite". The studied lithium-cesium-tantalum Boqueirão granitic pegmatite intruded meta-conglomerates of the Equador Formation during the late Cambrian (502.1 ± 5.8 Ma; $^{40}Ar/^{39}Ar$ plateau age of muscovite). The pegmatite exhibits a typical zona mineral pattern with four defined zones (Zone I: muscovite, tourmaline, albite, and quartz; Zone II K-feldspar (microcline), quartz, and albite; Zone III: perthite crystals (blocky feldspar zone); Zone IV massive quartz). Huge individual beryl, spodumene, tantalite, and cassiterite crystals are common as well. Microscopic examinations revealed that melt inclusions were entrapped simultaneously with fluid inclusions, suggesting the magmatic–hydrothermal transition. The magmatic–hydrothermal transition affected the evolution of the pegmatite, segregating volatile compounds (H_2O, CO_2, N_2) and elements that preferentially partition into a fluid phase from the viscous silicate melt. Fluid inclusion studies on microcline and associated quartz combined with microthermometry and Raman spectroscopy gave an insight into the P-T-X characteristics of entrapped fluids. The presence of spodumene without other $LiAl(SiO_3)_2$ polymorphs and constructed fluid inclusion isochores limited the magmatic–hydrotherma transition at the gem-bearing Boqueirão granitic pegmatite to the temperature range between 300 and 415 °C at a pressure from 1.8 to 3 kbar.

Keywords: gem-bearing pegmatite; fluid inclusions; P-T-X equilibria; spodumene; Ar/Ar dating

---

1. Introduction

Pegmatites, plutonic igneous rocks characterized by extremely coarse crystals with a systematically variable size, represent an important source of industrial minerals (feldspars, quartz, spodumene petalite), hi-tech mineral commodities (e.g., Li, Cs, Be, Nb, Ta, Sn), and gemstones. The most common pegmatite-hosted gem minerals are colored varieties of beryl (aquamarine, heliodor, and morganite) i-rich tourmaline (elbaite-rossmanite and liddicoatite), blue and sherry topaz, transparent varieties of podumene (kunzite), low-iron spessartine, and optical-grade quartz [1].

The Boqueirão granitic pegmatite, also known as the Alto da Cabeça pegmatite, is situated in the orthernmost part of the Serra das Queimadas Mountains, the State of Rio Grande do Norte, Northeast razil (Figure 1). The pegmatite is hosted by Borborema Pegmatitic Province (BPP), which represents ne of the world's most important sources of tantalum and beryllium, as well as of gemstones, including quamarine, morganite, and the high-quality turquoise-blue "Paraíba Elbaite" [2–5].

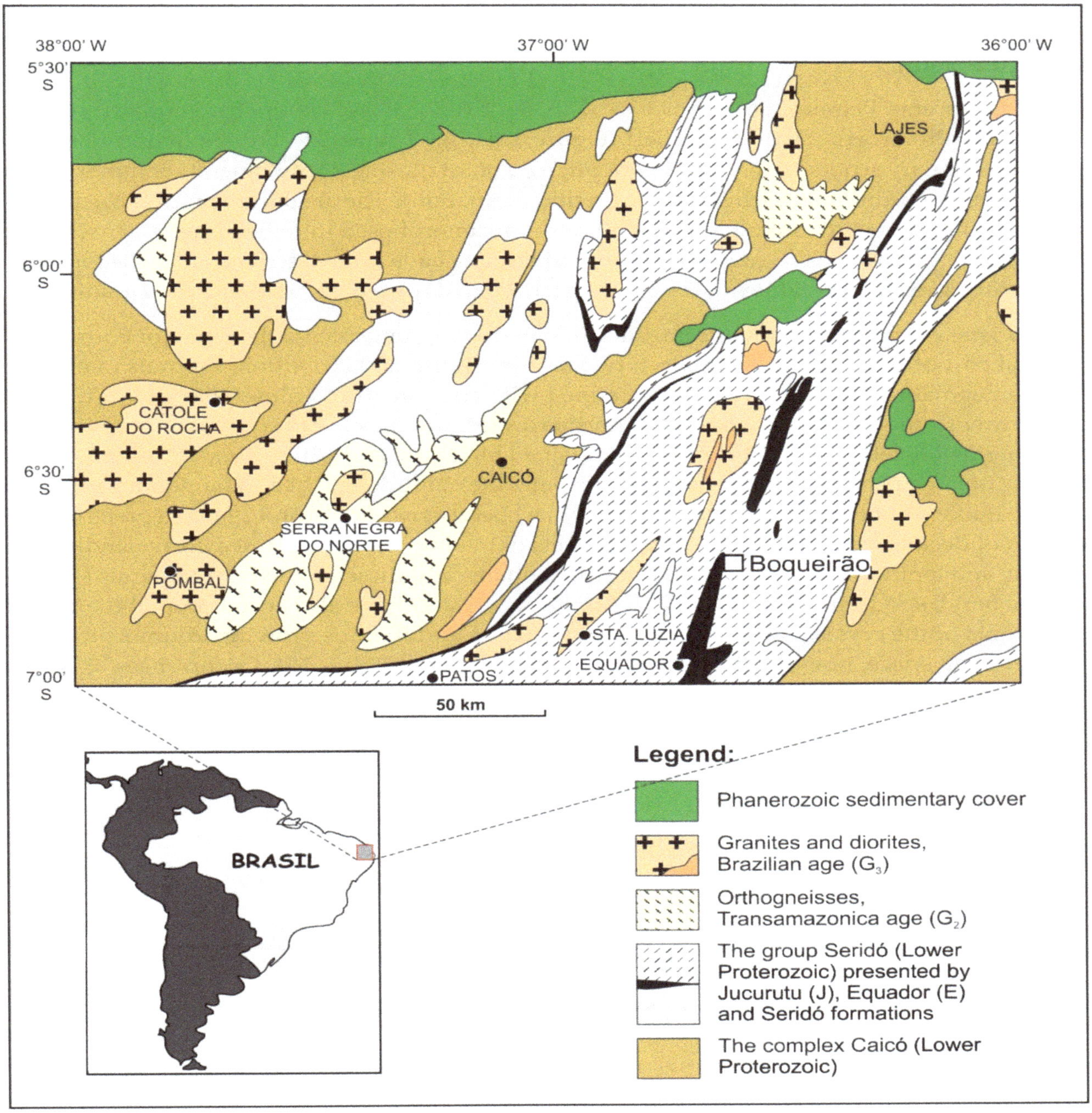

Figure 1. A simplified geological map of the Borborema Pegmatitic Province with the location of the Boqueirão granitic pegmatite, Parelhas, Rio Grande do Norte, Brazil. Adapted with permission from [6].

Mineralogical and geochemical features, characterized by a strong enrichment in numerous ncompatible elements, including Li, Rb, Cs, Be, Sn, Ta, Nb (with Ta > Nb), B, P, and F, classify the

Boqueirão granitic pegmatite into the lithium-cesium-tantalum (LCT) family [5–9]. According t previously published data, the BPP pegmatites were formed in a relatively narrow range of pressur (2.1 to 4 kbar) but within a wide range of temperatures (390–900 °C) [10–12].

This study brings new fluid inclusion and $^{40}Ar/^{39}Ar$ data from the gem-bearing Boqueirão graniti pegmatite. The fluid inclusion data combined with the calculated thermodynamic equilibria of th established mineral paragenesis shed light on the formation conditions of the Boqueirão graniti pegmatite, whereas the $^{40}Ar/^{39}Ar$ dating confirmed the late Cambrian age of pegmatite emplacemen A particular focus has been given to the recorded magmatic–hydrothermal transition and its potentia role in the evolution of the studied pegmatite.

2. Geological Setting

The Borborema Pegmatitic Province (6°–7°S, 36°15′–36°45′W) covers the southern part of th Meso- to Neo-Proterozoic Seridó Foldbelt (Figure 1; [13]). The Seridó Foldbelt comprises a basa volcano-sedimentary sequence (Jucurutu Formation), a quartzite–metaconglomerate complex (Equado Formation), and a turbidite–flysch sequence (Seridó Formation). The area is metamorphosed up to th upper amphibolite facies (Abukuma type) and retro-metamorphosed into the upper greenschist-facie grade [14]. Four generations of granite intrusions in the area have been described [15]. The formation o the pegmatites is related to granites of the late- to post-orogenic phase [16], labelled as G4 granites [15,17

The Boqueirão granitic pegmatite intruded into meta-conglomerates of the Equador Formation [17 The central part of the pegmatite exhibits a typical zonal mineral distribution. From its margin to th center, the pegmatite consists of (Figure 2): Zone I with comb-textured muscovite and/or tourmalin intergrown with medium-grained albite and quartz (Figure 3a–c); Zone II hosting homogeneou medium-grained K-feldspar (microcline) accompanied by quartz and albite; Zone III composed almos exclusively of large perthite crystals (blocky feldspar zone); and Zone IV, i.e., a monomineralic nucleu of massive milky and/or rose quartz [6,18]. The contact between Zones III and IV was a preferred site fo deposition of decimetric to metric, irregular pockets of medium- to fine-grained cleavelandite (albite muscovite, and lepidolite selvages with some phosphates and disseminated ore minerals (Figure 3c Individual beryl, spodumene, tantalite, and cassiterite crystals have also been found at the boundary with the roots of the crystals in the blocky feldspar zone and the tips growing idiomorphically int the former open space, now filled by the massive quartz core (Figure 3d). Occurrences of beryl an tourmaline mineralization at the boundaries between Zones II and III (Figure 3e,f), and spodumen and/or cassiterite in Zones I and II are common ([6] and references therein). The U-Pb dating o manganocolumbite and ferrocolumbite constrained the time of pegmatite emplacement between 50 and 515 Ma [19].

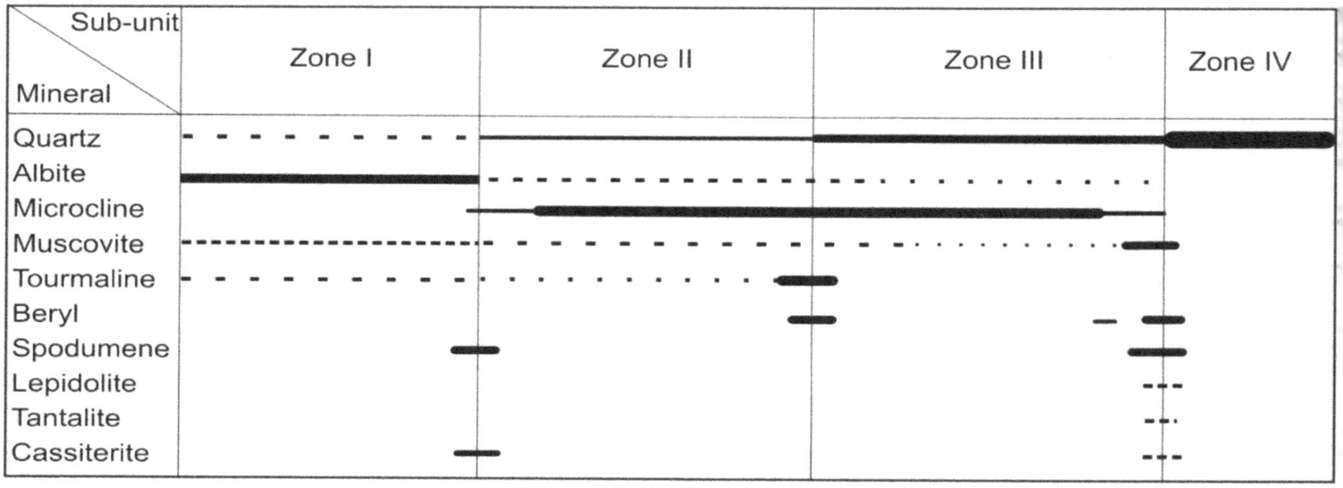

Figure 2. The paragenetic sequence of the gem-bearing Boqueirão granitic pegmatite.

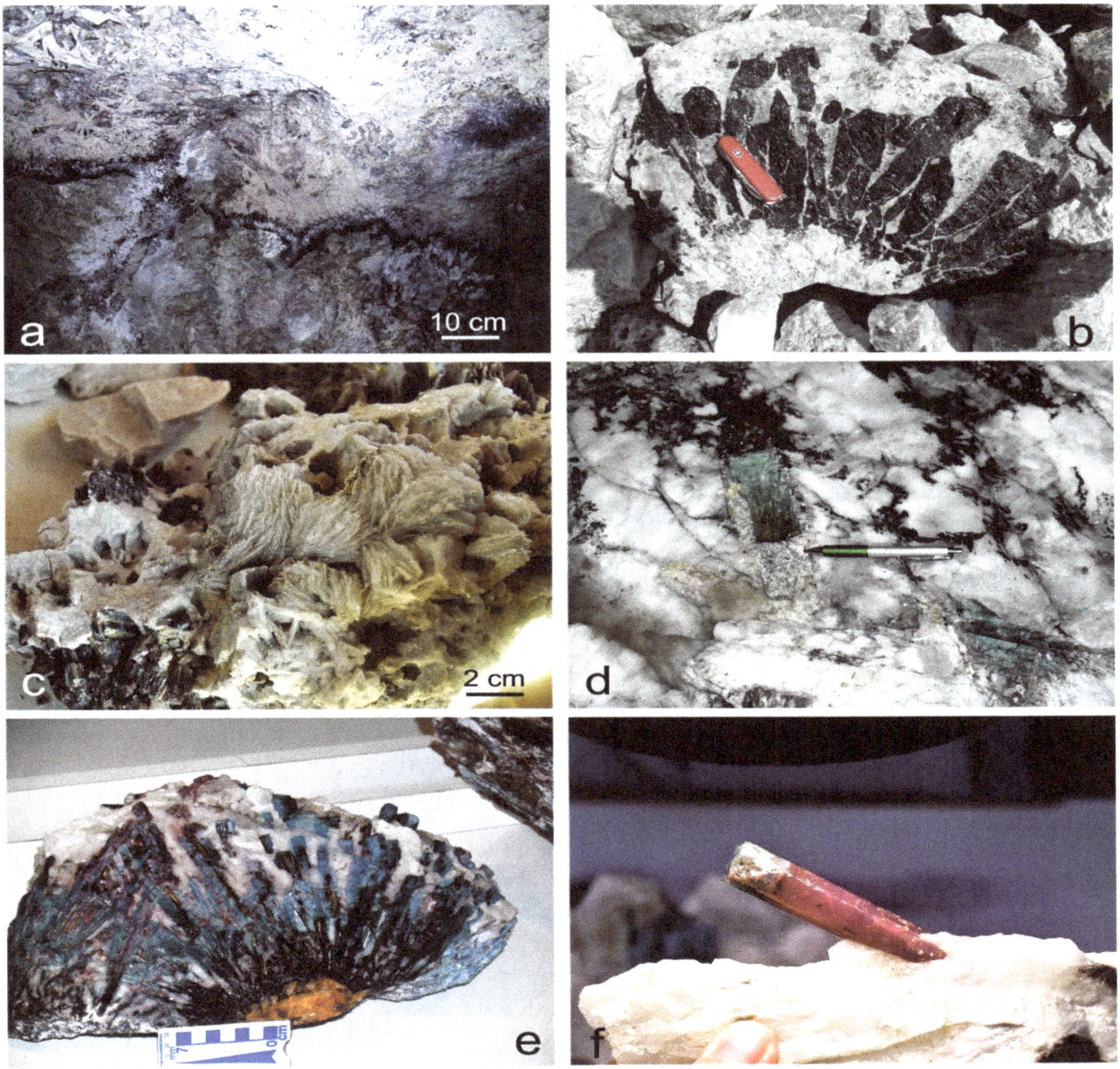

Figure 3. Macrophotographs showing: (**a**) A several-centimeters-thick tourmaline front, Zone I; (**b**) Comb-textured tourmaline intergrown within medium-grained albite and quartz, Zone I; (**c**) Mineral association composed of cleavelandite (albite), muscovite, and lepidolite; (**d**) Beryl crystals embedded within a lepidolite envelopment at the contact between Zones III and IV; (**e**) Transition from schorl to blue-colored tourmaline; (**f**) Red-colored tourmaline from the Boqueirão granitic pegmatite.

3. Materials and Methods

A fluid inclusion study was carried out on microcline and quartz crystals collected from Zone II. Muscovite grains gathered from Zone I were suitable for $^{39}Ar/^{40}Ar$ dating.

Petrographic and microthermometric measurements of fluid inclusions were performed at the University of Zagreb. Double polished, 0.1–0.3 mm thick, transparent mineral wafers were studied. Measurements were carried out on Linkam THMS 600 (Linkam Scientific Instruments Ltd., Tadworth, UK) stages mounted on an Olympus BX 51 (Olympus, Tokyo, Japan) using 10× and 50× Olympus long-working distance objectives. Two synthetic fluid inclusion standards (SYN FLINC; pure H_2O

and mixed H_2O–CO_2) were used to calibrate the equipment. The precision of the system was <2.0 °C for homogenization temperatures, and <0.2 °C in the temperature range between −60 and +10 °C. Microthermometric measurements were conducted on carefully defined fluid inclusion assemblages representing groups of inclusions that were trapped simultaneously. The fluid inclusion assemblage were identified based on petrography prior to heating and freezing. If all of the fluid inclusions within the assemblage showed similar homogenization temperatures, the inclusions were assumed to have trapped the same fluid and to have not been modified by leakage or necking; these fluid inclusions thu record the original trapping conditions [20–22]. The salinity of aqueous inclusions was calculated from the final ice melting temperature using the BULK computer program [23]. Calculations are based on purely empirical best–fits, with no fundamental thermodynamic modeling involved [24]. The salinity of aqueous-carbonic inclusions was calculated according to [25]. Isochores were calculated by the ISOC computer program [23].

Raman spectroscopy of fluid inclusions, used for the semiquantitative analysis of entrapped volatiles, was performed on a Dilor LabRAM instrument (Horiba, Kyoto, Japan). Investigations were carried out at Department of Mineralogy and Petrology, Montanuniversität Leoben. A laser beam was focused through an Olympus BX 40 microscope (Olympus, Tokyo, Japan) onto the fluid inclusion of interest. The objective lenses of 50× and 100× magnification, combined with a confocal optical arrangement, enable a spatial resolution in the order of a cubic micrometer. A frequency-doubled Nd-YAG green laser (532 nm, 100 mW) was employed.

The $^{40}Ar/^{39}Ar$ analysis of muscovite was carried out at the ARGONAUT laboratory at the Department of Geography and Geology, Paris-Lodron-University of Salzburg. Mineral concentrates of muscovite were packed in aluminum-foil, sealed in quartz vials, and irradiated in the MTA KFKI reactor (Budapest, Hungary) for 16 h. The neutron fluence was monitored with DRA1 sanidine standard for which a $^{40}Ar/^{39}Ar$ plateau age of 25.03 ± 0.05 Ma has been reported [26]. Analyses were performed using a defocused (~1.5 mm diameter) 25 W CO_2-IR laser operating at the wavelengths between 10.57 and 10.63 μm. Gas cleanup was performed using two Zr–Al SAES getters (Milan, Italy). Ar-isotopes were measured on the VG ISOTECHTM NG3600 mass spectrometer (Isotopx, Middlewich, UK) on an axial electron multiplier in a static mode. Intensities of the peaks were back-extrapolated over 16 measured intensities to the time of gas admittance either with a straight line or a curved fit. Intensities were corrected for system blanks, background, post-irradiation decay of ^{37}Ar, and interfering isotopes. Isotopic ratios, ages, and uncertainties for individual steps were calculated following the suggestions by [27] and [28] using decay factors reported by [29]. The calculation of the plateau age was carried out using ISOPLOT/EX [28].

To avoid $^{40}Ar/^{39}Ar$ dating of hydrothermally altered muscovite, prior to the $^{40}Ar/^{39}Ar$ analysis an aliquot of the muscovite sample was analyzed by applying the X-ray powder diffraction (XRD) technique. The XRD analysis was conducted at the University of Zagreb on a Philips PW 3040/60 X'Pert PRO powder diffractometer (45 kV, 40 μA), with CuK-monochromatized radiation (λ = 1.54056 Å) and θ–θ geometry. The area between 4 and 63° 2θ, with 0.02 steps, was measured with a 0.5 primary beam divergence.

4. Results

4.1. Petrography of Fluid Inclusions

Microscopic examinations, performed at the room temperature on double-side-polished microcline and quartz wafers from Zone II, distinguished four types of inclusions:

Type I. Two phase-aqueous fluid inclusions (FIs) are mostly irregular, but some of them show progressive formation of negative crystal forms. The degree of fill (F), around 0.9, is fairly uniform (Figure 4a–c).

Type II. Aqueous-carbonic FIs show mostly irregular forms. This type of inclusion is characterized by the presence of two immiscible liquid phases (L_1 and L_2) and a vapor (V) phase. The F value aries slightly around 0.7 (Figure 4a,d). L_1 represents an aqueous solution whereas L_2 is composed of quid CO_2.

Type III. Monophase elongated or slightly irregular gas inclusions (Figure 4e).

Type IV. Melt inclusions are mostly polyphase (Figure 4f), but they do not undergo any phase ransition in the temperature range between −180 and +600 °C.

All four types of inclusions occur together in numerous inclusion assemblages, reflecting melt-fluid immiscibility during crystallization of Zone II microcline and quartz.

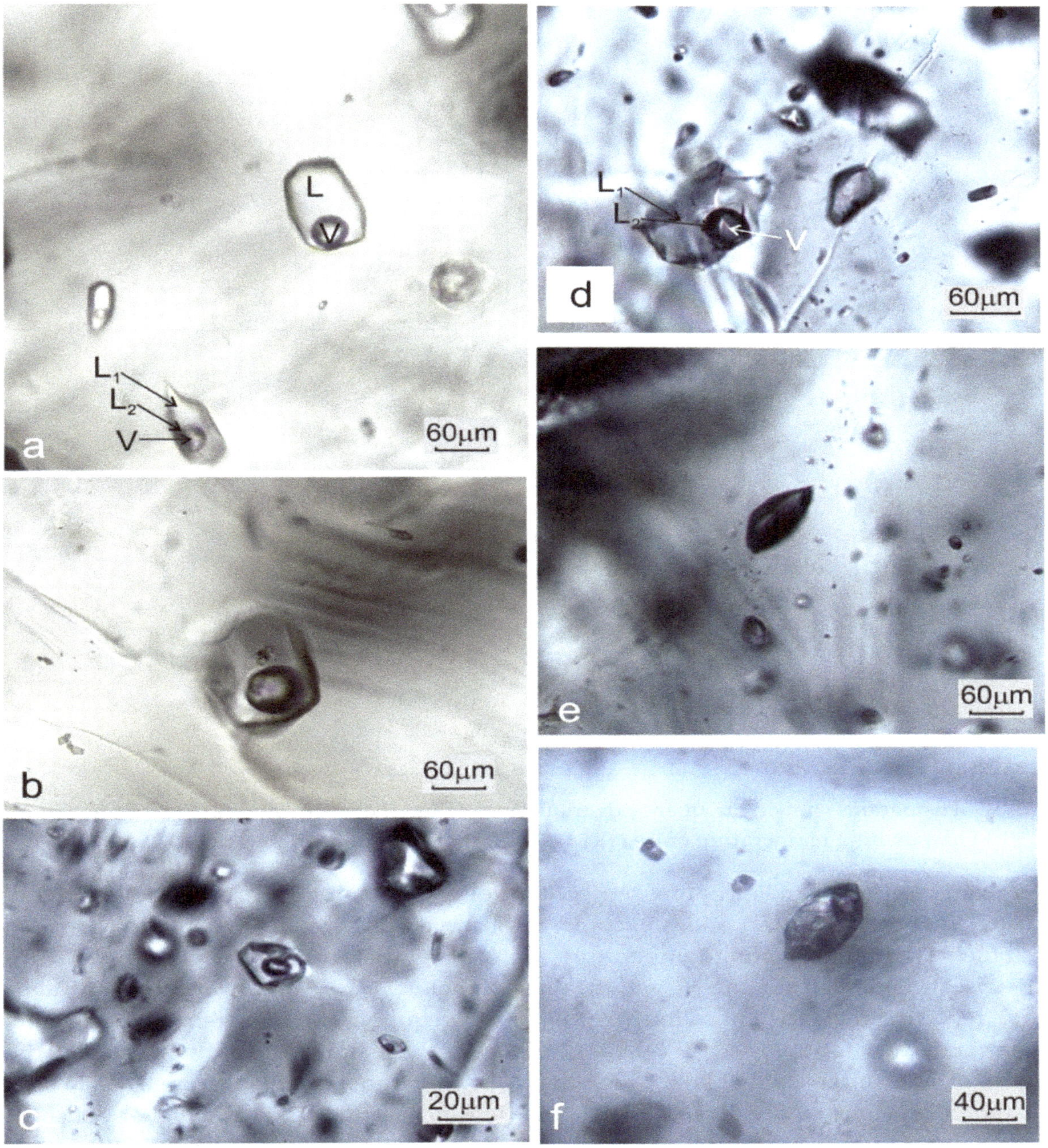

Figure 4. Macrophotographs of (**a**) coexisting two-phase, L + V, aqueous fluid inclusions and three-phase, $L_1 + L_2$ + V, aqueous-carbonic inclusions hosted by quartz; (**b**) two-phase, L + V, aqueous fluid inclusions seldom contain accidentally entrapped solid phases; (**c**) two-phase, L + V, aqueous fluid inclusions in microcline; (**d**) aqueous-carbonic inclusions in microcline; (**e**) a vapor-only inclusion hosted by microcline.

4.2. Microthermometry of Fluid Inclusions

Microthermometric data were collected from Type I and Type II inclusions hosted by microclin and quartz (Figure 5). Monophase fluid inclusions (Type III) as well as melt inclusions (Type IV) d not show any phase transition in the temperature range between −180 and +600 °C.

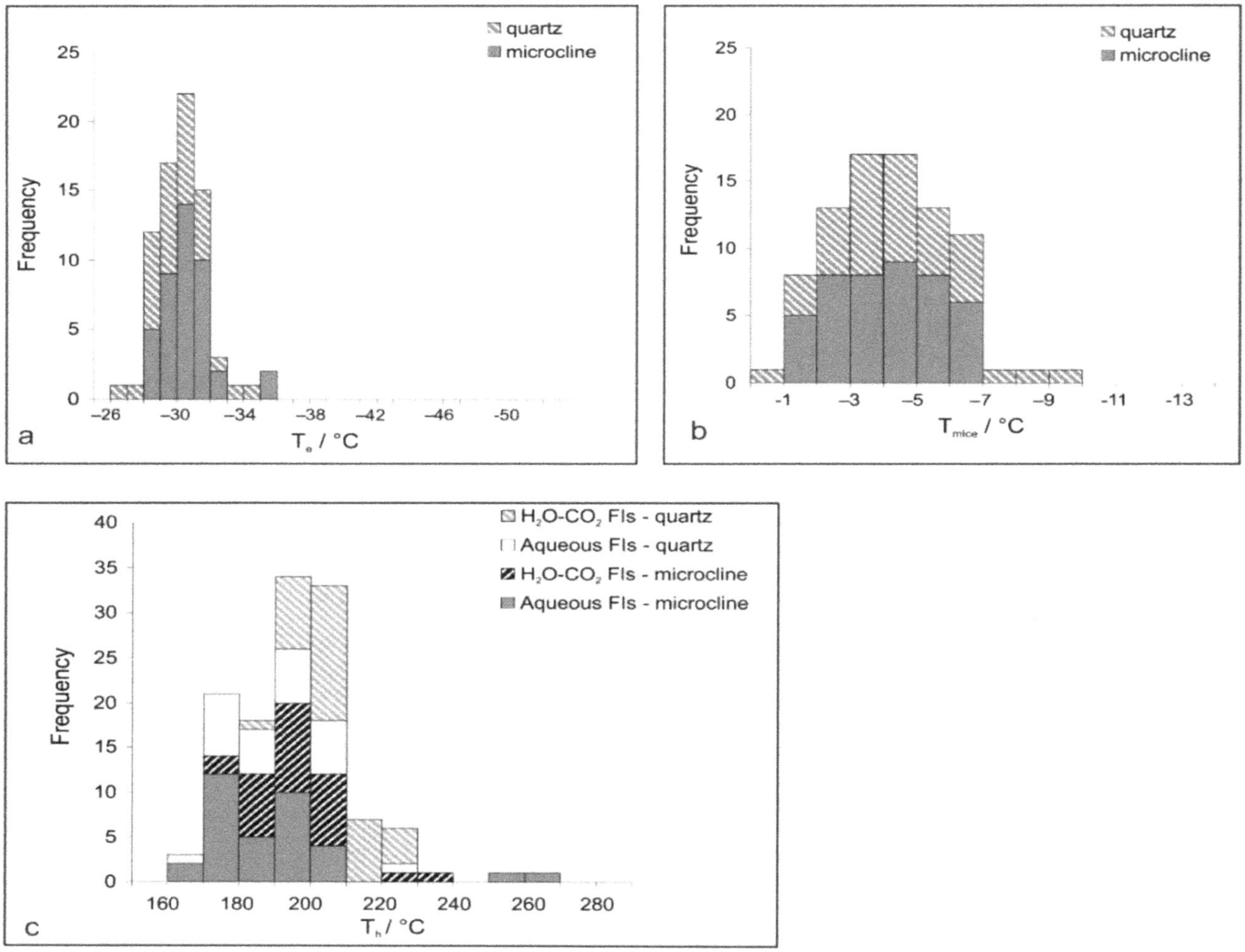

Figure 5. Histograms showing frequency distributions of fluid inclusion data: (**a**) Eutectic temperature (*Te*); (**b**) Final melting temperature of ice (T_{mIce}); (**c**) Homogenization temperature (T_h).

Microcline; Type I. A colorless frozen content of the aqueous FIs, formed at moderately lov temperatures (generally around −40 °C), could be observed only by distortion and shrinkage of th vapor bubble. The initial melting temperature (eutectic temperature, T_e) is observed in an interva between −25.2 and −32.7 °C (Figure 5a) and the final ice melting (T_{mIce}) occurs between −1.8 and −6.2 °C (Figure 5b), reflecting the apparent salinity of 3.1–9.5 wt.% equ. NaCl. Homogenization (T_h) follows th disappearance of the vapor phase at 170–220 °C (Figure 5c). The bulk fluid density, calculated utilizing the equation of state proposed by [30] through the BULK software [23], spans from 0.867 to 0.965 g/cm Type II carbonic-aqueous FIs are composed of a low-density gaseous phase and higher-density aqueou and carbonic liquid phases. The initial freezing of the aqueous phase was recorded around −40 °C and complete freezing occurs at around −100 °C. The first melting of solid CO_2 occurs mostly in th temperature range between −60.5 and −67.8 °C, which is assigned to the presence of other volatile (such as CH_4 and N_2; [31]). Melting of the aqueous part of inclusions was observed in a wide range o temperatures between −40 and −2 °C. The final melting of clathrate (T_{mClath}) spans between 8.0 and 9.0 °C and reflects salinities ranging from 2.5 to 4.0 wt.% equ. NaCl. Homogenization of the carbonic phas proceeds in two ways, L + V→V and L + V→L. Critical phenomena have not been observed. The data gather around 29 °C in either way of homogenization. Total homogenization (T_h) into the liquid phase occurred in the interval between 170 and 230 °C (Figure 5c). Bulk fluid densities, calculated according to the equation of state from [32] revised by [33], fall in the range between 0.808 and 0.893 g/cm^3.

Quartz, Type I: Measurements performed on aqueous FIs within quartz samples yielded microthermometric data that overlap with those gathered from microcline (Figure 5a–c). In contrast, Type II (carbonic-aqueous) FIs in quartz show some differences compared to the same type of FIs hosted by microcline. The initial melting of CO_2 occurs in the interval between −57.1 and −65.0 °C. The recorded T_{mClath} between 7.0 and 9.0 °C points to the salinity between 2.5 and 6.1 wt.% equ. NaCl. Homogenization of the carbonic phase proceeds mostly, at around 20 °C, into a liquid phase (Figure 5e). A critical phenomenon was observed only in one inclusion at +19.9 °C. The Th values have been recorded in an interval between 185 and 225 °C (Figure 5c).

4.3. Raman Spectroscopy of Fluid Inclusions

Raman spectroscopy measurements performed on aqueous FIs (Type I) within microcline and quartz samples recognized only the presence of water. The measurements of Type II FIs suggest CO_2 as the major non-H_2O volatile component (Figure 6a). In addition, variable amounts of N_2 were detected (Figure 6b). However, in several cases, the measured peak areas were too small for reliable quantification. A simple formula based on Placzek's polarizability theory was applied to derive quantitative molar fractions of species present in FIs [34–38]. The molar fraction of N_2 ranges up to 9 mole % (Table 1). No difference between quartz and microcline samples was observed.

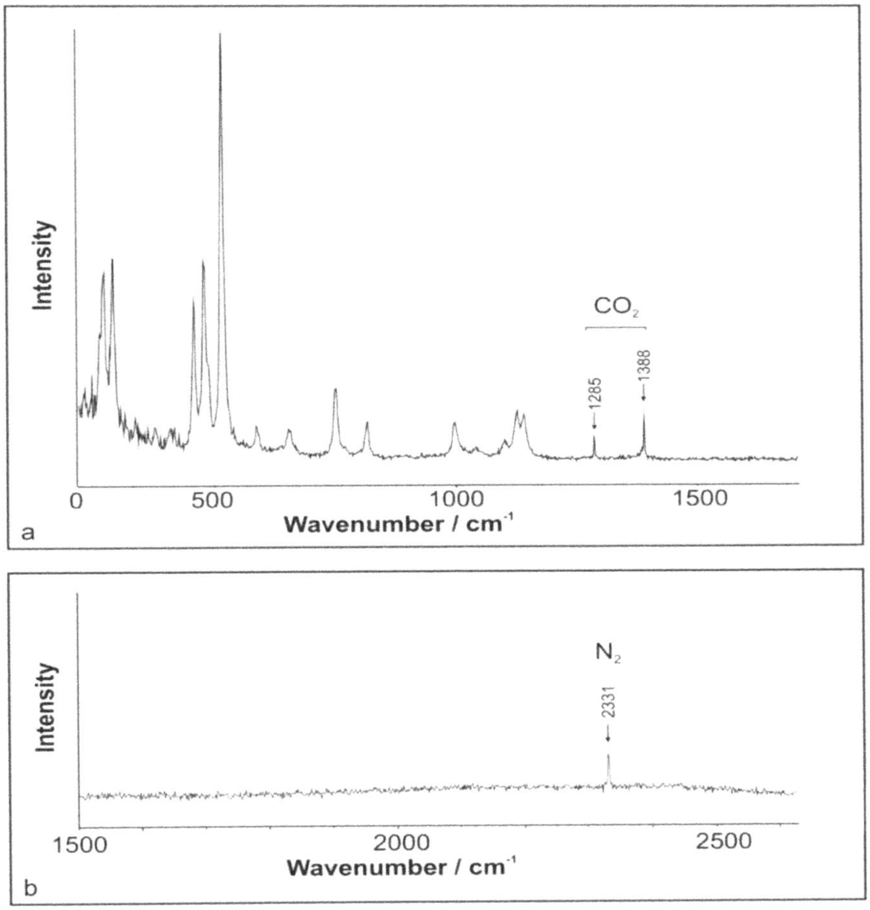

Figure 6. Raman spectra of the vapor bubble from an aqueous-carbonic inclusion reveal the presence of (**a**) CO_2 and (**b**) N_2.

Table 1. Microthermometric results, Raman data, and calculated bulk compositions of selected carbonic-aqueous fluid inclusions.

Sample	FI	Microthermometry (°C)				Raman (mol %)		Bulk Composition (mol %)			Density ($g \cdot cm^{-3}$)
		T_{mice}	T_{mclath}	T_{hCO2}	Mode	CO_2	N_2	H_2O	CO_2	N_2	
						Microcline					
m-1	1	−2.5	8.1	26.2	V	93.2	6.8	85.4	11.6	0.7	0.9366
m-1	2	−3.1	8.6	28.3	V	96.5	3.5	84.4	12.6	0.2	0.9576
m-1	3	−4.3	7.8	27.5	L	98.0	2.0	83.4	12.6	0.1	0.9691
m-1	4	−2.0	8.9	27.9	V	94.7	5.3	86.7	10.9	0.6	0.9253
m-2	1	−2.7	8.7	30.0	L	99.5	0.5	84.3	13.2	<0.05	0.9671
m-2	2	−3.3	9.2	28.0	L	97.8	2.2	84.2	12.7	0.2	0.9608
m-3	1	−4.0	8.1	28.6	V	92.1	7.9	83.6	12.1	0.6	0.9621
m-3	2	−2.9	7.6	29.1	V	99.9	0.1	84.1	13.2	<0.01	0.9694
m-3	3	−3.1	8.8	28.8	V	93.3	6.7	85.3	11.5	0.4	0.9359
						Quartz					
q-1	1	−4.0	8.2	20.1	V	93.4	6.6	83.7	12.0	0.8	0.9673
q-1	2	−3.1	8.9	18.5	V	99.7	0.3	82.8	14.4	<0.05	0.9950
q-1	3	−2.9	8.4	17.9	V	92.6	7.4	83.5	13.0	0.9	0.9749

4.4. The $^{40}Ar/^{39}Ar$ Age

A concentrate of a few relatively large flakes (<0.5 mm) of white mica (muscovite) free of any inclusions was selected for dating. The XRD pattern of the analysed muscovite is presented in Figure 7. The experimental results of $^{40}Ar/^{39}Ar$ dating are given in Table 2. The argon release pattern of the muscovite sample (Figure 8) shows a slightly disturbed U-shaped pattern with a plateau age of 502.5 ± 5.8 Ma, constituting together 88.9% of ^{39}Ar released (Steps 4 to 10). Low-energy Steps 1 to 3 yielded an addition of excess argon, whereas a significant increase in some ages (Steps 6 and 8) is attributed to an internally inhomogeneous distribution of argon in muscovite. The $^{37}Ar_{Ca}$ values are low (Table 2) and the variation of $^{37}Ar_{Ca}$ is small. In contrast, the chlorine-derived $^{38}Ar_{Cl}$ values are relatively high (Table 2), showing that muscovite did grow under some saline conditions. We consider the plateau age of 502.5 ± 5.8 Ma to be geologically significant and to date the cooling of pegmatite through the argon retention temperature, which is experimentally determined at 425 ± 25 °C in slowly cooling terranes [39].

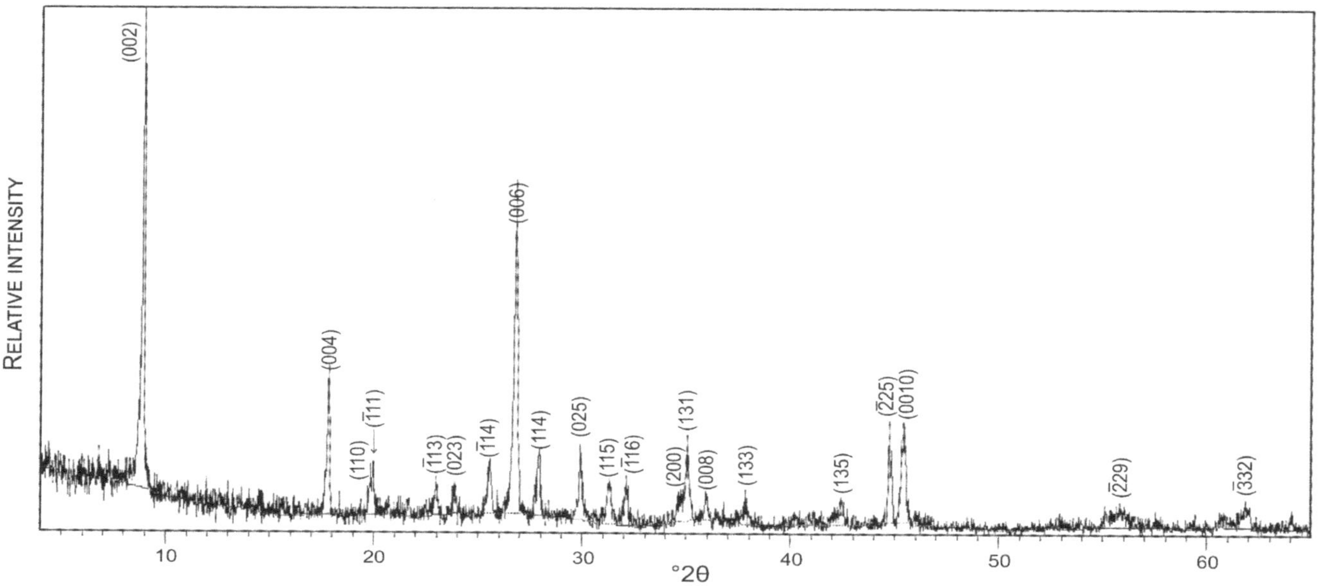

Figure 7. XRD pattern of muscovite dated by the of $^{40}Ar/^{39}Ar$ technique.

Table 2. The experimental results of $^{40}Ar/^{39}Ar$ analytical data.

J-Value:		0.01850	+/−	0.00019											
Step	^{36}Ar Meas.	$\pm\sigma_{36}$	^{37}Ar Decay Corr.	$\pm\sigma_{37}$	^{38}Ar Meas.	$\pm\sigma_{38}$	^{39}Ar Decay Corr.	$\pm\sigma_{39}$	^{40}Ar Meas.	$\pm\sigma_{40}$	$^{40}Ar^{*}/^{39}Ar_K$	$\pm\sigma$	$\%^{40}Ar^{*}$	$\%^{39}Ar$	Age (Ma)	±(Ma) 1-Sigma Abs.
1	62.9	9.3	56	25	120	7.9	7820	66.0	1.96×10^5	215	22.69	0.40	90.5	2.0	632.4	10.9
2	24.7	7.2	55	19	120	13	4070	26.2	9.59×10^4	183	21.74	0.54	92.4	1.1	610.0	13.9
3	56.1	12	52	29	360	14	33,100	68.6	6.27×10^5	253	18.44	0.11	97.3	8.6	529.6	5.5
4	17.7	8.7	46	22	940	15	90,500	119	1.59×10^6	916	17.49	0.04	99.6	23.5	505.7	4.6
5	99.8	1.3	37	25	1100	13	111,000	22.0	1.95×10^6	1340	17.24	0.04	98.4	28.9	499.5	4.6
6	27.8	18	63	32	130	11	9500	68.3	1.83×10^5	274	18.36	0.58	95.5	2.5	527.5	15.3
7	17.2	7.0	10	25	1100	18	108,000	142	1.88×10^6	1630	17.28	0.03	99.7	28.2	500.5	4.6
8	37.5	15	67	27	170	11	12,300	53.1	2.39×10^5	289	18.48	0.38	95.3	3.2	530.5	10.6
9	28.8	11	120	32	260	14	23,300	54.9	4.24×10^5	338	17.85	0.14	98.0	6.1	514.7	5.9
10	51.1	13	120	39	120	13	6060	43.0	1.26×10^5	187	18.25	0.67	88.0	1.6	524.7	17.3
Total	-	-	-	-	-	-	-	-	-	-	-	-	-	100.0	506.3	4.6

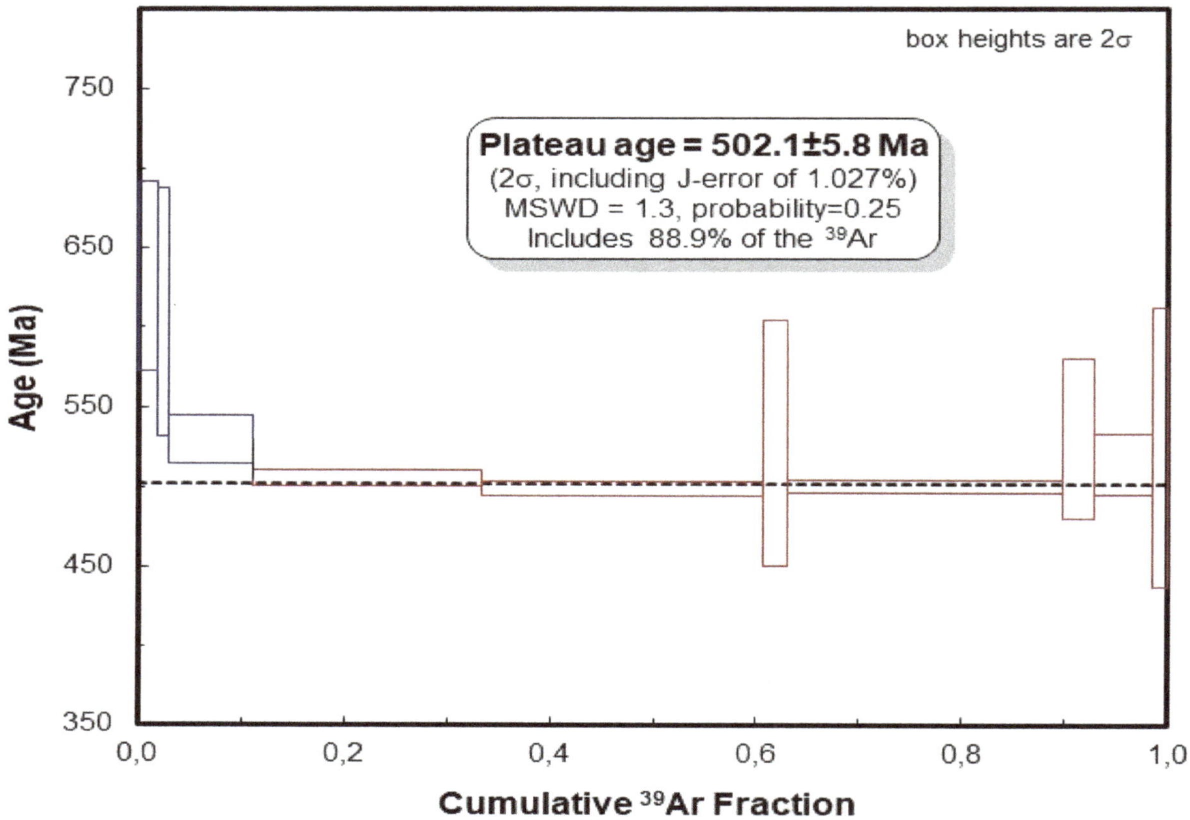

Figure 8. $^{40}Ar/^{39}Ar$ apparent age spectra of the coarse-grained muscovite sample. Laser energy increases from left to right. Vertical width of bars represents the 2σ error and includes the error of the J-value. Steps 4–10 are used for calculation of the plateau age.

5. Discussion

Fluid inclusions hosted by pegmatite crystals may provide a snapshot of P-T-X conditions at the time of their entrapment [40]. Numerous studies on the LCT-type of pegmatites worldwide have outlined the following types of inclusions: (1) melt inclusions; (2) saline aqueous fluid inclusions and (3) CO_2-enriched fluid inclusions [41–45]. Classical theories of pegmatite genesis emphasize the importance of volatiles during crystallization of granitic pegmatites from coexisting aluminosilicate melt and hydrous fluids [46,47]. According to the model proposed by [48], aluminosilicate melts can produce highly evolved pegmatitic liquids via continuous crystallization under particular kinetic conditions. The ubiquitous crystallization commenced by formation of schorl (tourmaline), which buffered the boron content in the hydrous silicic melt, produces a high amount of exsolution of hydrous fluid [48]. The "boron quenching" in turn was enhanced by consequent crystallization of elbaite. The sink of boron, due to tourmaline crystallization and removal of the Li-alkali borate fluxing component, caused the separation of silicate and aqueous fluids producing supersaturation of alkali aluminosilicates (albite, microcline, quartz) and oxides, and their massive growth. Experimental studies support the coexistence of alumino-silicate melt, hydrous fluid, and hydrosaline fluid during the late stage of magma evolution [49].

Fluid inclusion assemblages, composed of melt inclusions and aqueous and CO_2-bearing fluid inclusions in minerals from Zone II of the gem-bearing Boqueirão granitic pegmatite, indicate the magmatic–hydrothermal transition. Similar phenomena have been recorded in other granitic pegmatites worldwide [50,51]. Melt inclusions represent entrapped remains of the silicate-rich melt, whereas fluid inclusions contain the fluid phase exsolved during the crystallization process. Additionally, coexistence of aqueous and aqueous-carbonic inclusions as well as their overlapping homogenization emperatures reflect an immiscibility between the low-salinity, low-density CO_2-rich fluid phase nd the higher-salinity, higher-density fluid phase during the magmatic–hydrothermal transition. he recorded three-phase immiscibility (silicate melt–low salinity and low density carbonic-aqueous luid–moderate salinity and moderate density fluid) affected the fate of metals in the evolving pegmatite. he majority of lithophile metals stay in the silicate melt, but some elements preferentially enter mmiscible fluids. The partitioning is strongly controlled by the salinity of the exsolving fluids [52]. Pb, n, Ag, and Fe preferentially partition into fluids with a higher salinity, whereas Mo, B, As, Sb, and Bi refer low-salinity fluids. In contrast, Li and Sn do not show systematic variations in their partition oefficients with the salinity of the fluids [53]. Partitioning of P between silicate melts and exsolving luids strongly depends on pressure and temperature. Regardless, partition of flux elements (e.g., B, Li, nd P) into exsolving fluids together with the loss of H_2O during the magmatic–hydrothermal transition nay increase the viscosity of evolving silicate melts and affect textural features of pegmatites [54].

The lithium aluminosilicate phase diagram has been used as a petrogenetic grid for lithium-rich egmatites [55]. Spodumene and petalite are stable Li-phases in a quartz-saturated system up to temperature of 700 °C, which sets the upper limit on the crystallization conditions for Li-rich egmatites [56]. However, the Boqueirão granitic pegmatite is characterized by the presence of podumene as the only $LiAl(SiO_3)_2$ polymorph, which reflects the minimum formation pressure of .6 kbar (Figure 9).

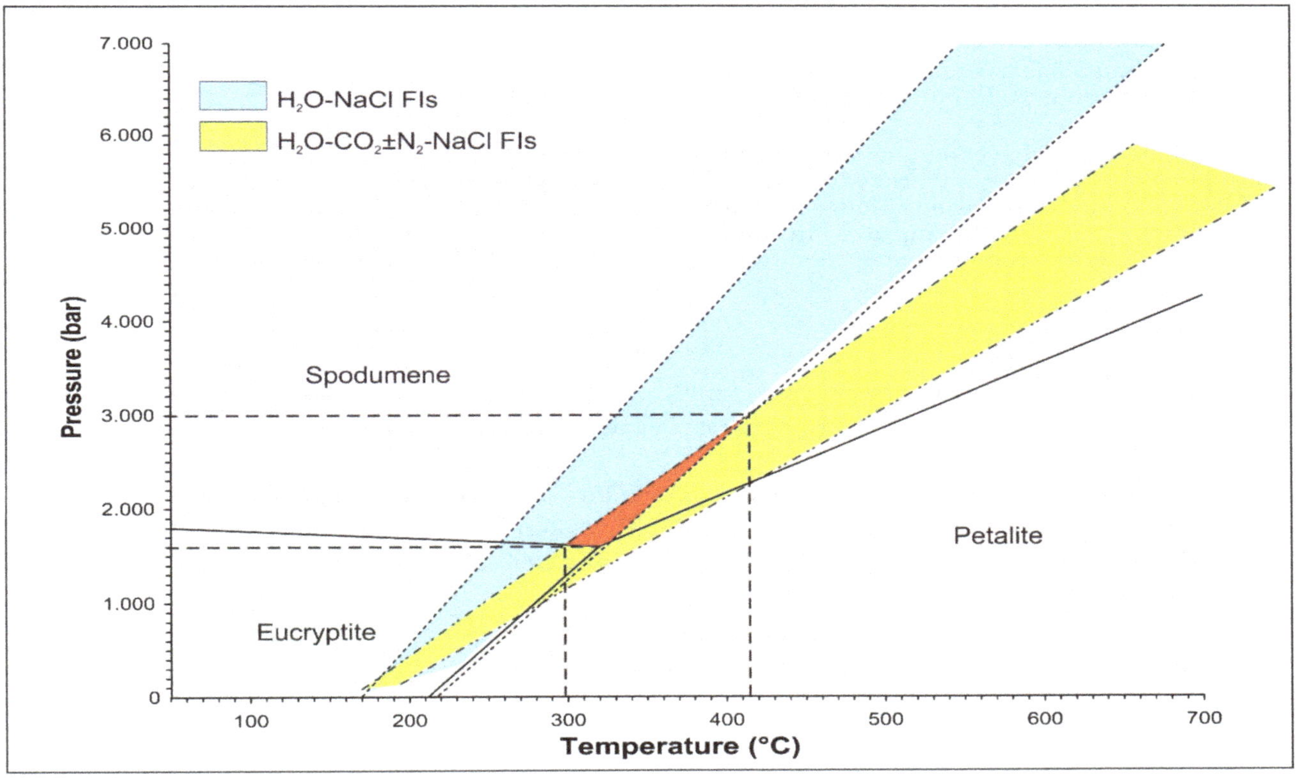

Figure 9. The isochores for aqueous (H_2O–NaCl FIs) and aqueous-carbonic (H_2O–CO_2 ± N_2-NaCl FIs) fluid inclusions extrapolated across the stability field of spodumene suggest a formation temperature in the range between 300 and 415 °C at a pressure from 1.8 to 3 kbar.

Isochores, constructed for the coexisting aqueous and aqueous-carbonic fluid inclusions (calculated ısing the equation of state proposed by [57] for the NaCl–H_2O system and [32] revised by [33] for the H_2O–CO_2–CH_4–N_2–NaCl system) extrapolated across the stability field of spodumene suggest that the nagmatic–hydrothermal transition associated with the formation of the Boqueirão granitic pegmatite occurred at a temperature from 300 to 415 °C and a pressure between 1.8 and 3 kbar (Figure 9).

According to the $^{40}Ar/^{39}Ar$ plateau age of muscovite, the Boqueirão granitic pegmatite crystallized simultaneously with late stage of magmatic activity in the BPP (511–500 Ma; e.g., [58]).

6. Conclusions

According to its mineralogical and geochemical characteristics, the Boqueirão granitic pegmatite has been classified as a member of the LCT pegmatite family, broadly widespread over the BPP. The pegmatite shows a zonal structure with significant enrichment on incompatible elements from its outer rim toward the massive quartz core.

Coexistence of melt inclusions and aqueous and CO_2-enriched fluid inclusions suggests the magmatic–hydrothermal transition that resulted with segregation of two liquid phases (low salinity and low density versus moderate salinity and moderate density) from the silicate melt. The fluid inclusion data together with the well-defined stability of $LiAl(SiO_3)_2$ polymorphs over the P-T area can be used as an indicator of formation conditions for this type of granitic pegmatite. The fluid inclusion data obtained from the Boqueirão granitic pegmatite accompanied by the P-T stability of spodumene revealed that the magmatic–hydrothermal transition occurred in the temperature range between 300 and 415 °C at a pressure ranging from 1.8 to 3 kbar.

The $^{40}Ar/^{39}Ar$ plateau age of muscovite, at 502.5 ± 5.8 Ma, sets the Boqueirão granitic pegmatite to the late stage of magmatic activity in the BPP.

Author Contributions: Conceptualization, S.S.P., L.P. and V.B.; Methodology, F.N., L.P., R.S. and S.S.P.; Formal Analysis, S.S.P. and S.B.Š.; Writing-Original Draft Preparation, all coauthors; Writing-Review & Editing, S.S.P., F.N. and R.S.; Visualization, S.S.P.; Funding Acquisition, S.S.P., L.P. and V.B.

Acknowledgments: Early versions of this paper greatly benefited from comments by David London and Rainer Thomas. Ronald J. Bakker is appreciated for help in obtaining the Raman spectroscopy data. Special thanks go to Panagiotis Voudouris and Vasilios Melfos for handling the manuscript as well as to the anonymous reviewers whose comments substantially improved the quality of this paper. The publication charges for this article have been funded by a grant from the publication fund of UiT The Arctic University of Norway.

References

1. London, D. Ore-forming processes within granitic pegmatites. *Ore Geol. Rev.* **2018**, *101*, 349–383. [CrossRef]
2. Pough, M.W. Simpsonite and the Northern Brazilian pegmatite region. *Bull. Geol. Soc. Am.* **1945**, *56*, 505–514. [CrossRef]
3. Beurlen, H. The mineral resources of the Borborema Province in Northeastern Brazil and its sedimentary cover: A review. *J. Am. Earth Sci.* **1995**, *8*, 365–376. [CrossRef]
4. Magyar, M.J. Tantalum. In *Mineral Commodity Summaries 2007*; U.S. Geological Survey: Reston, VA, USA, 2007; pp. 164–165.
5. Beurlen, H.; Barreto, S.; Martin, R.; Melgarejo, J.; Da Silva, M.R.R.; Souza Neto, J.A. The Borborema pegmatite province. NE Brazil revisited. *Estud. Geológicos* **2009**, *19*, 62–66.
6. Beurlen, H.; Da Silva, M.R.R.; Thomas, R.; Soares, D.R.; Olivier, P. Nb–Ta–(Ti–Sn) oxide mineral chemistry as tracer of rare-element granitic pegmatite fractionation in the Borborema Province, Northeastern Brazil. *Miner. Depos.* **2008**, *43*, 207–228. [CrossRef]
7. Černy, P. Rare element granitic pegmatites. Part 1: Anatomy and internal evolution of pegmatite deposits. *Geosci. Can.* **1991**, *18*, 49–67.
8. Černy, P.; Ercit, T.S. The Classification of Granitic Pegmatites Revisited. *Can. Miner.* **2005**, *43*, 2005–2026. [CrossRef]
9. Soares, D.R.; Beurlen, H.; Barreto, S.D.B.; Da Silva, M.R.R.; Ferreira, A.C.M. Compositional variation of tourmaline-group minerals in the borborema pegmatite province, northeastern brazil. *Can. Miner.* **2008**, *46*, 1097–1116. [CrossRef]
10. Beurlen, H.; da Silva, M.R.; de Castro, C. Fluid inclusion microthermometry in Be–Ta–(Li–Sn)-bearing pegmatites from the Borborema Province, northeast Brazil. *Chem. Geol.* **2001**, *173*, 107–123. [CrossRef]

11. Thomas, R.; Davidson, P.; Beurlen, H. Tantalite-(Mn) from the Borborema Pegmatite Province. northeastern Brazil: Conditions of formation and melt- and fluid-inclusion constraints on experimental studies. *Miner. Depos.* **2011**, *46*, 749–759. [CrossRef]
12. Strmić Palinkaš, S.; Wegner, R.; Čobić, A.; Palinkaš, L.A.; Barreto, S.D.B.; Váczi, T.; Bermanec, V. The role of magmatic and hydrothermal processes in the evolution of Be-bearing pegmatites: Evidence from beryl and its breakdown products. *Am. Mineral.* **2014**, *99*, 424–432. [CrossRef]
13. Scorza, E.P. Província Pegmatítica da Borborema. DNPM. *Div Geol Min Boletim. Rio de Janeiro* **1944**, *112*, 57.
14. Lima, E.S. Metamorphism and Tectonic Evolution in the Seridó Region. Northeastern Brazil. Ph.D. Thesis, Universty of California Los Angeles, Los Angeles, CA, USA, 1986; p. 215.
15. Jardim de Sá, E.F.; Legrand, J.M.; McReath, I. Estratigrafia de rochas granitóides na Região do Seridó (RN-PB) com base em critérios estruturais. *Rev. Bras. Geoc.* **1981**, *11*, 50–57. [CrossRef]
16. Araújo, M.N.C.; Silva, F.C.A.; Jardim, E.F. Pegmatite emplacement in the Seridó Belt. Northeastern Brazil: Late stage tectonics of the Brasiliano Orogen. *Gondwana Res.* **2001**, *4*, 75–85.
17. DA Silva, M.R.R.; Höll, R.; Beurlen, H. Borborema Pegmatitic Province: Geological and geochemical characteristics. *J. S. Am. Earth Sci.* **1995**, *8*, 355–364. [CrossRef]
18. Johnston, J.R. Beryl–tantalite pegmatites of Northeastern Brazil. *Geol. Soc. Am. Bull.* **1945**, *56*, 1015–1070. [CrossRef]
19. Baumgartner, R.; Romer, R.L.; Möritz, R.; Sallet, R.; Chiaradia, M. Columbite-tantalite-bearing granitic pegmatites from the serido belt, northeastern brazil: Genetic constraints from U-Pb dating and Pb isotopes. *Can. Miner.* **2006**, *44*, 69–86. [CrossRef]
20. Goldstein, R.H. Fluid inclusions in sedimentary and diagenetic systems. *Lithos* **2001**, *55*, 159–193. [CrossRef]
21. Goldstein, R.H.; Reynolds, T.J. Systematics of fluid inclusions in diagenetic minerals. *Soc. Sediment. Geol. Short Course* **1994**, *31*, 199.
22. Bodnar, R.J. Introduction to fluid inclusions. In *Short Course, Fluid Inclusions: Analysis and Interpretation*; Mineralogical Association of Canada: Ottawa, ON, Canada, 2003; Volume 32, pp. 1–8.
23. Bakker, R.J.; Package, F. Computer programs for analysis of fluid inclusion data and for modelling bulk fluid properties. *Chem. Geol.* **2003**, *194*, 3–23. [CrossRef]
24. Naden, J. *Calcic Brine: A Microsoft Excel 5.0 Add-in for Calculating Salinities from Microthermometric Data in the System $NaCl-CaCl_2-H_2O$*; Brown, P.E., Hagemann, S.G., Eds.; PACROFI VI: Madison, WI, USA, 1996; pp. 97–98.
25. Collins, P.L. Gas hydrates in CO_2-bearing fluid inclusions and the use of freezing data for estimation of salinity. *Econ. Geol.* **1979**, *74*, 1435–1444. [CrossRef]
26. Wijbrans, J.R.; Pringle, M.S.; Koopers, A.A.; Schveers, R. Argon geochronology of small samples using the Vulkaan argon laserprobe. *Proc. K. Ned. Akad. Van Wet.* **1995**, *98*, 185–218.
27. Mcdougall, I.; Harrison, M.T. *Geochronology and Thermochronology by the $^{40}Ar/^{39}Ar$ Method*, 2nd ed.; Oxford University Press: Oxford, UK, 1999; p. 26.
28. Ludwig, K.R. *Users Manual for ISOPLOT/Ex 3.75c: A Geochronological Toolkit for Microsoft Excel*; Special Publication; Berkeley Geochronology Center: Berkeley, CA, USA, 2012.
29. Steiger, R.H.; Jäger, E. Subcommission on geochronology: Convention on the use of decay constants in geo- and cosmochronology. *Earth Planet. Sci. Lett.* **1977**, *36*, 359–362. [CrossRef]
30. Zhang, Y.G.; Frantz, J.D. Determination of the homogenization temperatures and densities of supercritical fluids in the system $NaCl-KCl-CaCl_2-H_2O$ using synthetic fluid inclusions. *Chem. Geol.* **1987**, *64*, 335–350. [CrossRef]
31. Burrus, R.C. Hydrocarbon fluid inclusions in studies of sedimentary diagenesis. In *Fluid Inclusions: Applications to Petrology*; Short Course Notes; Mineralogical Association of Canada: Toronto, ON, Canada, 1981; Volume 6, pp. 138–156.
32. Bowers, T.S.; Helgeson, H.C. Fortran programs for generating fluid inclusion isochores and fugacity coefficients for the system H_2O-CO_2-NaCl at high pressures and temperatures. *Comput. Geosci.* **1985**, *11*, 203–213. [CrossRef]
33. Bakker, R.J. Adaptation of the Bowers and Helgeson (1983) equation of state to the $H_2O-CO_2-CH_4-N_2-NaCl$ system. *Chem. Geol.* **1999**, *154*, 225–236. [CrossRef]
34. Schrötter, H.W.; Klöckner, H.W. Raman scattering cross-sections in gases and liquids. In *Raman Spectroscopy of Gases and Liquids*; Weber, A., Ed.; Springer: Berlin, Germany, 1979; pp. 123–166.

35. Wopenka, B.; Pasteris, J.D. Limitation to quantitative analysis of fluid inclusions in geological samples b laser Raman microprobe spectroscopy. *Appl. Spectrosc.* **1986**, *40*, 144–151. [CrossRef]
36. Wopenka, B.; Pasteris, J.D. Raman intensities and detection limits of geochemically relevant gas mixtures fo a laser Raman microprobe. *Anal. Chem.* **1987**, *59*, 2165–2170. [CrossRef]
37. Dubessy, J.; Poty, B.; Ramboz, C. Advances in C-O-H-N-S fluid geochemistry based on micro-Rama spectroscopic analysis of fluid inclusions. *Eur. J. Mineral.* **1989**, *1*, 517–534. [CrossRef]
38. Burke, E.A. Raman microspectrometry of fluid inclusions. *Lithos* **2001**, *55*, 139–158. [CrossRef]
39. Harrison, T.M.; Célérier, J.; Aikman, A.B.; Hermann, J.; Heizler, M.T. Diffusion of ^{40}Ar in muscovit *Geochim. Cosmochim. Acta* **2009**, *73*, 1039–1051. [CrossRef]
40. Roedder, E. Fluid inclusions. *Rev. Mineral.* **1984**, *12*, 644.
41. Thomas, A.; Spooner, E. The volatile geochemistry of magmatic H_2O-CO_2 fluid inclusions from the Tanc zoned granitic pegmatite, southeastern Manitoba, Canada. *Geochim. Cosmochim. Acta* **1992**, *56*, 49–6 [CrossRef]
42. London, D. Geochemical features of peraluminous granites. pegmatites. and rhyolites as sources of lithophil metal deposits. Magmas, Fluids and Ore Deposits. In *Magmas, Fluids, and Ore Deposits*; Short Cours Handbook; Mineralogical Association of Canada: Nepean, ON, Canada, 1995; Volume 23, pp. 175–202.
43. Webster, J.D.; Thomas, R.; Rhede, D.; Förster, H.-J.; Seltmann, R. Melt inclusions in quartz from a evolved peraluminous pegmatite: Geochemical evidence for strong tin enrichment in fluorine-rich an phosphorus-rich residual liquids. *Geochim. Cosmochim. Acta* **1997**, *61*, 2589–2604. [CrossRef]
44. Thomas, R.; Webster, J.D.; Heinrich, W. Melt inclusions in pegmatite quartz: Complete miscibility betwee silicate melts and hydrous fluids at low pressure. *Contrib. Mineral. Petrol.* **2000**, *139*, 394–401. [CrossRef]
45. Sirbescu, M.-L.C.; Nabelek, P.I. Crystallization conditions and evolution of magmatic fluids in the Harne Peak Granite and associated pegmatites, Black Hills, South Dakota—Evidence from fluid inclusion *Geochim. Cosmochim. Acta* **2003**, *67*, 2443–2465. [CrossRef]
46. Jahns, R.H.; Burnham, C.W. Experimental studies of pegmatite genesis: I. A model for the derivation an crystallization of granitic pegmatites. *Econ. Geol.* **1969**, *64*, 843–864. [CrossRef]
47. Veksler, I.V. Liquid immiscibility and its role at the magmatic–hydrothermal transition: A summary c experimental studies. *Chem. Geol.* **2004**, *210*, 7–31. [CrossRef]
48. London, D. Magmatic-hydrothermal transition in Tanco Pegmatite. *Am. Mineral.* **1986**, *71*, 76–395.
49. Veksler, I.V.; Thomas, R.; Schmidt, C. Experimental evidence of three coexisting immiscible fluids in syntheti granitic pegmatite. *Am. Mineral.* **2002**, *87*, 775–779. [CrossRef]
50. Sirbescu, M.-L.C.; Nabelek, P.I. Crustal melts below 400 °C. *Geology* **2003**, *31*, 685–688. [CrossRef]
51. Kaeter, D.; Barros, R.; Menuge, J.F.; Chew, D.M. The magmatic–hydrothermal transition in rare-elemen pegmatites from southeast Ireland: LA-ICP-MS chemical mapping of muscovite and columbite–tantalit *Geochim. Cosmochim. Acta* **2018**, *240*, 98–130. [CrossRef]
52. Webster, J.D. The exsolution of magmatic hydrosaline chloride liquids. *Chem. Geol.* **2004**, *210*, 33–4 [CrossRef]
53. Zajacz, Z.; Halter, W.E.; Pettke, T.; Guillong, M. Determination of fluid/melt partition coefficients b LA-ICPMS analysis of co-existing fluid and silicate melt inclusions: Controls on element partitionin *Geochim. Cosmochim. Acta* **2008**, *72*, 2169–2197. [CrossRef]
54. Nabelek, P.I.; Whittington, A.G.; Sirbescu, M.L.C. The role of H_2O in rapid emplacement an crystallization of granite pegmatites: Resolving the paradox of large crystals in highly undercoole melts. *Contrib. Mineral. Petrol.* **2010**, *160*, 313–325. [CrossRef]
55. London, D.; Burt, D.M. Lithium aluminosilicate occurrences in pegmatites and the lithium aluminosilicat phase diagram. *Am. Mineral.* **1982**, *67*, 483–493.
56. London, D. Experimental phase equilibria in the system $LiAlSiO_4$-SiO_2-H_2O: A petrogenetic grid fo lithium-rich pegmatites. *Am. Mineral.* **1984**, *69*, 995–1004.
57. Potter, R.W., II. Pressure corrections for fluid-inclusion homogenization temperatures based on the volumetri properties of the system NaCl-H_2O. *J. Res. U. S. Geol. Surv.* **1977**, *5*, 603–607.
58. Corsini, M.; De Figueiredo, L.L.; Caby, R.; Féraud, G.; Ruffet, G.; Vauchez, A. Thermal history o the Pan-African/Brasiliano Borborema Province of northeast Brazil deduced from ^{40}Ar/^{39}Ar analysi *Tectonophysics* **1998**, *285*, 103–117. [CrossRef]

6

Gem-Quality Green Cr-Bearing Andradite (var. Demantoid) from Dobšiná, Slovakia

Ján Štubňa [1,*], Peter Bačík [2], Jana Fridrichová [2], Radek Hanus [3], Ľudmila Illášová [1], Stanislava Milovská [4], Radek Škoda [5], Tomáš Vaculovič [6] and Slavomír Čerňanský [7]

1 Gemmological Institute, Faculty of Natural Sciences, Constantine the Philosopher University in Nitra, Nábrežie mládeže 91, 949 74 Nitra, Slovakia; lillasova@ukf.sk

2 Department of Mineralogy and Petrology, Faculty of Natural Sciences, Comenius University in Bratislava, Ilkovičova 6, Mlynská dolina, 842 15 Bratislava, Slovakia; peter.bacik@uniba.sk (P.B.); fridrichova.jana2@gmail.com (J.F.)

3 Gemological Laboratory of e-gems.cz, 110 00 Prague, Czech Republic; kakt@centrum.cz

4 Earth Science Institute of the Slovak Academy of Sciences, Ďumbierska 1, 974 01 Banská Bystrica, Slovakia; milovska@savbb.sk

5 Department of Geological Sciences, Masaryk University, Kotlářská 2, 611 37 Brno, Czech Republic; rskoda@sci.muni.cz

6 Central European Institute of Technology, Masaryk University, Kamenice 5, 625 00 Brno, Czech Republic; tomas.vaculovic@ceitec.muni.cz

7 Department of Environmental Ecology, Faculty of Natural Sciences, Comenius University in Bratislava, Ilkovičova 6, Mlynská dolina, 842 15 Bratislava, Slovakia; slavomir.cernansky@uniba.sk

* Correspondence: janstubna@gmail.com

Abstract: Andradite, variety demantoid, is a rare gem mineral. We describe gem-quality garnet crystals from serpentinized harzburgites from Dobšiná, Slovakia which were faceted. Both the andradite samples were transparent, with a vitreous luster and a vivid green color. They were isotropic with refractive indices >1.81. The measured density ranged from 3.82 to 3.84 $g \cdot cm^{-3}$. Andradite var. demantoid appeared red under Chelsea filter observation. Both samples contained fibrous crystalline inclusions with the typical "horsetail" arrangement. The studied garnet had a strong Fe^{3+} dominant andradite composition with 1.72–1.85 apfu Fe^{3+}, Cr^{3+} up to 0.15 apfu, Al^{3+} 0.03 to 0.04 apfu, V^{3+} up to 0.006 apfu substituted for Fe^{3+}, Mn^{2+} up to 0.002 apfu, and Mg up to 0.04 apfu substituted for Ca. Raman spectrum of garnet showed three spectral regions containing relatively strong bands: I—352–371 cm^{-1}, II—816–874 cm^{-1}, and III—493–516 cm^{-1}. The optical absorption spectrum as characterized by an intense band at 438 nm and two broad bands at 587 and 623 nm and last one at 861 nm, which were assigned to Fe^{3+} and Cr^{3+}. Transmission was observed in the ultraviolet spectral region (<390 nm), near the infrared region (700–800 nm), and around 530 nm in the green region of visible light, resulting in the garnet's green color.

Keywords: andradite; demantoid; gemstone; Raman spectroscopy; UV-Vis-NIR spectroscopy; X-ray fluorescence spectroscopy; gem-quality; garnet

---

1. Introduction

Slovakia has only a few occurrences which produce gem-quality minerals. For a long time in history, precious opal from Červenica-Dubník was the only mineral considered as a gemstone [1–7]. Only a few Slovak minerals, including quartz var. smoky quartz [8,9], sphalerite [8,10], pyrite [8], garnet var. almandine [8], hematite [8], and corundum var. sapphire [11–13] are suitable for faceting.

Andradite garnet, $Ca_3Fe_2Si_3O_{12}$, is a rock-forming mineral typically found in metamorphic rocks such as skarns and rodingites. The andradite crystal structure consists of alternating SiO_4 tetrahedra and FeO_6 octahedra, and Ca with eight-fold coordination. Several studies have been carried ou on the crystal chemistry and the thermodynamic properties of andradite garnets [14–22]. Green to yellow-green andradite garnets, demantoid variety, represent one of the most appreciated and precious gemstones among the garnet-group minerals, due to their color, brilliance, and rarity [23–25 Gem-quality andradite-garnet sources occur in Russia [26–29], Italy [30], Iran [31–33], Pakistan [25,34–37 Namibia [38–41], Madagascar [42–47], Afganistan [24], Switzerland [24], Canada [48], Mexico [49], and China [50].

Our goal is a mineralogical and crystal-chemical investigation of andradite var. demantoid from a serpentinite body in Dobšiná, Slovakia using various analytical methods.

Geological Setting

The serpentinized harzburgites from Dobšiná are part of the Jurassic mélange complex (Meliata Unit). This inner Western-Carpathian unit forms small nappes over the Central Western Carpathians [51]. The Meliata Ocean formed after the continental break up in the Anisian [52] probably as a back-arc basin, due to northward subduction of the Paleotethys oceanic crust beneath the Eurasian continent [53]. The back-arc basin (BAB) basalts and mid-ocean ridge basalts (MORB were dominant in the Meliatic oceanic rift [54,55]. The association of eupelagic and deepwate pelagic sediments, basalts, dolerite dikes, gabbros, and serpentinized mantle peridotites found in the Meliata Unit clearly indicates an oceanic suite [56,57]. The geochemistry of the basalts reveals a normal mid-ocean ridge basalts (N-MORB) [(La/Sm) n = 0.6–0.9] composition [54,55]. Traces of obducted oceanic crust (blocks of basalts alternating with radiolarite layers) were also identified from deepwater sedimentary breccias with a radiolaritic matrix [58]. The Neotethyan Meliata Oceanic back-arc basin closed in the Late Jurassic, according to $^{40}Ar–^{39}Ar$ ages of white mica from the blueschists and U/Pb LA-ICP-MS age of metamorphic-metasomatic perovskite [59,60]. The Meliatic subduction–accretionary wedge [61] was incorporated into the Cretaceous orogenic wedge of the Central Western Carpathians [62] between 130 and 50 Ma dated by $^{40}Ar–^{39}Ar$ method on white micas [63] in the form of mostly small nappe-sheeted bodies, thrust over the Gemeric and Veporic units of the Central Western Carpathians [64,65]. The zircon (U-Th)/He data recorded three evolutionary stages: (i) cooling through the ~180 °C isotherm at 130–120 Ma related to the startling collapse of the accretionary wedge, following exhumation of the high-pressure slices in the Meliatic accretionary wedge; (ii) postponed exhumation and cooling of some fragments through the ~180 °C isotherm from 115 to 95 Ma due to ongoing collapse of this wedge; and (iii) cooling from 80 to 65 Ma, postdating the thrusting (~100–80 Ma) during the Late Cretaceous compression related to formation of the Centra Western Carpathians orogenic wedge [66].

The Dobšiná quarry (48°49.622′ N, 20°21.977′ E) (Figure 1) is in a tectonic fragment of the Meliata tectonic unit overlying the Gemeric tectonic unit. The locality is famous for the occurrence of yellow-green Cr-rich andradite (demantoid) in serpentinite veins [67–69]. Andradite-bearing serpentinites are part of the Late Jurassic (to Early Cretaceous) serpentinite–blueschist mélange complex. The mélange complex contains talc–phengite–glaucophane schists, marbles, metaconglomerates, blueschists of magmatic and sedimentary origin, and serpentinites. All rocks form decimeter- to 100 m-sized fragments within the Late Jurassic very low-temperature greenschist-facies metasediments [70].

Partly-preserved magmatic structures of harzburgite consist of olivine (~70 vol. %), orthopyroxene (~25 vol. %), accessory spinel (~4 vol. %), and rare clinopyroxene (~1 vol. %). Magmatically corroded and partly dissolved porphyric orthopyroxene (±clinopyroxene) is enclosed in olivine matrix with spinel. Both pyroxenes show exsolution lamellae, of clinopyroxene in orthopyroxene or orthopyroxene in clinopyroxene, respectively. Orthopyroxene often forms sigmoidally rotated porphyroclasts, including clinopyroxene exsolution lamellae. The serpentinized harzburgite contains serpentinite group minerals (antigorite replaced by chrysotile and lizardite), Cr-Fe spinel, perovskite pyrophanite, Ti andradite, magnetite, and hematite, and rarely carbonates with talc, quartz, and characteristic relics of olivine, orthopyroxene, clinopyroxene, and spinel [70,71].

Figure 1. Locations of andradite var. demantoid source (small point—capital town of region) (CZ—Czech Republic, AT—Austria, HU—Hungary, UA—Ukraine, PL—Poland).

. Materials and Methods

.1. *Sample Description*

Two pieces of transparent andradite var. demantoid were separated from the host rock and aceted into round brilliant cuts with dimensions of 2.80 × 2.80 × 1.64 mm (0.11 carat) and 1.71 × 1.74 × 1.14 mm (0.025 carat) dimensions (Figure 2).

Figure 2. Andradite var. demantoid from Dobšiná, Slovakia (0.11 and 0.025 carat). Photo by J. Štubňa.

.2. *Analytical Methods*

The faceted samples were examined by standard gemmological methods at the Gemmological nstitute, Faculty of Natural Sciences, Constantine the Philosopher University in Nitra, to determine heir optical properties (optical character and refractive indices), hydrostatic specific gravity, ultraviolet luorescence, and microscopic features. The refractive indices were measured with a refractometer Kruss and Presidium Refractive Index Meter II. A Kern ABT 120-5DM with extension KERN ABT A01 hydrostatic balance was used to determine the density of the gemstones. The ultraviolet fluorescence was investigated with a short (254 nm) and long (366 nm) wavelength ultraviolet lamp. We used a Chelsea filter.

The chemical composition of andradite garnet was measured using a Cameca SX100 electron microprobe operated in wavelength-dispersion mode at the Masaryk University, Brno, Czech Republic, under the following conditions: accelerating voltage 15 kV, beam current 20 nA, and beam diameter 5 μm. The samples were analyzed with the following standards: fluorapatite (PKα), sanidine (SiKα, AlKα, KKα), titanite (TiKα), chromite (CrKα), vanadinite (VKα), YAG (YLα), almandine (FeKα), spessartine (MnKα), forsterite (Mg Kα), gahnite (ZnKα), wollastonite (CaKα), albite (NaKα) and topaz (FKα). Lower detection limits of the measured elements varied between 200–400 ppm except for Na (530–570 ppm), Fe (670–730 ppm), Zn (880–900 ppm), and Y (470–500 ppm). The measured content of Y, Mn, Zn, Na, and K was always below the detection limit. Analytical times were 15 to 40 s during measurement depending on the expected concentration of the element in the mineral phase. Major elements were measured using shorter times, whereas longer times were applied for elements with a low concentration. The andradite chemical formula was calculated based on 12 anions assuming all Fe as ferric.

Instrumentation for laser ablation inductively coupled plasma mass spectrometry (LA-ICP-MS) at the Department of Chemistry, Masaryk University, Brno, consists of a UP 213 laser ablation system (New Wave Research, Inc., Fremont, CA, USA) and Agilent 7500 CE ICP-MS spectrometer (Agilent Technologies, Santa Clara, CA, USA). A commercial Q-switched Nd-YAG laser ablation device worked at the 5th harmonic frequency corresponding to 213 nm wavelength. The ablation device was equipped with a programmable XYZ-stage to move the sample along a programmed trajectory during ablation. Target visual inspection and photographic documentation were achieved by the built-in microscope/CCD-camera system. A sample was enclosed in the SuperCell and was ablated by the laser beam, which was focused onto the sample surface through a quartz window. The ablation cell was flushed with a helium carrier gas which transported the laser-induced aerosol to the inductively coupled plasma (1 L·min^{-1}). Sample argon gas flow was admixed with the helium carrier gas flow after the laser ablation cell to 1.6 L·min^{-1} total gas flow. NIST SRM 610 silicate glass reference material was used to optimize the gas flow rates, the sampling depth, and MS electrostatic lens voltage in LA-ICP-MS conditions. This provided maximum signal-to-noise ratio and minimum oxide formation (ThO^+/Th^+ count ratio 0.2%, U^+/Th^+ counts ratio 1.1%). Laser ablation required a 100 μm laser spot diameter, 8 J·cm^{-2} laser fluence, and a 20 Hz repetition rate. The fixed sample position during laser ablation enabled 60 s hole-drilling duration for each spot. All element measurements were normalized on ^{28}Si in the investigated andradite. Analyses were gained from 10 spots in a line across the crystal.

Raman analyses were performed on the LabRAM-HR Evolution (Horiba Jobin-Yvon, Kyoto, Japan) spectrometer system with a Peltier cooled CCD detector and Olympus BX-41 microscope (Department of Geological Sciences, Masaryk University). Raman spectra excited by blue diode laser (473 nm) in the range of 50–4000 cm^{-1} were collected from each stone using 50× objectives. The acquisition time of 10 s per frame and 2 accumulations were used to improve the signal-to-noise ratio.

Optical absorption spectra of cut stones in the region (400–750 nm) were measured with the GL Gem SpectrometerTM at room temperature in the Gemmological Institute of Constantine the Philosopher University, Nitra. Both Raman and UV/Vis/NIR spectra were processed in Seasolve PeakFit 4.1.12 software. Raman and absorption bands were fitted by Lorentz function with the automatic background correction and Savitsky-Golay smoothing.

3. Results

Both the andradite samples were transparent, with a vitreous luster and a yellow-green color which defines the andradite variety. They were all isotropic with refractive indices >1.81. The measured density ranged from 3.82 to 3.84 g·cm^{-3}, and the gemstones were inert in shortwave and longwave ultraviolet radiation. Andradite var. demantoid appeared red under Chelsea filter observation, which is typical for Cr-bearing gemstones. Both samples contained fibrous crystalline inclusions with the typical "horsetail" arrangement (Figure 3).

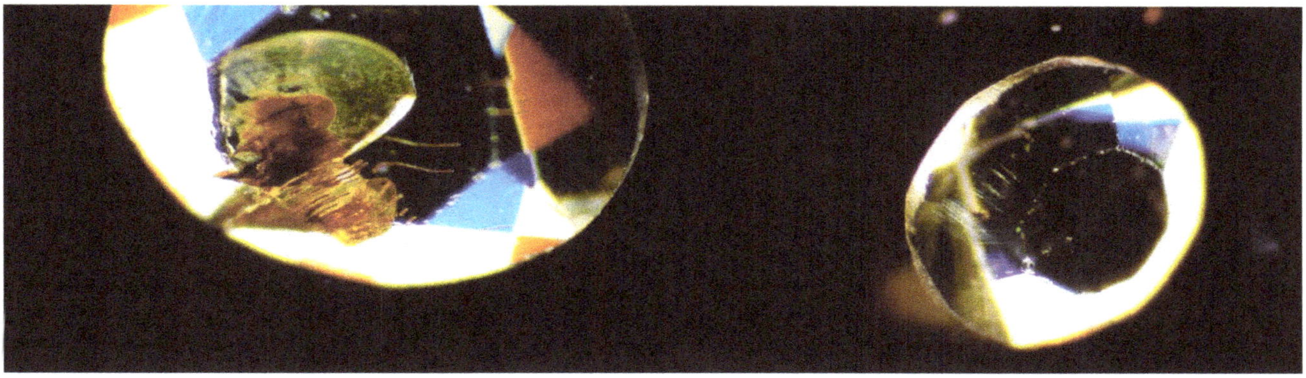

Figure 3. Fibrous crystalline inclusions of serpentine group minerals in andradite var. demantoid from Dobšiná, Slovakia. Photomicrograph by J. Štubňa; magnified 50×.

The studied garnet had a strongly Fe^{3+} dominant andradite composition with 1.72–1.85 apfu Fe^{3+} (Table 1 and Figure 4). Only Cr^{3+} (up to 0.15 apfu), Al^{3+} (0.03 to 0.04 apfu), and Mg (up to 0.04 apfu) substituted for Fe^{3+}. From the trace elements, Cr was the most abundant, along with moderate to minor concentrations (in relative decreasing order of abundance) of V, Ti, and Mn. The other trace elements measured (Sc, Ge, As, Y, and Gd) were near or below their respective detection limit (Table 2). There was no specific chemical zoning of crystal observed in major or trace elements.

Table 1. Electron microprobe analyses of demantoids, first provided in wt % of elements and then recalculated on the basis of 12 anions.

	1	2	3	4	5	6	7	8
P_2O_5	0.02	0.02	0.02	0.02	0.04	0.00	0.02	0.02
SiO_2	35.21	35.41	35.38	35.33	35.47	35.43	35.74	35.72
TiO_2	0.05	0.05	0.04	0.03	0.00	0.03	0.08	0.04
Al_2O_3	0.33	0.34	0.31	0.32	0.40	0.34	0.37	0.32
V_2O_3	0.09	0.07	0.07	0.08	0.07	0.09	0.06	0.05
Cr_2O_3	1.24	1.50	2.31	2.14	1.54	1.34	0.97	0.63
Fe_2O_3	29.15	29.15	27.69	28.34	28.64	28.82	29.16	29.65
MgO	0.20	0.23	0.20	0.17	0.34	0.23	0.20	0.21
CaO	33.88	33.80	33.74	33.72	33.85	33.85	33.74	33.65
F	0.08	0.11	0.09	0.11	0.09	0.08	0.09	0.10
O=F	−0.04	−0.05	−0.04	−0.06	−0.04	−0.04	−0.05	−0.05
Total	100.19	100.63	99.80	100.20	100.40	100.16	100.38	100.34
P^{5+}	0.002	0.001	0.001	0.001	0.003	0.000	0.001	0.001
Si^{4+}	2.975	2.977	2.993	2.981	2.984	2.990	3.003	3.004
F^-	0.021	0.028	0.023	0.030	0.023	0.022	0.024	0.026
Σ	2.997	3.006	3.018	3.012	3.010	3.012	3.029	3.031
Ti^{4+}	0.003	0.003	0.003	0.002	0.000	0.002	0.005	0.003
Al^{3+}	0.033	0.034	0.031	0.032	0.040	0.034	0.037	0.032
V^{3+}	0.006	0.005	0.005	0.006	0.005	0.006	0.004	0.003
Cr^{3+}	0.083	0.100	0.155	0.143	0.102	0.089	0.064	0.042
Fe^{3+}	1.845	1.837	1.757	1.793	1.806	1.823	1.840	1.873
Σ	1.969	1.979	1.950	1.975	1.953	1.954	1.950	1.952
Mg^{2+}	0.025	0.029	0.026	0.022	0.043	0.028	0.025	0.026
Ca^{2+}	3.067	3.044	3.059	3.049	3.052	3.060	3.037	3.032
Σ	3.092	3.073	3.084	3.071	3.095	3.089	3.063	3.058

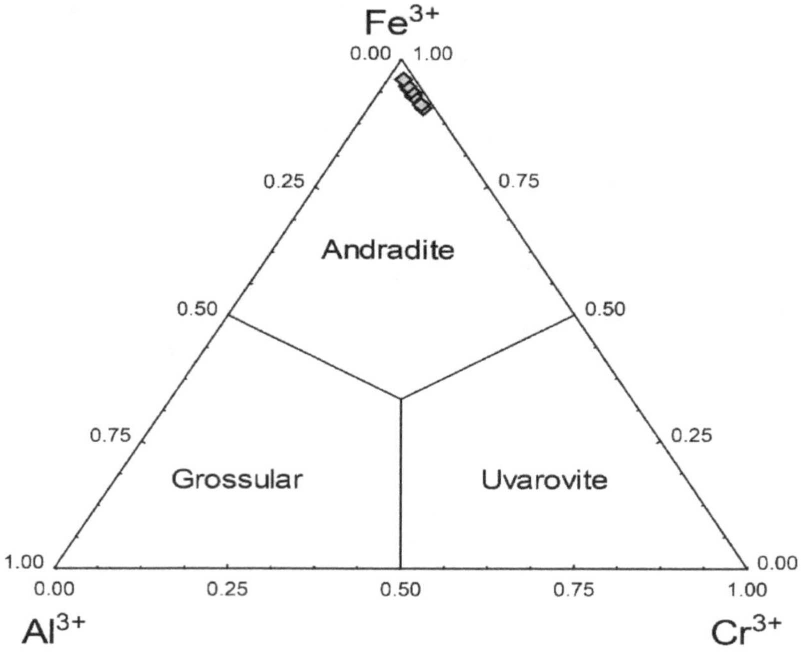

Figure 4. Ternary classification diagram of garnets.

Table 2. The trace-elements contents (in ppm) in demantoid from Dobšiná measured by laser ablation inductively coupled plasma mass spectrometry (LA-ICP-MS). (LOD—limit of detection).

	K	Sc	Ti	V^{51}	Cr^{53}	Mn	Ge^{72}	As^{75}	Y^{89}	Mo^{95}	Gd^{157}
Min	392	16	103	305	3814	103	4	11	2	0	0
Max	787	40	389	572	13331	224	13	22	5	1	2
Average	523	25	244	444	9199	143	9	16	3	1	1
LOD	48.7	2.1	11.4	0.7	9.2	7.8	3.8	9.1	0.1	0.1	0.1

From the calculated crystal-chemical formulae, it is obvious that *Z* site occupancy by Si is slightly lower than three. It could be filled to three by Al or Fe^{3+}, but their content is sufficient only to accommodate the *Y* site. Even after adding all possible octahedral cations, the *Y*-site sum is always slightly below two which, however, could result from the analytical error. However, the small deficiency at the *Z* site can be justified by the presence of F. The *X* site is almost uniformly occupied by Ca with a very small admixture of Mg.

Raman spectrum of garnet (Figure 5 and Table 3) showed three spectral regions containing relatively strong bands: I—352–371 cm^{-1}, II—816–874 cm^{-1}, III—493–516 cm^{-1}. The other bands exhibited relatively lower intensities. Measured bands were assigned to various molecular motions (Table 3): translation of SiO_4—T(SiO_4), rotation of SiO_4—R(SiO_4), translation of YO_6 (Y = Fe^{3+} Cr^{3+})—T(M), symmetric band ν_2, asymmetric stretching frequency ν_3, asymmetric bending frequency ν_4 of SiO_4 group. Our spectrum is consistent with published andradite spectra [72], except the 343 cm^{-1} band, which is not present in published data. Only bands with similar Raman shift are in almandine and pyrope spectra associated with rotation of SiO_4 [72].

The optical absorption spectrum (Figure 6 and Table 4) is characterized by small absorption features below 370 nm, an intense band at 438 nm with shoulders at 470 and 488 nm, and two broad bands at 587 and 623 nm, and 861 nm with a shoulder at 964 nm. These spectral features were al assigned to Fe^{3+} and Cr^{3+}. Transmission was observed in the near the infrared region (700–800 nm) and around 530 nm in the green region of visible light resulting in the garnet's green color.

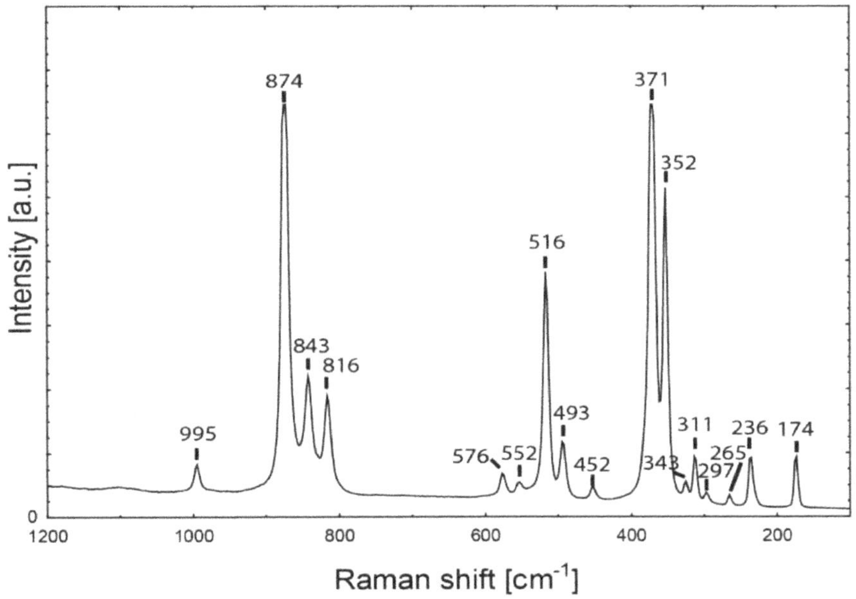

Figure 5. Raman spectrum of the andradite var. demantoid from Dobšiná, Slovakia.

Table 3. Raman shifts (cm^{-1}) of Raman-active bands in the studied sample. Assignment of band modes was made according to Reference [72].

Symmetry	Assignment	Demantoid Dobšiná	Andradite [72]
$T_{2g} + T_{1u}$	T(M)	174	173
$T_{2g} + T_{1u}$	T(M)	236	235
T_{2g}	$T(SiO_4)_{mix}$	265	264
E_g	$T(SiO_4)$	297	296
$T_{2g} + T_{1u}$	$T(SiO_4)$	311	311
?	?	343	-
E_g	ν_2	352	352
A_{1g}	$R(SiO_4)$	371	370
$T_{2g} + T_{1u}$	ν_2	452	452
E_g	ν_2	493	494
A_{1g}	ν_2	516	516
$T_{2g} + T_{1u}$	ν_4	552	553
E_g	ν_4	576	576
$T_{2g} + T_{1u}$	ν_3	816	816
$T_{2g} + T_{1u}$	ν_3	843	842
E_g	ν_3	874	874
$T_{2g} + T_{1u}$	ν_3	995	995

Table 4. The optical spectral bands (nm) in the studied sample. Assignment of bands was made according to Reference [22] and references therein.

Chromophores	Demantoid Dobšiná
Fe^{3+}	438
Fe^{3+}	470
Cr^{3+}	488
Fe^{3+}	587
Cr^{3+}	623
Fe^{3+}	861
Fe^{3+}	964

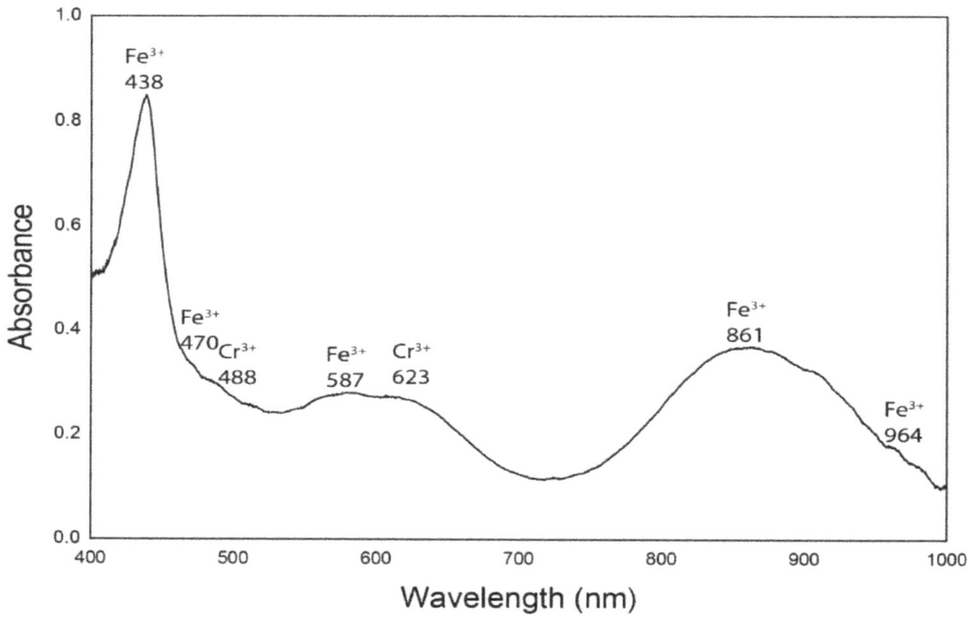

Figure 6. Optical absorption spectrum of the andradite var. demantoid from Dobšiná, Slovakia.

4. Discussion and Conclusions

Chromium-bearing garnets are relatively rare and are associated with specific geologica environments and geochemical conditions. In the Western Carpathians, only a few occurrence have been described, including Pezinok–Rybníček [73], Čierna Lehota [74], and Dobšiná [67].

The garnet from Dobšiná occurs as the three following types [67]. The first type is present a relatively evenly dispersed minute grains in the rock. It originated preferentially at the expens of pseudomorphs after rhombic pyroxene. The garnet also forms lenticular or elongated layer 1 to 2 mm thick and several centimeters long. The second-type of garnet occurs as infilling of chrysotil veinlets. The veinlets' thickness is up to 3 mm. The third type of garnet forms monomineral fills i 0.3 to 3 mm-thick fissures. The garnet crystals display usually euhedral habit. However, only th third type of garnets are emerald-green color. All three garnet types occur in pale-colored, pale-gree varieties of serpentinite.

Andradite color differences are caused by the variable Cr content. The studied garnet belongs t the third type which is generally enriched in Cr compared to the other two types. In some crystals o andradite, the determined value exceeded 6 wt % Cr_2O_3 [67], but the studied crystals only contain up t 2.3 wt % Cr_2O_3. However, this content is still significantly higher than in other demantoid occurrence around the world [21,30,47]. Moreover, demantoid from Dobšiná is enriched in V (up to 0.09 wt % V_2O_3 from EMPA, 305–572 ppm from LA-ICP-MS) compared to other localities with V usually belo the detection limit of EMPA and below 270 ppm determined by LA-ICP-MS [21,30,47]. The As conten between 10–22 ppm (although very close to the detection limit, were below detection limit on othe localities [21]) and absence of REE (except very low Y), which are usually between 1–90 ppm [21] coul also be a fingerprint feature for Dobšiná demantoid.

Raman mode analysis predicts the existence of 25 active Raman modes in all garnet species $3A_{1g}$ $8E_g$, and $14T_{2g}$. The vibrational spectra of garnet clearly possess two different groups of vibrationa modes. The first group extends from 816 to 995 cm^{-1} in the Raman spectrum and possesses three stron bands and one weak one. In the garnet structure, these modes are commonly assigned to the ν_1 Si–O stretching motions within SiO_4 groups [72,75–78]. The second region extends from 174 to 576 cm^{-1} and possesses three strong and ten weak bands. The bands at 352, 452, 493, 516, 552, and 576 cm^{-1} are commonly assigned to the Si–O bending motions within SiO_4 groups [72]. The bands at 343 an

371 cm^{-1} can be assigned to rotation of the $[SiO_4]^{4-}$ tetrahedron $R[SiO_4]^{4-}$ [72]. The bands at 174, 236, 265, 297, and 311 cm^{-1} can be assigned to translation modes of the tetrahedra and octahedra [72].

The optical absorption spectra of studied garnet revealed a few absorption and transmission regions. The absorption feature at 438 nm and its shoulders are attributed to octahedral Fe^{3+}, although a further contribution from Cr^{3+} cannot be excluded [22,50,75,79–82], but this was not very significant due to the low Cr content. The weak absorption band at 488 nm is related to Fe^{3+} [30,47]. Two bands at 587 and 623 nm belong to octahedral Fe^{3+} and Cr^{3+}, respectively [22,50,75,79–82]. Presumed contribution of Cr^{3+} to the band at 438 nm can be assigned, according to the Tanabe–Sugano diagram [83–85], to the dominant $^4A_{2g} \rightarrow {}^4T_{2g}$ transition in Cr^{3+}. The second $^4A_{2g} \rightarrow {}^4T_{1g}$ transition in Cr^{3+} was at 623 nm but was overlapping with the Fe^{3+} absorption band at 587 nm. Moreover, a broad band at about 861 nm was due to octahedral Fe^{3+} [22,50,75,79–82].

Author Contributions: Concept of study, J.Š., R.H., J.F., and P.B.; methodology, J.Š., R.H., J.F., P.B, S.M., R.Š., T.V.; formal analysis, S.M., R.Š., T.V., and P.B.; formal investigation, J.Š., L'.I., J.F., and P.B.; compilation and writing of the manuscript, J.Š., J.F., and P.B.; review and editing of manuscript, J.Š., L'.I., J.F., S.Č., and P.B.

References

1. Kaličiak, M.; Ďud'a, R.; Burda, P.; Kaličiaková, E. Structural geological characteristics of the Dubník opal deposits. *Zborník Vychodoslovenského Múzea v Košiciach* **1976**, *17*, 7–22.
2. Rondeau, B.; Fritsch, E.; Guiraud, M.; Renac, C. Opals from Slovakia ("Hungarian" opals): A re-assessment of the conditions of formation. *Eur. J. Mineral.* **2004**, *16*, 789–799. [CrossRef]
3. Gaillou, E.; Fritsch, E.; Aguilar-Reyes, B.; Rondeau, B.; Post, J.; Barreau, A.; Ostroumov, M. Common gem opal: An investigation of micro- to nano-structure. *Am. Mineral.* **2008**, *93*, 1865–1873. [CrossRef]
4. Gaillou, E.; Delaunay, A.; Rondeau, B.; Bouhnik-le-Coz, M.; Fritsch, E.; Cornen, G.; Monnier, C. The geochemistry of gem opals as evidence of their origin. *Ore Geol. Rev.* **2008**, *34*, 113–126. [CrossRef]
5. Kiefert, L.; Karampelas, S. Use of the Raman spectrometer in gemmological laboratories: Review. *Spectrochim. Acta-Part A Mol. Biomol. Spectrosc.* **2011**, *80*, 119–124. [CrossRef] [PubMed]
6. Caucia, F.; Marinoni, L.; Leone, A.; Adamo, I. Investigation on the gemological, physical and compositional properties of some opals from Slovakia ("Hungarian" opals). *Periodico di Mineralogia* **2013**, *82*, 251–261.
7. Caucia, F.; Ghisoli, C.; Marinoni, L.; Bordoni, V. Opal, a beautiful gem between myth and reality. *Neues Jahrbuch fur Mineralogie Abhandlungen* **2013**, *190*, 1–9. [CrossRef]
8. Ďud'a, R. Gemmological and genetic classification of precious and trim stones of Slovakia. *Mineral. Slovaca* **1987**, *19*, 353–362. (In Slovakian)
9. Fridrichová, J.; Bačík, P.; Illášová, L'.; Kozáková, P.; Škoda, R.; Pulišová, Z.; Fiala, A. Raman and optical spectroscopic investigation of gem-quality smoky quartz crystals. *Vib. Spectrosc.* **2016**, *85*, 71–78. [CrossRef]
10. Fridrichová, J.; Bačík, P.; Malíčková, I.; Illášová, L'.; Pulišová, Z. Optical-spectroscopic study of gem sphalerite from Banská Štiavnica and Hodruša-Hámre. *Esemestník* **2017**, *6*, 13–15. (In Slovakian)
11. Uher, P.; Sabol, M.; Konečný, P.; Gregáňová, M.; Táborský, Z.; Puškelová, L'. Sapphire from Hajnáčka Cérová Highland, southern Slovakia. *Slovak Geol. Mag.* **1999**, *5*, 273–280.
12. Uher, P.; Sabol, M.; Konečný, P.; Gregáňová, M.; Táborský, Z.; Puškelová, L'. Zafír vo výplni vrchnopliocénneho maaru pri Hajnáčke. *Miner. Slovaca* **2001**, *33*, 307–308.
13. Uher, P.; Giuliani, G.; Szakáll, S.; Fallick, A.; Strunga, V.; Vaculovič, T.; Ozdín, D.; Gregáňová, M. Sapphires related to alkali basalts from the Cerová Highlands, Western Carpathians (southern Slovakia): Composition and origin. *Geol. Carpath.* **2012**, *63*, 71–82. [CrossRef]
14. Quareni, S.; De Pieri, R. La struttura dell'andradite. Atti. e Memorie dell'. *Accademia Patavina di Science Lettere ed Arti* **1966**, *78*, 151–170.
15. Novak, G.A.; Gibbs, G.V. The crystal chemistry of the silicate garnets. *Am. Mineral.* **1971**, *56*, 791–825.
16. Hazen, R.M.; Finger, L.W. High-pressure crystal chemistry of andradite and pyrope: Revised procedure for high-pressure diffraction experiments. *Am. Mineral.* **1989**, *74*, 352–359.
17. Lager, G.A.; Armbruster, T.; Rotella, F.J.; Rossman, G.R. OH substitution in garnets: X-ray and neutron diffraction, infrared, and geometric-modeling studies. *Am. Mineral.* **1989**, *74*, 840–851.

18. Armbruster, T.; Geiger, C.A. Andradite crystal chemistry, dynamic X-site disorder and structural strain in silicate garnets. *Eur. J. Mineral.* **1993**, *5*, 59–71. [CrossRef]
19. Armbruster, T.; Birrer, J.; Libowitzky, E.; Beran, A. Crystal chemistry of Ti-bearing andradites. *Eur. J. Mineral.* **1998**, *10*, 907–921. [CrossRef]
20. Pavese, A.; Diella, V.; Pischedda, V.; Merli, M.; Bocchio, R.; Mezouar, M. Pressure-temperature equation of state of andradite and grossular, by high-pressure and -temperature powder diffraction. *Phys. Chem. Miner.* **2001**, *28*, 242–248. [CrossRef]
21. Bocchio, R.; Adamo, I.; Diella, V. Trace-element profiles in gem-quality green andradite from classic localities. *Can. Mineral.* **2010**, *48*, 1205–1216. [CrossRef]
22. Adamo, I.; Gatta, G.D.; Rotiroti, N.; Diella, V.; Pavese, A. Green andradite stones: Gemmological and mineralogical characterisation. *Eur. J. Mineral.* **2011**, *23*, 91–100. [CrossRef]
23. Balčiūnaitė, I.; Kleišmantas, A.; Norkus, E. Chemical composition of rare garnets, their colours and gemmological characteristics. *Chemija* **2015**, *26*, 18–24.
24. O'Donoghue, M. *Gems*, 6th ed.; Butterworth-Heinemann: Oxford, UK, 2006; p. 936.
25. Adamo, I.; Bocchio, R.; Diella, V.; Caucia, F.; Schmetzer, K. Demantoid from Balochistan, Pakistan: Gemmological and Mineralogical Characterization. *J. Gemmol.* **2015**, *34*, 428–433. [CrossRef]
26. Phillips, W.R.; Talantsev, A.S. Russian demantoid, czar of the garnet family. *Gems Gemol.* **1996**, *32*, 100–111. [CrossRef]
27. Krzemnicki, M.S. Diopside needles as inclusions in demantoid garnet from Russia: A Raman microspectrometric study. *Gems Gemol.* **1999**, *35*, 192–195. [CrossRef]
28. Laurs, B.M. A new find of demantoid at a historic side in Kladovka, Russia. *Gems Gemol.* **2003**, *39*, 54–55.
29. Mayerson, W.M. Lab Notes: Large demantoid of exceptional color. *Gems Gemol.* **2006**, *42*, 261–262.
30. Adamo, I.; Bocchio, R.; Diella, V.; Pavese, A.; Vignola, P.; Prosperi, L.; Palanza, V. Demantoid from Val Malenco, Italy: Review and update. *Gems Gemol.* **2009**, *45*, 280–287. [CrossRef]
31. Douman, M.; Dirlam, D. Update on demantoid and cat´s-eye from Iran. *Gems Gemol.* **2004**, *40*, 67–68.
32. Du Toi, G.; Mayerson, W.; van Der Bogert, C.; Douman, M.; Befi, R.; Koivula, J.I.; Kiefert, L. Demantoid from Iran. *Gems Gemol.* **2006**, *42*, 131.
33. Karampelas, S.; Gaillou, E.; Fritsch, E.; Douman, M. Les grenats andradites-démantoïde d'Iran: Zonage de couleur et inclusions. *Rev. Gemmol.* **2007**, *160*, 14–20.
34. Fritz, E.A.; Laurs, B.M. Gem News International: Andradite from Balochistan, Pakistan. *Gems Gemol.* **2007**, *43*, 373.
35. Milisenda, C.C.; Henn, U.; Henn, J. Demantoide aus Pakistan. *Gemmol. Z. Dtsch. Gemmol. Ges.* **2001**, *50*, 51–56.
36. Quinn, E.P.; Laurs, B.M. Demantoid from northern Pakistan. *Gems Gemol.* **2005**, *41*, 176–177.
37. Palke, A.C.; Pardieu, V. Gem News International: Demantoid from Baluchistan Province in Pakistan. *Gems Gemol.* **2014**, *50*, 302–303.
38. Lind, T.; Henn, U.; Bank, H. New occurrence of demantoid in Namibia. *Aust. Gemmol.* **1998**, *20*, 75–79.
39. Cairncross, B.; Bahmann, U. The Erongo Mountains, Namibia. *Mineral. Rec.* **2006**, *37*, 361–470.
40. Koller, F.; Niedermayar, G.; Pintér, Z.; Syabó, C. The demantoid garnets of the Green Dragon Mine (Tubussis, Erongo region, Namibia). *Acta Mineral.-Petrogr.* **2012**, *7*, 72.
41. Kaiser, H.; Enzersdorf, M. Demantoid aus der Green Dragon Mine, Erongo Region, Namibia. *Gemmo News* **2012**, *32*, 5.
42. Danet, F. Gem News International: New discovery of demantoid from Ambanja, Madagascar. *Gems Gemol.* **2009**, *45*, 218–219.
43. Rondeau, B.; Mocquet, B.; Lulzac, Y.; Fritsch, E. Les nouveaux grenats démantoïde d'Ambanja, Province d'Antsiranana, Madagascar. *Le Règne Mineral* **2009**, *90*, 41–45.
44. Rondeau, B.; Fritsch, E.; Mocquet, B.; Lulyac, Y. Ambanja (Madagascar)—New source of gem demantoid garnet. *InColor* **2009**, *11*, 16–20.
45. Praszkier, T.; Gajowniczek, J. Demantoide aus Antetezambato auf Madagascar. *Miner. Welt* **2010**, *21*, 32–41.
46. Pezzotta, F. Andradite from Antetezambato, North Madagascar. *Mineral. Rec.* **2010**, *41*, 209–229.
47. Pezzotta, F.; Adamo, I.; Diella, V. Demantoid and topazolite from Antetezambato, northern Madagascar: Review and new data. *Gems Gemol.* **2011**, *47*, 2–14. [CrossRef]
48. Wilson, B.S. Coloured gemstones from Canada. *Rocks Miner.* **2009**, *85*, 24–43. [CrossRef]
49. Ostrooumov, M. Mexican demantoid from new deposits. *Gems Gemol.* **2015**, *51*, 450–452.

0. Guobin, L.; Xu, K.; Lin, Z. On the Genesis of Demantoid from Xinjiang, China. *Chin. J. Geochem.* **1986**, *5*, 381–390. [CrossRef]
1. Mock, R.; Sýkora, M.; Aubrecht, R.; Ožvoldová, L.; Kronome, B.; Reichwalder, P.; Jablonský, J. Petrology and stratigraphy of the Meliaticum near the Meliata and Jaklovce Villages, Slovakia. *Slovak Geol. Mag.* **1998**, *4*, 223–260.
2. Kozur, H. The Evolution of the Meliata-Hallstatt ocean and its significance for the early evolution of the Eastern Alps and Western Carpathians. *Palaeogeogr. Palaeoclimatol. Palaeoecol.* **1991**, *87*, 109–135. [CrossRef]
3. Stampfli, G.M. The Intra-Alpine terrain: A Paleotethyan remnant in the Alpine Variscides. *Eclogae Geol. Helv.* **1996**, *89*, 13–42.
4. Ivan, P. Relics of the Meliata Ocean crust: Geodynamic implications of mineralogical, petrological and geochemical proxies. *Geol. Carpath.* **2002**, *53*, 245–256.
5. Faryad, S.W.; Spišiak, J.; Horváth, P.; Hovorka, D.; Dianiška, I.; Józsa, S. Petrological and geochemical features of the Meliata mafic rocks from the sutured Triassic Oceanic Basin, Western Carpathians. *Ofioliti* **2005**, *30*, 27–35.
6. Kozur, H.; Mock, R. New paleogeographic and tectonic interpretations in the Slovakian Carpathians and their implications for correlations with the Eastern Alps and other parts of the Western Tethys, Part II: Inner Western Carpathians. *Miner. Slovaca* **1997**, *29*, 164–209.
7. Kozur, H.; Mock, R. Erster Nachweis von Jura der Meliata-Einheit der südlichen Westkarpaten. *Geologisch-Paläontologische Mitteilungen Innsbruck* **1985**, *13*, 223–238.
8. Putiš, M.; Radvanec, M.; Hain, M.; Koller, F.; Koppa, M.; Snárska, B. 3-D analysis of perovskite in serpentinite (Dobšiná quarry) by X-ray micro-tomography. In *Proceedings of the Petros Symposium*; Ondrejka, M., Šarinová, K., Eds.; Univerzita Komenského: Bratislava, Slovakia, 2011; pp. 33–37.
9. Dallmeyer, R.D.; Neubauer, F.; Handler, R.; Fritz, H.; Müller, W.; Pana, D.; Putiš, M. Tectonothermal evolution of the internal Alps and Carpathians: $^{40}Ar/^{39}Ar$ mineral and whole rock data. *Eclogae Geol. Helv.* **1996**, *89*, 203–277.
0. Putiš, M.; Yang, Y.-H.; Koppa, M.; Dyda, M.; Šmál, P. U/Pb LA-ICP-MS age of metamorphic-metasomatic perovskite from serpentinized harzburgite in the Meliata Unit at Dobšiná, Slovakia: Time constraint of fluid-rock interaction in an accretionary wedge. *Acta Geol. Slovaca* **2015**, *7*, 63–71.
1. Faryad, S.W.; Henjes-Kunst, F. Petrological and K-Ar and 40Ar/39Ar age constraints for the tectonothermal evolution of the high-pressure Meliata unit, Western Carpathians (Slovakia). *Tectonophysics* **1997**, *280*, 141–156. [CrossRef]
2. Plašienka, D. Cretaceous tectonochronology of the central western Carpathians, Slovakia. *Geol. Carpath.* **1997**, *48*, 99–111.
3. Putiš, M.; Frank, W.; Plašienka, D.; Siman, P.; Sulák, M.; Biroň, A. Progradation of the Alpidic Central Western Carpathians orogenic wedge related to two subductions: Constrained by $^{40}Ar/^{39}Ar$ ages of white micas. *Geodin. Acta* **2009**, *22*, 31–56. [CrossRef]
4. Dal Piaz, G.; Martin, S.; Villa, I.M.; Gosso, G.; Marschalko, R. Late Jurassic blueschist facies pebbles from the Western Carpathians orogenic wedge and paleostructural implications for Western Tethys evolution. *Tectonics* **1995**, *14*, 874–885. [CrossRef]
5. Putiš, M.; Gawlick, H.J.; Frisch, W.; Sulák, M. Cretaceous transformation from passive to active continental margin in the Western Carpathians as indicated by the sedimentary record in the Infratatric Unit. *Int. J. Earth Sci.* **2008**, *97*, 799–819. [CrossRef]
6. Putiš, M.; Danišík, M.; Ružička, P.; Schmiedt, I. Constraining exhumation pathway in an accretionary wedge by (U-Th)/He thermochronology—Case study on Meliatic nappes in the Western Carpathians. *J. Geodyn.* **2014**, *81*, 80–90. [CrossRef]
7. Fediuková, E.; Hovorka, D.; Greguš, J. Compositional zoning of andradite from serpentinite at Dobšiná (West Carpathias). *Věst. Ústr. Úst. Geol.* **1976**, *51*, 339–345.
8. Hovorka, D.; Jaroš, J.; Kratochvíl, M.; Mock, R. The Mesozoic ophiolites of the Western Carpathians. *Krystalinikum* **1984**, *17*, 143–157.
9. Hovorka, D.; Ivan, P.; Jaroš, J.; Kratochvíl, M.; Reichwalder, P.; Rojkovič, I.; Spišiak, J.; Turanová, L. *Ultramafic rocks of the Western Carpathians, Czechoslovakia*; Geological Institute of Dionýz Štúr Publishers: Bratislava, Czechoslovakia, 1985; p. 258.

70. Putiš, M.; Koppa, M.; Snárska, B.; Koller, F.; Uher, P. The blueschist-associated perovskite-andradite-bearin serpentinized harzburgite from Dobšiná (the Meliata Unit), Slovakia. *J. Geosci.* **2012**, *57*, 221–240. [CrossRe
71. Putiš, M.; Yang, Y.-H.; Vaculovič, T.; Koppa, M.; Li, X.-H.; Uher, P. Perovskite, reaction product of harzburgite with Jurassic-Cretaceous accretionary wedge fluids (Western Carpathians, Slovakia): Evidenc from the whole-rock and mineral trace element data. *Geol. Carpath.* **2016**, *67*, 133–146. [CrossRef]
72. Hofmeister, A.M.; Chopelas, A. Vibrational spectroscopy of end-member silicate garnets. *Phys. Chem. Mine* **1991**, *17*, 503–526. [CrossRef]
73. Uher, P.; Kováčik, M.; Kubiš, M.; Shtukenberg, A.; Ozdín, D. Metamorphic vanadian-chromian silicat mineralization in carbon-rich amphibole schists from the Malé Karpaty Mountains, Western Carpathian Slovakia. *Am. Mineral.* **2008**, *93*, 63–73. [CrossRef]
74. Bačík, P.; Uher, P.; Kozakova, P.; Števko, M.; Ozdín, D.; Vaculovič, T. Vanadian and chromian garnet- an epidote-supergroup minerals in metamorphosed Paleozoic black shales from Cierna Lehota, Strazovsk vrchy Mountains, Slovakia: Crystal chemistry and evolution. *Mineral. Mag.* **2018**, *82*, 889–911. [CrossRef]
75. Peters, T. A water-bearing andradite from the Totalp serpentine (Davos, Switzerland). *Am. Mineral.* **196** *50*, 1482–1486.
76. Passaglia, E.; Rinaldi, R. Katoite, a new member of the $Ca_3Al_2(SiO_4)_3$-$Ca_3Al_2(OH)_{12}$ series and a nev nomenclature for the hydrogrossular group of minerals. *Bulletin de la Société Française de Minéralogie e de Cristallographie* **1984**, *107*, 605–618. [CrossRef]
77. Galuskina, I.O.; Galuskin, E.V. Garnets of the hydrogrossular-"hydroandradite"-"hydroschorlomite" serie *Spec. Pap. Mineral. Soc. Pol.* **2003**, *22*, 54–57.
78. Galuskin, E. *Minerals of the Vesuvianite Group from the Achtarandite Rocks (Wiluy River, Yakutia)*; University c Silesia Publishing House: Katowice, Poland, 2005; p. 192.
79. Moore, R.K.; White, W.B. Electronic spectra of transition metal ions in silicate garnets. *Can. Mineral.* **197** *11*, 791–811.
80. Manning, P.G. Optical absorption spectra of Fe^{3+} in octahedral and tetrahedral sites in natural garnet *Can. Mineral.* **1972**, *11*, 826–839.
81. Amthauer, G. Crystal chemistry and colour of chromium bearing garnets. *Neues Jahrbuch für Mineralog Abhandlungen* **1976**, *126*, 158–186.
82. Stockton, C.M.; Manson, D.V. Gem andradite garnets. *Gems Gemol.* **1983**, *19*, 202–208. [CrossRef]
83. Tanabe, Y.; Sugano, S. On the absorption spectra of complex ions II. *J. Phys. Soc. Jpn.* **1954**, *9*, 766–77 [CrossRef]
84. Tanabe, Y.; Sugano, S. On the absorption spectra of complex ions I. *J. Phys. Soc. Jpn.* **1954**, *5*, 753–76 [CrossRef]
85. Tanabe, Y.; Sugano, S. On the absorption spectra of complex ions III. *J. Phys. Soc. Jpn.* **1956**, *11*, 864–87 [CrossRef]

7

Gemstones of Greece: Geology and Crystallizing Environments

Panagiotis Voudouris [1,*], Constantinos Mavrogonatos [1], Ian Graham [2], Gaston Giuliani [3], Alexandre Tarantola [4], Vasilios Melfos [5], Stefanos Karampelas [6], Athanasios Katerinopoulos [1] and Andreas Magganas [1]

1 Faculty of Geology and Geoenvironment, National and Kapodistrian University of Athens, GR-15784 Athens, Greece
2 PANGEA Research Centre, School of Biological, Earth and Environmental Sciences, University of NSW, 2052 Sydney, Australia
3 CRPG/CNRS, Université Paul Sabatier, GET/IRD et Université de Lorraine, 54501 Vandœuvre cedex, France
4 GeoRessources, Faculté des Sciences et Technologies, Université de Lorraine, CNRS, F-54506 Vandœuvre-lès-Nancy, France
5 Faculty of Geology, Aristotle University of Thessaloniki, 54124 Thessaloniki, Greece
6 Bahrain Institute for Pearls & Gemstones (DANAT), WTC East Tower, P.O. Box 17236 Manama, Bahrain
* Correspondence: voudouris@geol.uoa.gr

Abstract: In the Hellenides Orogen, minerals of various gem quality occur in various rock types from mainly four tectono-metamorphic units, the Rhodope, Pelagonian, and the Attico-Cycladic massifs, and the Phyllites-Quartzites unit of Crete Island. In crystalline rocks, gemstones are related to both regional metamorphic-metasomatic processes (e.g., gem corundums, Mn-andalusite, thulite/clinothulite, spessartine, titanite, jadeite), and to the formation of late alpine-type fissures, such as, for example, quartz, albite, adularia and titanite. The Tertiary (and Mesozoic) magmatic-hydrothermal environments provide gem-quality sapphire, beryl, garnet, vesuvianite, epidote, fluorite, and SiO_2 varieties. The supergene oxidation zone of the Lavrion deposit hosts gem-quality smithsonite and azurite. Coloration in the studied material is either due to various chromophore trace elements present in the crystal structure, or due to inclusions of other mineral phases. Future modern exploration methods combined with gemological investigations (such as treatment and faceting of selected stones), are necessary in order to obtain a better knowledge of the gemstone potential of Greece for its possible exploitation.

Keywords: gemstones; corundum; beryl; jadeitite; garnet; quartz varieties; Greece

---

1. Introduction

By definition, gems are materials used for adornment or decoration that must be relatively rare, hard, and tough enough (shock resistant) to resist "normal" wear and withstand corrosion by skin contact and cosmetics [1]. Gems have been prized for thousands of years for their color, luster, transparency, durability, and high value-to-volume ratio [1,2]. A natural gem is one that has been fashioned (or faceted) after having been found in nature, even if it later undergoes treatment processes. Gem materials cover a large variety of products found in the jewelry market today. Most natural gems are single crystals of natural minerals, although others are amorphous (some varieties of opal, natural glass), some are solid solutions (garnets, peridot, etc.), others are rocks (jade, lapis), and some are composed partly or wholly of organic materials (amber, pearl, coral, etc.) [1].

According to Groat [3,4] and Stern et al. [5], any mineral or stone beautiful enough to be sought, mined, and sold for its beauty alone is a gemstone. Among the most important gemstones are diamond, ruby (red-colored corundum) and sapphire (blue-colored corundum), emerald and other gem forms of the mineral beryl, chrysoberyl, tanzanite (blue zoisite), tsavorite (green grossular garnet), "paraiba-type" tourmaline (copper-colored "neon"-blue tourmaline), topaz and jadeite-"jade" (jadeitite) [4]. More common gem materials include, among others, amber, silica gems, feldspar, tourmaline, spinel, garnets, zircon etc. It is not the mineral itself that makes a gemstone, it is the characteristics of a specific sample [2].

Understanding the geological conditions that give rise to gem deposits is of great importance because it can provide guidelines for exploration [2,3,6]. Specific geological conditions include the availability of uncommon major constituents, the presence of adequate chromophores and open space in order to form crystals of sufficient size and transparency [2], and the absence of dramatic post-growth events (e.g., fracturing).

Traditionally, Greece has not been regarded as a source country for gemstones [7]. Known gem material included rubies from Xanthi [8], sapphire and beryl from Naxos Island [9], red-colored spessartine from Paros Island [10], smithsonite from the oxidation zone of the Lavrion mines [11] and the green quartz variety (prase) from the Serifos skarn [12]. Exploration work in Greece over the last three decades has resulted in the discovery of new occurrences of mineral megacrysts many of them being gem-quality [13–19]. The gemstones of Greece are set within diverse geological settings the study of which will increase our knowledge on the conditions necessary for their crystallization and concentration into economically-viable deposits. The aim of this work is to review all available information on the geology, mineralogy, geochemistry, and fluid characteristics involved in the formation of various gemstones in Greece. The occurrences of gemstones in Greece are presented in Figures 1 and 2 and described with respect to their geological framework and the conditions of their formation.

2. Materials and Methods

Analytical methods included optical and electron microscopy, X-ray powder diffraction studies and electron microprobe analyses, in the Section of Mineralogy-Petrology at University of Athens and at the Institute of Mineralogy and Petrology, University of Hamburg. X-ray powder diffraction measurements were obtained using a SIEMENS D-500 diffractometer with Cu tube and Co filter. For the mineral analyses, a scanning electron microscope JEOL JSM-5600 combined with an energy dispersive X-ray microanalysis system Oxford Link Isis 300 system were used. Thin and thin-and-polished sections of mineral samples and host rocks were studied by a JEOL JSM 5600 scanning electron microscope equipped with back-scattered imaging capabilities, respectively, at the Department of Mineralogy and Petrology at the University of Athens (Greece). Quantitative analyses were carried-out at the Institute of Mineralogy and Petrology, University of Hamburg using a Cameca-SX 100 WDS. Analytical conditions were as follows: accelerating voltage of 20 kV, a beam current of 20 nA and counting time of 20 s for Al, Si, Ca, Fe and Mg, 60 s for Mn and 120 s for Cr. The X-ray lines used were: AlKα, SiKα, TiKα, FeKα, MnKα, MgKα, CrKα, and CaLα. The standards used were: andradite (for Si, Ca and Fe) and synthetic Al_2O_3 (for Al), MnTiO3 (for Mn and Ti), Fe_2O_3 (for Fe), Cr_2O_3 (for Cr), and MgO (for Mg). Corrections were applied using the PAP online program [20].

LA-ICP-MS analyses presented in this study are from Voudouris et al. [18]. The analyses were conducted at the CODES ARC Centre of Excellence in Ore Deposits of the University of Tasmania, Australia, and the Institute of Mineralogy, University of Münster, Germany. For analytical conditions see Voudouris et al. [18].

Stable isotope analyses were performed at the Stable Isotope and Atmospheric Laboratories, Department of Geology, Royal Holloway, University of London (London, UK). The oxygen isotope composition of quartz was obtained using a CO_2 laser fluorination system similar to that described by Mattey [21]. Each mineral, separate or standard, is weighed at 1.7 mg ± 10%. These were loaded into the 16-holes of a nickel sample tray, which was inserted into the reaction chamber and then evacuated. The oxygen was released by a 30W Synrad CO_2 laser in the presence of BrF_5 reagent. The yield of oxygen was measured as a calibrated pressure based on the estimated or known oxygen content of the mineral being analyzed. Low yields result in low $\delta^{18}O$ values for all mineral phases, so accurate yield calculations are essential. Yields of >90% are required for most minerals to give satisfactory $\delta^{18}O$ values. The oxygen gas was measured using a VG Isotech (now GV Instruments) Optima dual inlet isotope ratio mass spectrometer (IRMS). All values are reported relative to Vienna Standard Mean Ocean Water (V-SMOW). The data are calibrated to a quartz standard (Q BLC) with a known $\delta^{18}O$ value of +8.8‰ V-SMOW from previous measurements at the University of Paris-6 (France). This has been further calibrated for the RHUL laser line by comparison with NBS-28 quartz. Each 16-hole tray contained up to 12 sample unknowns and 4 of the Q BLC standard. For each quartz run a small constant daily correction, normally less than 0.3‰, was applied to the data based on the average value for the standard. Overall, the precision of the RHUL system based on standard and sample replicates is better than ±0.1‰.

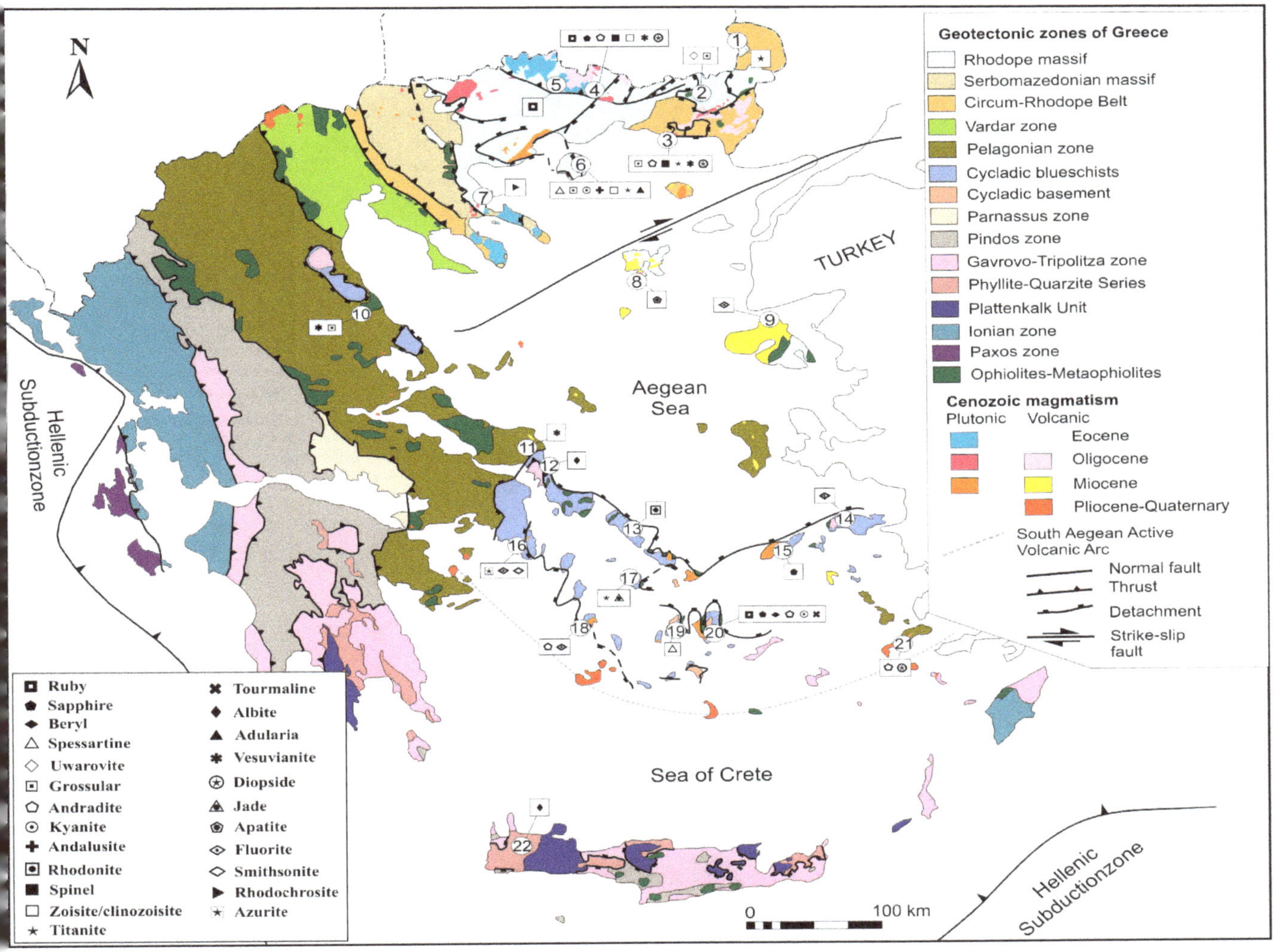

Figure 1. Occurrences of various gemstones in metamorphic and igneous rocks of Greece (Geological map of Greece modified after Ottens and Voudouris [17]): 1. Therapio/Evros, 2. Vyrsini/Evros, 3. Maronia/Rhodopi 4. Gorgona-Kimmeria/Xanthi, 5. Paranesti/Drama, 6. Trikorfo/Thassos, 7. Stratoni-Olympiada/Chalkidiki, 8. Fakos/Limnos Island, 9. Megala Therma/Lesvos Island, 10. Larissa, 11. Kymi/Evia Island, 12. Krieza-Koskina/Evia Island, 13. Petalo/Andros Island, 14. Ampelos/Samos Island, 15. Pounta/Ikaria Island, 16. Lavrion/Attica, 17. Kampos/Syros Island, 18. Agia Marina-Avessalos/Serifos Island, 19, Thapsana/Paros Island, 20. Kinidaros-Kavalaris/Naxos Island, 21. Kardamena/Kos Island, 22. Lakkoi/Crete Island.

3. Geological Setting

The Hellenides constitute part of the Alpine-Himalayan Orogen and formed when Apulia collided with Europe during the Late Cretaceous to Tertiary. They are subdivided into several units: the Rhodope Massif, Servo-Macedonian Massif, Vardar Zone, Pelagonian Zone (Internal Hellenides) and the External Hellenides built-up by Mesozoic and Cenozoic rocks [22–25] (Figures 1 and 2). The Hellenides can be considered an accretionary orogen, resulted from thrusting and SW-verging nappe-stacking of the Rhodopes, Pelagonia and Adria continental blocks, and closure of the Vardar and Pindos oceanic domains of the Neotethys [22]. A Permo-Carboniferous igneous event (known from the Pelagonian Zone, the Rhodope Massif, the Attico-Cycladic Zone, Peloponnese and Crete) documents an active continental margin evolution in the Precambrian-Silurian basement of the Hellenides. Final collision between Europe and Pelagonia at the end of the Cretaceous closed the Neotethys Ocean along the Vardar Suture Zone, as evidenced by obducted Jurassic ophiolites on the Pelagonia continental block [26].

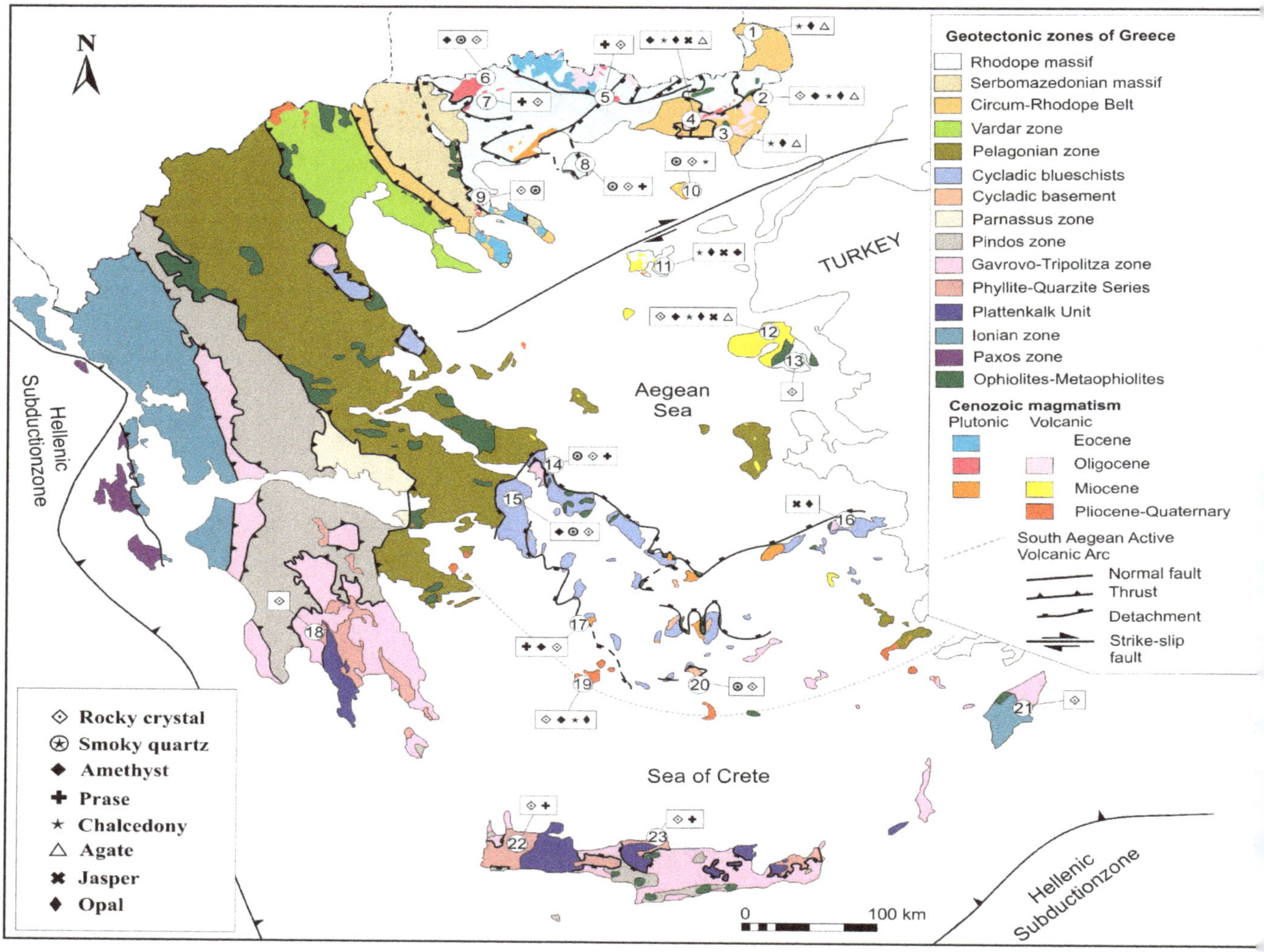

Figure 2. Occurrences of gem-quality silica varieties in metamorphic and igneous rocks of Greece (Geological map of Greece modified after Ottens and Voudouris [17]): 1. Dikea/Evros, 2. Kornofolia/Evros, 3. Aetochori/Evros 4. Sapes/Rhodopi, 5. Kimmeria/Xanthi, 6. Dassoto/Drama, 7. Cresti/Drama, 8. Trikorfo/Thassos, 9. Stratoni-Olympiada/Chalkidiki, 10. Fengari/Samothraki, 11. Moudros-Roussopouli/Limnos Island, 12. Megala Therma/Lesvos Island, 13. Geras/Lesvos, 14. Krieza-Koskina/Evia Island, 15. Penteli Mt/Attika, 16. Ampelos/Samos Island, 17. Avessalos/Serifos Island, 18 Taygetos,. 19. Chondro Vouno-Vani/Milos Island, 20. Mylopotas/Ios Island, 21. Rhodes Island, 22. Prases/Crete Island, 23. Agia Pelagia/Crete Island.

Shortening and syn-orogenic exhumation of HP-LT rocks occurred during the late Cretaceous-Eocene, before an acceleration of slab retreat changed the subduction regime and caused the collapse of the Hellenic mountain belt and thinning of the Aegean Sea from the middle Eocene/late Oligocene to the present [22]. During this post-orogenic episode, large-scale detachments formed, which exhumed metamorphic core complexes in a back-arc setting. Tertiary magmatism in the Aegean region occurred mostly in a post-collisional setting behind the active Hellenic subduction zone [26]. The Pliocene to recent volcanic rocks in the active Aegean volcanic arc formed as a consequence of active subduction beneath the Hellenic trench. In the Hellenides Orogen, gemstones occur in various rock types of mainly three tectono-metamorphic units, the Rhodope- and the Attico-Cycladic massifs and the Phyllites-Quartzites unit of Crete Island.

4. Formation Environments of Gems: An Overview

4.1. *Metamorphic-Metasomatic*

In crystalline rocks, two groups of gemstones are distinguished: those formed during regional metamorphism-metasomatism and those associated with late alpine-type fissures [14–16,18,19,27–29] (Figures 1 and 2). The first group includes ruby, sapphire, jadeite "jade" (jadeitite), Mn-andalusite, spessartine, Mn-grossular, Mn-zoisite/clinozoisite, Fe-Mn-bearing kyanite and rhodonite. Corundums in the Xanthi-Drama areas (Rhodope massif) and Naxos-Ikaria Islands (Attico-Cycladic massif), are hosted in pargasitic schists, marbles and metabauxites. Host rocks are Carboniferous orthogneisses, metapelites and skarns for Mn-andalusite, spessartine and thulite at Thassos Island, Permian - Middle Triassic schists for rhodonite in Andros Island (Attico-Cycladic massif, [30]) and orthogneisses and marbles in Paros Island [10]. Jadeitites, eclogites and blueschists are host rocks for jadeite "jade" and titanite at Syros Island.

Alpine-type fissure minerals [31,32] in Greece are related to tension gashes formed under brittle-ductile to ductile conditions during the retrograde, late-stage exhumation of metamorphic core complexes [14,33,34]. The alpine-type fissures in Greece contain gem-quality quartz and albite together with adularia, chlorite, epidote, actinolite, hematite, muscovite, rutile, tourmaline and pyrite. Host lithologies are ortho- and paragneisses in the Rhodope Massif (Drama, Thassos Island) and Attico-Cycladic Massif (Penteli Mt, Evia, Ios Islands), amphibolites (Evros, Thassos, Evia and Andros Islands) and finally phyllites (Lesvos Island) and metaquartzites (Crete Island). Quartz in both transparent smoky and colorless (rock crystal) varieties were found in all the above localities and/or host rocks—deep violet amethyst occurs in orthogneisses at Drama area [34]. Albite crystals of gem-quality are abundant in metabasites from Crete Island and orthogneisses from Evia Island.

4.2. *Magmatic-Hydrothermal*

The Tertiary (and Mesozoic) magmatic-hydrothermal environments in Greece (granitoids, pegmatites, skarns, non-skarn carbonate-replacement deposits, volcanic and ophiolitic rocks) also provide gem-quality material of beryl, corundum, garnet, vesuvianite, diopside, epidote, titanite, spinel, fluorite, rhodochrosite, quartz varieties, and silica microcrystalline species (Figures 1 and 2). Miarolitic cavities and quartz veins cross-cutting granitoids host gem-quality quartz varieties and apatite. Samothraki and Limnos Islands in the northern Aegean, and Kimmeria/Xanthi, Maronia/Rhodopi in the mainland of northern Greece and Tinos Island in the Cyclades are the most important localities. The Greek pegmatites are mostly deficient in miarolitic cavities, and thus contain only matrix embedded mineral crystals. On Naxos Island, blue sapphires are associated with granite pegmatites intruding ultramafic lithologies (plumasites). Blue beryl crystals (var. aquamarine) are typical constituents of Naxos pegmatites both within the migmatitic domes as well as in pegmatites cross-cutting the surrounding metamorphic rocks. Black tourmaline crystals occur in pegmatites near Nevrokopi-Drama and at Naxos Island.

The skarns of Serifos, Kimmeria/Xanthi and Drama, are characterized by abundant quartz. Epidote crystals are associated with quartz at Lefkopetra, near Kimmeria within metasomatized granodiorite and gneiss-hosted exoskarn bodies. Vesuvianite occurs at Kimmeria and Maronia exoscarns, and garnet is a major constituent of the Greek skarns and represented by several gemmy varieties at Maronia, Kimmeria, Kresti, Kos and Serifos. Titanite and spinel occur in the Maronia endo- and exoskarns respectively. The carbonate-replacement Pb-Zn-Au-Ag deposits of Stratoni, Olympias, Lavrion and Serifos host a large variety of non-metallic minerals of gem-quality of both hypogene and supergene origin. Primary, gem-quality minerals from the Chalkidiki mines include: quartz and rhodochrosite. The Lavrion and Serifos deposits are well-known localities for fluorite. Rodingitized gabbros and dolerites at Evros, Evia, Larissa and Othrys Mesozoic ophiolites (Rhodope massif and Pelagonian zone) host gem-quality hessonite and vesuvianite [17].

Hydrothermal-altered volcanic rocks throughout Greece host several silica varieties most of them in gem-quality. Amethyst and chalcedony occur at Kornofolia/Evros area, in Sapes-Rhodope region and in Lesvos and Milos Islands in epithermal veins accompanying calcite, and/or adularia and barite [35]. Opal occurs in several varieties and colors at Lesvos and Limnos Islands being a constituent of fossilized wood, as well as at various localities at Evros and Milos, Limnos and Lesvos Island and is considered to be part of either silica sinters or steam-heated alteration zones. Fluorite megacrysts occur at Samos and Lesvos Islands, in both cases in the form of monomineralic fluorite veins, cross-cutting epithermally-altered silicified zones and propylitically altered lavas, respectively.

5. Mineralogy of Gems

5.1. Corundum

Greece contains gem corundums mostly within the Rhodope (Xanthi and Drama areas) and Attico-Cycladic (Naxos and Ikaria Islands) tectono-metamorphic units (Figures 1 and 3). In the Xanthi area (Gorgona-Stirigma localities) sapphire deposits are stratiform, occurring within marble layers alternating with eclogitic amphibolites [18,36,37] (Figure 3a). Rubies in the Paranesti-Drama area are restricted to boudinaged lenses of pargasitic schists alternating with amphibolites and gneisses [28,29]. Both occurrences are oriented parallel to the UHP-HP Nestos suture zone. Sapphire from Xanthi marbles is of pink, orange, purple to blue color, usually of tabular or barrel-shaped euhedral form and reaches sizes of up to 4 cm (Figure 3b,c). In some cases, blue corundum alters to spinel. The corundums from Xanthi are transparent with very clear parting and fine cracks. Blue sapphires are zoned, with alternating deep blue and colorless domains. This zoning, or irregular color distribution in the sapphires, is attributed to different Fe and Ti contents in the crystals [18]. The Xanthi sapphires are associated with calcite, dolomite, brown or blue spinel, margarite, and nickeloan tourmaline.

In the Paranesti/Drama area, rubies are associated with kyanite and pargasitic hornblende, and are rimmed by margarite, muscovite, chlorite and chromian spinel. Ruby crystals, ranging in size up to 5 cm and of pale pink to deep red color [28], are mainly flat tabular and less commonly prismatic and barrel-shaped (Figure 3d). The Paranesti rubies are opaque to transparent exhibiting clear parting and lamellar twinning.

On Naxos Island, about 2 km East and Southeast of Kinidaros, blue sapphires up to 3 cm are associated with granite pegmatites intruding ultramafic lithologies (plumasites), occurring either within the pegmatites themselves or the surrounding metasomatic reaction zones [18,37,38] (Figure 3e). Within the plumasites, colorless to blue, purple, and pink corundum may occur either as isolated crystals within the plagioclase matrix, and/or associated with tourmaline and phlogopite (Figure 3f). In the blackwalls that developed at the contacts between the pegmatites and the meta-peridotite country rock purple and pink sapphires are enveloped by phlogopite. The Naxos plumasite sapphire crystals are barrel-shaped, display macroscopic color zoning from a blue core surrounded by a white rim or as a blue-zoned outer rim surrounding a colorless core, or pink cores to purple rims [18]. They are transparent with inner fractures. In the southern part of the Island, close to Kavalaris Hill, a rock ermed "corundite" by Feenstra and Wunder [39], which was formed by the dissociation of former iasporites, in meta-bauxites during prograde regional metamorphism, is composed almost entirely of lue corundum. The corundum from this locality does not occur in well-shaped crystals, it is mostly nassive, and translucent to opaque, appropriate only to be cut as cabochon.

Figure 3. Field and hand specimens photographs demonstrating gem corundum and beryl occurrences/crystals from Greece: (**a**) Corundum-bearing marbles alternating with amphibolites along the Nestos suture zone, Gorgona/Xanthi area; (**b**,**c**) Pink to purple sapphires within Xanthi marbles; (**d**) Ruby within pargasite schist from Paranesti/Drama; (**e**) Ultrabasic bodies rimming the migmatite core of Naxos Island. Plumasite formation took place after metasomatic reaction between ultrabasites and pegmatites, (**f**) Blue and colorless sapphires within desilicated pegmatite (plumasite) from Kinidaros, Naxos Island; (**g**) Beryl-bearing pegmatite penetrating migmatite at Kinidaros, Naxos Island, (**h**,**i**) Blue beryl (var. aquamarine) associated with orthoclase and muscovite within pegmatite from Kinidaros, Naxos Island. Photographs 3**c**,**d**,**i** are courtesy of Berthold Ottens.

On Ikaria Island, blue sapphires occur in the metabauxites of Atheras Mt, hosted within marble which lie on top of gneisses. The corundum megacrysts fill together with margarite, extensiona fissures and networks of veins discordant to the metabauxite foliation [18,37]. The corundums ar deep-blue in color, tabular to well-shaped, reach sizes up to 4 cm and are accompanied by Fe-chlorit hematite, rutile and diaspore. The corundum contains inclusions of ilmenite, hematite, ulvospine rutile and zircon.

The LA-ICP-MS results for averages and ranges for chromophores and genetic indicator element (Fe, Cr, Ti, V, Ga and Mg) are listed in Table 1. Colorless to blue sapphires from Gorgona/Xanthi displa significant variations in their Ti content which reflects their zoned coloration. Bluish areas display hig Ti values up to 6462 ppm, and pink varieties are characterized by much less Ti (up to 810 ppm). Iro content reaches values up to 1339 ppm in both colorless and blue areas and up to 1782 ppm in the pin varieties. Pink corundums are characterized by significantly higher Cr (up to 1082 ppm) concentration compared to the colorless/blue areas (Cr < 298 ppm). Maximum values for V and Ga content of u to 227 and 121 ppm, respectively are fixed regardless of the color. Mg values (up to 536 ppm) ar generally higher in the colorless to blue grains, compared to the pink grains (<65 ppm). Rubies fro Paranesti/Drama display very high Cr contents (up to 15347 ppm), and Fe (up to 4348 ppm). The T Mg, Ga and V values are low (up to 148, 31, 24 and 6 ppm, respectively).

Table 1. Chromophores and key trace elements (Mg and Fe) LA-ICP-MS analyses (ppm) of the Greek corundum crystals (data from Voudouris et al. [18]).

Sample (Number of Analyses)	Locality	Color	ppm	Mg	Ti	V	Cr	Fe	Ga
			aver	17	41	2.9	9142	26,558	16
Dr1a-b (n = 21)	Paranesti/Drama	red	min	8	8	2	2431	1799	12
			max	31	148	5.7	15,347	4384	24
			aver	322	4255	78	84	1003	89
Go1a-b (n = 26)	Gorgona/Xanthi	blue-colorless	min	78	1007	23	35	363	29
			max	601	6462	207	251	1339	121
			aver	28	102	20	104	1291	47
Go5a (n = 8)	Gorgona/Xanthi	pink-purple	min	13	30	5	4	962	9
			max	49	205	60	298	1782	100
			aver	35	509	65	297	424	78
Go5b (n = 14)	Gorgona/Xanthi	pink	min	17	39	47	105	294	72
			max	53	810	79	1082	494	87
			aver	15	1263	116	223	4326	90
Ik1a (n = 24)	Ikaria isl.	blue	min	5	267	31	103	2710	68
			max	54	4508	164	313	7324	114
			aver	58	594	22	214	3400	60
Nx1-4 (n = 77)	Naxos isl./Kinidaros	blue-colorless	min	16	10	9	1	1377	42
			max	208	3222	42	851	6361	184
			aver	64	520	22	43	4677	63
Nx1b (n = 14)	Naxos isl./Kinidaros	purple	min	39	124	15	36	3324	55
			max	199	848	31	48	6670	76
			aver	52	181	20	428	3096	45
Nx5b (n = 12)	Naxos isl./Kinidaros	pink	min	17	60	14	274	2138	37
			max	133	348	43	548	4716	52
			aver	5	462	52	262	3706	87
Nx5a (n = 8)	Naxos isl./Kavalaris	blue	min	2	238	43	227	3301	84
			max	14	784	65	339	4268	90

Sapphires from Naxos Island are characterized by high Fe concentrations (up to 6678 ppm), related to blue-colored domains. Ti is considerably lower (up to 966 ppm), with the exception of one colorless to blue corundum, where values up to 3222 ppm were detected. Chromium in blue to colorless varieties is low and reaches values of up to 851 ppm in grains with pink and purple hues. Vanadium content in the majority of the samples measured varies in the range of 20–40 ppm. Mg and Ga values range between 2–208 ppm and 42–184 ppm, respectively. The higher Ga concentrations occur in the metabauxite-hosted sapphires. Sapphires from Ikaria display high Fe and Ti values (up to 7324 ppm and 4508 ppm, respectively). Cr, Mg and V values reach maximum values of up to 313, 54 and 64 ppm, respectively. Gallium shows elevated concentrations, up to 114 ppm. Figure 4 distinguishes different primary sources for the studied corundums based on the chromophore and genetic indicator elements. In the (Cr + V)/Ga versus Fe/Ti diagram (Figure 4a, [40,41]), the majority of the samples plot in the field of metamorphic corundum, exhibiting a large variation in Fe/Ti ratios.

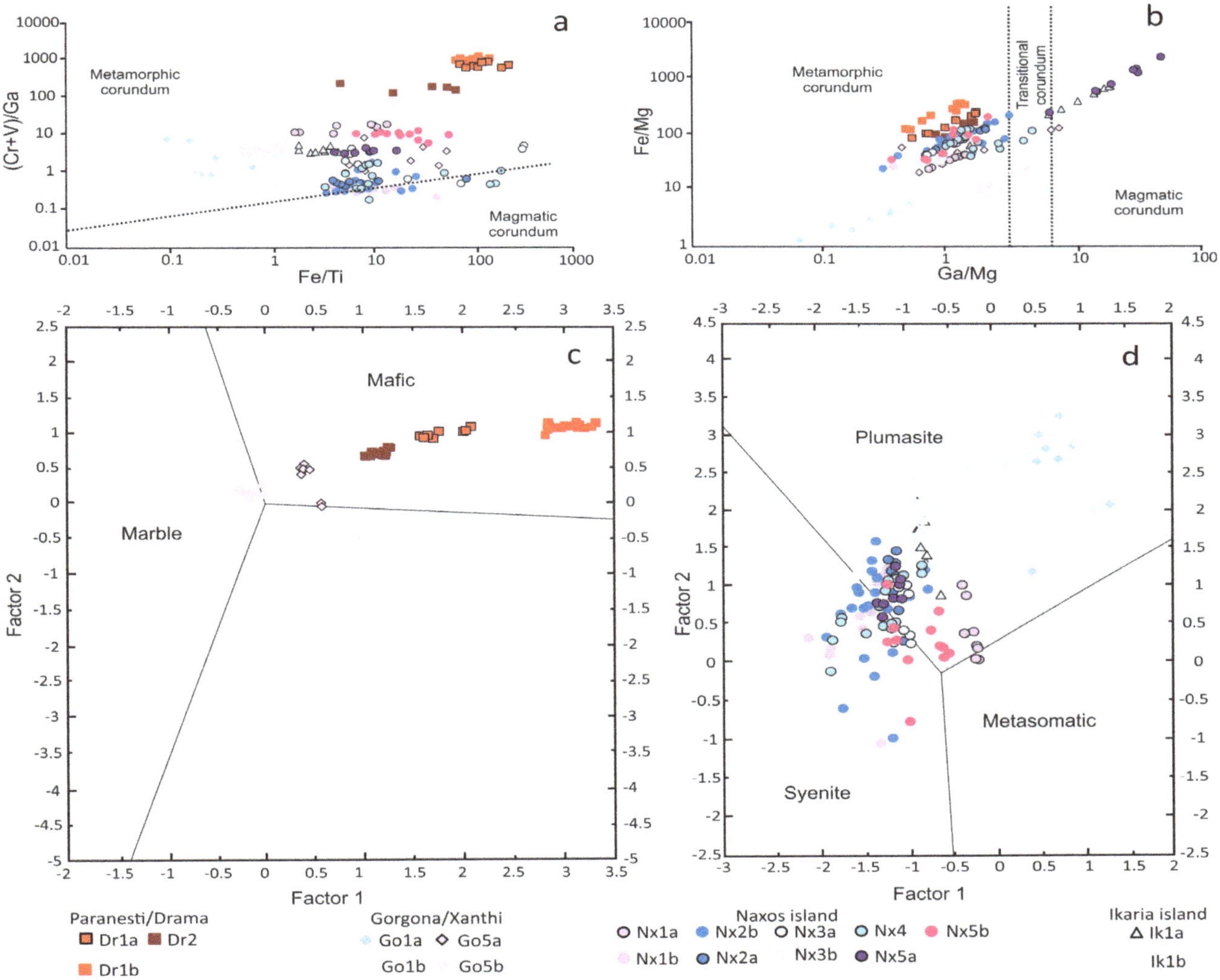

Figure 4. Greek corundum LA-ICP-MS analyses plotted on (**a**) (Cr+V)/Ga versus Fe/Ti discrimination diagram separating the fields for magmatic and metamorphic corundums (adapted from Sutherland et al. [40] and Harris et al. [41]; (**b**) Fe/Mg versus Ga/Mg discrimination separating the fields for magmatic, transitional and metamorphic corundums (adapted from Peucat et al. [42], Sutherland et al. [40] (both **a** and **b** are from Voudouris et al. [18]); (**c**) plot of Greek rubies (and Xanthi pink sapphires) within mafic rocks and marble field in discriminant factors diagram adapted after Giuliani et al. [43]; (**d**) plot of Greek sapphires within plumasite and syenite field in discriminant factors diagram adapted after Giuliani et al. [43].

Rubies from Paranesti/Drama show high Cr/Ga and Fe/Ti ratios, followed by the pink and purpl sapphires from Naxos Island. In the Fe/Mg versus Ga/Mg plot (Figure 4b, [40,42]), most sample plot in the metamorphic corundum field, except for sapphires from the metabauxites of Naxos anc Ikaria Islands. A few blue sapphires from Naxos Island and pink sapphires from Gorgona/Xanth plot in the area of transitional corundum. In the discriminant factor diagrams of Giuliani et al. [43 the Paranesti rubies within the pargasitic schists, plot in the "mafic" field and pink corundums (e.g sapphires because Fe > Cr + V), from Xanthi in the "marble" field. In our study we considered pin corundum for which Cr + V < Fe as sapphires. From Figure 4c we can observe that the marble-hostec pink corundum from Xanthi (sample GO5b) plots in the marble area, while those of Go5a (pink purple plot in the metasomatite field, suggesting that for sapphires the "marble" field (as defined for rubies may extend to the right into the "metasomatic" field. Only part of the plumasitic sapphires from Naxo plot in the "plumasitic" field, the rest plotting in the "syenite" field (Figure 4d).

5.2. Beryl

The Greek pegmatites are mostly deficient in miarolitic cavities, and thus only contair matrix-embedded mineral crystals. Blue beryl crystals (var. aquamarine) up to 5cm long are typical constituents of Naxos pegmatites both within the migmatitic dome as well as in pegmatite cross-cutting the surrounding metamorphic rocks (Figure 3g–i).

5.3. Jadeitite

Jadeitite, a rock that is also termed jadeite "jade", consists almost entirely of the pyroxene minera jadeite. Its green color is attributed to iron substituting for aluminum in the jadeite [$Na(Al,Fe^{3+})Si_2O_6$ crystal structure [4]. Jadeitite together with omphacitite occurs within the Kampos mélange, Syros Island in contact relationship or enveloping eclogite [27,44,45] as shown in Figure 5a,b. A productior of polished jade axe heads on Syros took place at least since the Neolithic period [46].

Figure 5. Field photographs (**a**) and hand specimens (**b**) of jadeitite from Kampos, Syros Island.

5.4. Al_2SiO_5 Polymorphs

Kyanite in blue, green, yellowish to orange colored crystals up to 20 cm long, is found in quartz ± feldspar boudins intercalated within metapelites at Trikorfo, Thassos Island (Figure 6a–d) [17,19,47] Orange kyanite also occurs in association with spessartine, and muscovite. Distinct dark blue color zoning is a common feature in the centre of some light blue kyanite; this color zoning can also be observed in some yellow kyanite. Microprobe analyses for orange kyanite indicate MnO up to 0.1 wt. % contributing to the coloration of the crystals [19] and up to 1.5 wt. % FeO. In a similar occurrence on Naxos Island in the Cyclades, kyanite forms up to 10 cm blue-colored crystals in quartz lenses and veins.

Mn-Andalusite (formerly viridine) in dark green-colored euhedral to subhedral crystals up to 7 cm is found in quartz-feldspar boudins (Figure 6e,f) at Trikorfo, Thassos Island. Mn-andalusite alsc occurs in orange kyanite-mica schists or in association with muscovite. Microprobe data revealed Mn_2O_3 contents, up to 2.90 wt. % (Table 2). Mn^{3+} substituting for Al^{3+} in the mineral formula is responsible for the dark green color [19,47].

.5. *Epidote Group Minerals—Zoisite*

Mn-poor zoisite (var. thulite) and Mn-poor clinozoisite in calc-silicate layers at Trikorfo, Thassos Island, are intergrown with Mn-grossular and quartz (Figure 6g–i). They form light pink to red colored translucent subhedral to euhedral crystals up to 10cm long. Available electron microprobe data revealed low Fe_2O_3 content (1.5 wt. %–2 wt. %) and very low Mn_2O_3 values from 0.15 wt. %–0.21 wt. %, but enough to be responsible for the pink-red coloration [19,47]. Deep green colored epidote crystals, up to 10 cm long associated with quartz and garnets at Lefkopetra, near Kimmeria, Xanthi, within metasomatized granodiorite and gneiss-hosted exoskarn bodies (Figure 6j–l) are partly of gem-quality.

Figure 6. Field and hand specimens photographs demonstrating occurrences and crystals of gem-quality Al_2SiO_5-polymorphs and epidote-group minerals of Greece: (**a**–**d**) Blue, green to orange kyanite within metapelites and quartz veins at Trikorfo, Thassos Island; (**e**,**f**) Mn-rich andalusite (var. viridine) in quartz from Trikorfo, Thassos Island; (**g**–**i**) Pink-red, Mn-bearing zoisite/clinozoisite from Trikorfo, Thassos Island. (**j**) Epidote-bearing skarn of Lefkopetra, Kimmeria/Xanthi. Granodiorite is at the front of the photo; (**k**,**l**) Green epidote crystals associated with andradite from Lefkopetra; Photographs 6b,d,e,f,i,l are courtesy of Berthold Ottens.

.6. *Garnets*

Spessartine and Mn-grossular are crystallized within andalusite-kyanite schists and in the thulite-bearing calc-silicate layers of Trikorfo respectively [17,19,47]. Spessartine from quartz-muscovite veins within mica schists forms translucent, orange-colored euhedral crystals up to 1 cm and o gem-quality (Figure 7a,b). Electron microprobe analyses of spessartine indicate MnO contents u to 42.9 wt. % (Table 2). Mn-bearing grossular from the calc-silicate layers form euhedral yellowis crystals of gem-quality in close association with pink-to red colored Mn-zoisite and quartz. Mn(content ranges from 1.9 wt. %–2.4 wt. % (Table 2).

Figure 7. Field and hand specimens photographs demonstrating gem garnet occurrences/crystals of Greece: (**a**) alternating marbles and metapelites at Trikorfo, Thassos Island. Metapelites host spessartine together with other Mn-bearing silicates; (**b**) spessartine crystals associated with muscovite within quartz matrix; (**c**) spessartine within braunite-phlogopite bearing skarn from Thapsana. Paros Island; (**d**) Maronia skarn at the interface between marbles and the Maronia pluton (half right part of the photograph); (**e**) grossular crystals in contact with the Maronia monzogabbro; (**f**) partly transparent pale green grossular crystals from the Maronia skarn; (**g**) andradite-bearing wollastonitic skarn in contact with marbles, Kimmeria, Xanthi; (**h**) Green andradite and calcite from Kimmeria, Xanthi; (**i**) yellow andradite within amphibolite, Kimmeria, Xanthi; (**j**) Panoramic view of skarn at Dikeon Mt, southern Kos Island; (**k**) orange andradite and diopside from the Dikeon Mt skarn, Kos Island; (**l**) red-brown andradite with green quartz from Avessalos, Serifos Island. Photographs 7**f,h,i,l** are courtesy of Berthold Ottens.

Deep red colored spessartine crystals up to 3 cm at Paros Island/Cyclades, are related to Mn-bearin skarn occurrences at Thapsana (Figure 7c). Spessartine at Thapsana contains up to 34.4 wt. % MnO an is associated with rhodonite and Mn-oxides [48]. Garnet is also a major constituent and represented b several varieties at Maronia, Kimmeria, Kresti, Kos and Serifos skarns. The Maronia skarn (Figure 7d) includes dark green-colored Ti-Cr andradite-grossular up to 1 cm, which postdates crystallization of earlier black-colored schorlomite-uvarovite-kimzeyite solid solutions [13]. Late pale green, brown to orange grossular-andradite solid solutions reach sizes up to 5cm (Figure 7e,f). Ti-bearing brown grossular-andradite contains up to 4.6 wt. % TiO_2. Transparent specimens with shiny dodecahedron faces are of gem-quality. A wide spectrum of colors in andradite-grossular garnets also occurs at Kimmeria where dark green-, brown-, yellow- to orange-colored garnets (up to 3 cm) occur (Figure 7g–i). Orange-colored grossular contains up to 1.15 wt. % MnO, whereas dark green-colored andradite-grossular solid solutions are characterized by elevated Ti contents (up to 2.86 wt. % TiO_2; Table 2).

Red-brown Mn-bearing andradite-grossular is present at Kresti in Drama (with up to 3.6 wt. % MnO) and at Kos Island (with up to 1.1 wt. % MnO and 1 wt.% TiO_2, Table 2), where they reach spectacular sizes of up to 20 cm (Figure 7j,k). The Serifos andradites are famous due to their zonal growth with colors ranging from deep brown to orange. The Agia Marina and Avessalos areas at Serifos are characterized by splendid occurrences of red-brown, locally transparent andraditic garnets in massif garnetitic skarns (Figure 7l). The garnets (up to 5 cm in size) accompany quartz and hematite in hedenbergitic skarn. Rodingitized gabbros at Evros, Evia, Larissa and Othrys Mesozoic ophiolites host gem-quality orange to brown grossular (hessonite), vesuvianite and diopside crystals up to 2 cm long [17,49].

5.7. Quartz

Quartz is found in an enormous variety of forms and colors from alpine-type fissures, granite-hosted miarolitic cavities and/or quartz veins, skarn-carbonate replacement deposits and volcanic-hosted epithermal environments [14,17,35,50,51]. Famous localities of gem-quality alpine-type quartz include the Attica and Drama districts, as well as Evia, Ios, Thassos and Crete Islands (Figures 2 and 8a–c). Crystals reach sizes up to 40 cm. Smoky quartz, black quartz (morion), rock crystal, amethyst, chloritized quartz and rutilated quartz are among the varieties found throughout Greece.

Quartz crystals up to 50 cm, in colorless, smoky and black varieties (morion) fill miarolitic cavities in aplitic granites at Samothraki Island (Figure 8d). Colorless quartz crystals (up to 10 cm) also occur in quartz veins cross-cutting granitoids at Samothraki, Kimmeria, Maronia/Rhodopi and Tinos Island.

Quartz is a very common mineral in the skarns of Serifos, Kimmeria/Xanthi and Kresti/Drama. Combinations of amethyst and green quartz (prase) forming sceptre growths at Serifos are worldwide unique specimens. Green-colored (due to actinolite inclusions) quartz crystals, are also developed within wollastonite skarn, at both Kimmeria and Kresti.

The Avessalos area at Serifos Island is one of the best sites in the world in respect to its green quartz (prase) [12,13,52]. The area is characterized by a garnetitic and hedenbergitic skarn and by the development of huge geodes filled by prograde and retrograde skarn minerals. At Neroutsika location of Avessalos, two forms of prase occur (Figure 8e,f): the first variety refers to very deep green colored crystals accompanied by hematite roses. The second variety refers to double-colored crystals of prase-amethyst. The transition between these two crystals is abrupt within the same crystal, where prase occurs at the base and amethyst at the top of the crystal. The amethysts are transparent and of gem-quality [15,16]. In the southern part of the Avessalos area, rare combinations of scepter-shaped crystals contain both prase and amethyst. These crystals commonly comprise a lower prase part that evolves upwards into amethyst and further upwards again into prase. These alternations can be remarked even within a single crystal [13]. Gem-quality quartz, in crystals up to 60 cm also occurs at the carbonate-replacement deposits of Chalkidiki [53].

Table 2. Representative EPMA of various gem-quality minerals from Greece: Andradite-grossular, Kimmeria (1–2); Andradite-grossular, Maronia (3–4); Andradite-Grossular, Kos Island (5); Grossular, Thassos (6–8); Spessartine, Thassos (9–10); Zoisite, Thassos (11–12); Mn-andalusite, Thassos (13–14); Orange kyanite, Thassos (15–16); Blue kyanite, Thassos (17).

Wt%	1	2	3	4	5	6	7	8	9*	10*	11	12	13*	14*	15*	16*	17*
SiO_2	39.77	37.73	38.22	39.03	38.05	38.97	39.05	39.31	35.71	34.31	39.23	39.11	36.08	36.07	36.40	36.68	36.71
Al_2O_3	20.21	6.02	18.79	19.60	8.17	21.74	21.88	22.20	20.78	23.40	32.71	32.11	58.98	58.24	61.90	61.29	61.18
MgO	bd	0.23	0.64	0.39	0.19	0.14	0.17	0.12	0.31	0.41	0.02	0.10	0.11	0.16	bd	bd	0.01
FeO	3.68	19.28	5.62	4.40	19.00	1.03	1.08	0.61	0.08	0.52	-	-	-	-	0.76	1.52	1.49
Fe_2O_3	-	-	-	-	-	-	-	-	-	-	1.55	2.01	1.93	2.14	-	-	-
Cr_2O_3	0.05	0.73	bd	0.09	-	0.03	bd	0.04	-	-	-	-	-	-	-	-	-
MnO	1.15	0.10	0.20	0.07	1.14	1.90	2.18	2.42	42.87	40.40	-	-	-	-	0.08	0.11	0.02
Mn_2O_3	-	-	-	-	-	-	-	-	-	-	0.15	0.21	2.50	2.89	-	-	-
CaO	34.55	33.59	34.92	35.40	32.10	36.00	35.76	35.39	0.55	0.93	24.78	24.73	0.02	bd	bd	bd	0.03
Na_2O	-	0.26	-	-	bd	0.01	0.01	bd	bd	bd	bd	bd	0.00	0.02	0.04	bd	0.04
TiO_2	0.03	2.86	1.10	0.13	0.92	0.42	0.25	0.12	0.09	0.05	0.02	0.04	0.01	0.02	0.02	0.01	bd
Total	99.39	100.54	99.49	99.11	99.57	100.23	100.37	100,21	100.39	100.02	98.44	98.31	99.64	99.52	99.20	99.62	99.48
Formulae						24 (O)					8 cations				5 (O)		
Si	6.076	6.032	5.856	5.974	6.094	5.872	5.876	5.920	5.880	5.619	2.970	2.970	0.993	0.996	0.995	1.001	1.003
Al	3.640	1.134	3.392	3.536	1.542	3.860	3.880	3.940	4.033	4.517	2.918	2.874	1.913	1.895	1.994	1.973	1.971
Mg	0.000	0.054	0.148	0.088	0.046	0.032	0.038	0.026	0.076	0.100	0.003	0.011	0.005	0.007	0.007	0.000	0.000
Fe^{2+}	0.276	0.554	0.078	0.086	0.496	0.000	0.000	0.000	0.011	0.071	0.000	0.000	0.000	0.000	0.017	0.034	0.033
Fe^{3+}	0.196	2.024	0.642	0.478	2.048	0.296	0.312	0.190	0.000	0.000	0.088	0.115	0.040	0.045	0.000	0.000	0.000
Cr^{3+}	0.003	0.092	0.000	0.010	-	0.004	0.000	0.002	-	-	-	-	-	-	-	-	-
Mn^{2+}	0.148	0.014	0.026	0.010	0.154	0.242	0.278	0.308	5.979	5.604	0.000	0.000	0.000	0.000	0.002	0.003	0.000
Mn^{3+}	0.000	0.000	0.000	0.000	-	-	-	-	0.000	0.000	0.010	0.000	0.052	0.061	0.000	0.000	0.000
Ca	5.658	5.752	5.732	5.804	5.508	5.812	5.764	5.710	0.097	0.163	2.010	2.013	0.001	0.000	0.000	0.000	0.000
Na	0.000	0.011	0.000	0.000	-	0.000	0.000	0.000	0.000	0.000	0.000	0.000	0.000	0.002	0.002	0.000	0.011
Ti	0.004	0.344	0.126	0.014	0.110	0.048	0.028	0.014	0.011	0.006	0.001	0.002	0.000	0.000	0.000	0.000	0.000

(-) = not analyzed; bd = below detection. Analyses marked with (*) are from Tarantola et al. [19] (this volume).

Amethyst is the main quartz variety of gem-quality in the volcanic rock-hosted epithermal nvironments in Greece. It occurs at Kornofolia/Evros area, in Sapes/Rhodopi region and in Lesvos nd Milos Islands in epithermal veins accompanying calcite, and/or adularia and barite [13,35]. At apes, amethyst occurs in massive form within crustiform banded quartz-chalcedony veins, however avities in the veins may host deep purple crystals up to 2cm. In Milos and Kornofolia, well-developed methyst crystals (up to 2 cm) occur as open space filling in the centre of veins cross-cutting propylitic nd sericitic altered volcanics (Figure 8g,h). At Megala Therma, Lesvos Island, amethyst is prismatic, p to 10 cm in length and displays sceptre and window growths and shares similarities to the Mexican nd Sardinian occurrences.

Figure 8. Hand specimens illustrating the gem quartz crystals of Greece: (**a**,**b**) smoky quartz and amethyst from Dassoto, Drama; (**c**) smoky quartz from Koskina, Evia Island; (**d**) morion from Samothraki Island; (**e**) green quartz from Avessalos, Serifos Island; (**f**) amethyst on green quartz from Avessalos, Serifos Island; (**g**) amethyst on barite from Chondro Vouno, Milos Island; (**h**) amethyst from Kornofolia, Evros district. Photographs 8e,f are courtesy of Berthold Ottens.

The oxygen isotopic composition of quartz from the various geological environments has been analyzed in hand-picked crystals and the results are presented in Table 3. Quartz from the alpine-type fissures yield isotopic $\delta^{18}O$ values between 9.4‰ and 23.7‰ (Figure 9).

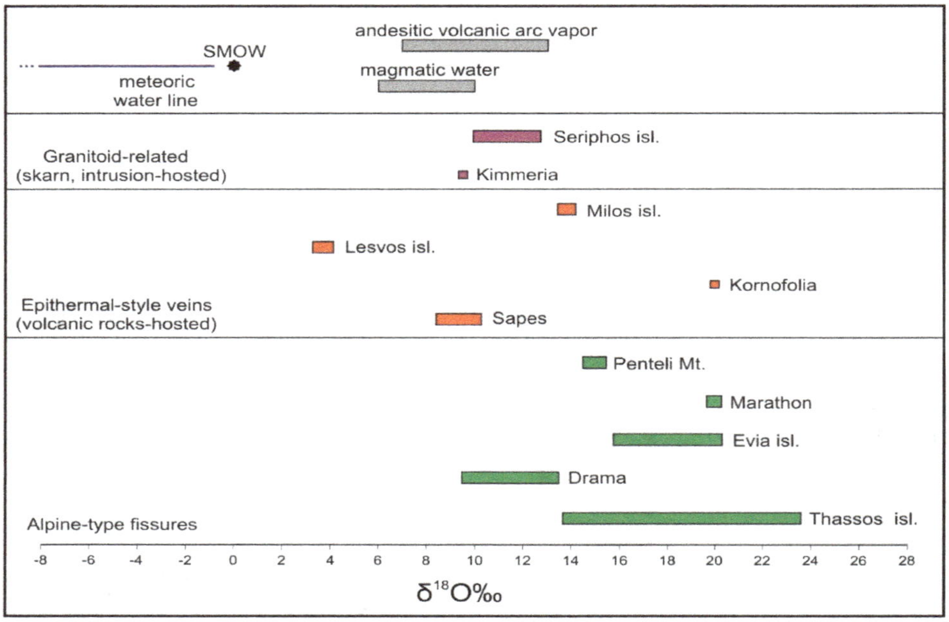

Figure 9. $\delta^{18}O$ values of quartz varieties from Greece. The values for magmatic water and andesitic volcanic arc vapor are from Taylor [54] and Giggenbach [55], respectively.

Table 3. Oxygen isotope composition of quartz from various geological settings in Greece (δ values in ‰ relative to SMOW).

Sample	Locality	Type	$\delta^{18}O_{Qtz}$ (‰)
		Alpine-type fissures	
TH1	Thassos Island	Colorless Qtz, Tessin habit	18.44
TH2	Thassos Island	Smoky Qtz, Tessin habit	13.77
TH3	Thassos Island	Colorless Qtz, skeletal habit	18.82
TH4	Thassos Island	Colorless Qtz, sceptre, prismatic habit	21.41
TH5	Thassos Island	Smoky Qtz, sceptre, prismatic habit	23.66
TH6	Thassos Island	Smoky Qtz, Tessin habit	14.56
TH7	Thassos Island	Choritized Qtz, Tessin habit	18.11
TH8	Thassos Island	Smoky Qtz, Tessin habit	14.66
KN1	Drama	Amethyst, prismatic habit	13.38
KN2	Drama	Amethyst, prismatic habit	13.36
KN3	Drama	Morion, Tessin habit	10.63
KN4	Drama	Morion, Tessin habit	10.83
AL1	Evia Island	Colorless Qtz, prismatic habit	16.80
AL2	Evia Island	Smoky Qtz, prismatic habit	20.77
AL2B	Evia Island	Smoky Qtz, prismatic habit	15.74
AL3	Evia Island	Smoky Qtz, prismatic habit	16.04
AL4	Evia Island	Smoky Qtz, prismatic habit	18.56
AL5	Evia Island	Colorless Qtz, prismatic habit	20.15
MA1	Marathon, Attica	Amethyst, prismatic habit	20.74
MA2	Marathon, Attica	Smoky Qtz, prismatic habit	19.43
PEN1	Penteli Mt, Attica	Colorless Qtz, prismatic habit	15.37
PEN2	Penteli Mt, Attica	Smoky Qtz, prismatic habit	14.14

Table 3. *Cont.*

Sample	Locality	Type	$\delta^{18}O_{Qtz}$ (‰)
		Epithermal veins (volcanic rock-hosted)	
SH1	Sapes, Rhodopi	Smoky Qtz	10.49
SH2	Sapes, Rhodopi	Amethyst	8.24
KIR1	Kirki, Evros	Amethyst	19.10
SF2	Soufli, Evros	Amethyst	20.54
LS2	Lesvos Island	Amethyst	3.33
LS3	Lesvos Island	Colorless Qtz	2.68
LS5	Lesvos Island	Colorless Qtz, Muzo habit	4.13
M1	Milos Island	Amethyst	14.12
M2	Milos Island	Amethyst	13.41
		Granitoid-related (skarn, intrusion-hosted)	
KIM1	Kimmeria, Xanthi	Colorless Qtz	9.49

In the absence of fluid inclusion temperature results, the composition of the fluid in equilibrium with the quartz crystals on the basis of the fractionation equilibrium equation of Sharp et al. [56], cannot be estimated. However, the available data, especially for sceptered quartz crystals from the Rhodope area (Dassoto and Thassos Island), indicate an increase in $\delta^{18}O$ values from the basal Tessin-habit parts of the crystals, towards the amethystine or clear and prismatic-habit sceptres on the upper parts of the crystal. Melfos and Voudouris [57] suggested that smoky quartz and amethyst at Dassoto were formed under different conditions due to mixing of carbonic metamorphic fluids with meteoric waters during the exhumation of the Rhodope core complex. The presence of hematite and pyrite within smoky quartz indicates fS_2/fO_2 conditions at the hematite/pyrite buffer, which were followed by more oxidizing conditions necessary for trivalent Fe to be incorporated into the quartz structure and to the crystallization of amethyst [57].

Similarly, the Thassos colorless and smoky quartz sceptre, display $\delta^{18}O$ values between 21.4 and 23.66‰, heavier than the basal Tessin habit quartz crystals (values from 13.8‰ to 18.8‰, Table 3), thus suggesting involvement of two different fluid types in the crystallization of quartz. This is in accordance to the fluid inclusion results of Bitte et al. [58], who suggested for the quartz crystals from Thassos Island, a continuous event at ductile-brittle to brittle conditions from 360 to 170 °C and at pressures lower than 2 MPa. Carbonic-rich fluids are replaced by surficial fluids, derived from evaporated sea-water during the final period of exhumation.

Smoky, chloritized and colorless quartz crystals in the Attico-Cycladic massif (Evia Island, Penteli Mt including Marathon area) display a similar range of $\delta^{18}O$ values (from 14.1 to 20.8‰) to those from Thassos (Table 3, Figure 9). According to Tarantola et al. [59], metamorphic and basinal fluids were trapped in Evia Island quartz crystals during exhumation of the Attic-Cycladic metamorphic complex. Mineral inclusions of biotite and chlorite in quartz indicate continuous crystallization from 430 to 250 °C.

Volcanic-rock hosted amethysts yield isotopic $\delta^{18}O$ values between 3.3‰ and 20.5‰ (Figure 9; see also Voudouris et al. [35]. Those from Kassiteres-Sapes show $\delta^{18}O$ values of 8.2 to 10.5‰. The amethyst at Kornofolia, Evros district, yielded the highest $\delta^{18}O$ values (20.5‰) and those from Lesvos the lowest $\delta^{18}O$ value of all amethysts analysed (3.3‰). Finally, the Milos amethyst have $\delta^{18}O$ values of 13.4 and 14.1‰. Based on fluid inclusion data, Voudouris et al. [35] suggested that most $\delta^{18}O$ values correspond to a mixing of magmatic and oceanic (and/or meteoric) water, with the highest magmatic component in Kornofolia and the lowest in Lesvos Island.

Quartz from a granite-hosted vein at Kimmeria, Xanthi yielded a $\delta^{18}O$ value of 9.49‰. The isotopic signature of prase at Serifos skarn is comparable to those of quartz samples from different assemblages of the skarn zone at Serifos, but amethyst displays a significantly lighter isotopic signature [52].

5.8. Chalcedony and Other Silica Rich Varieties

Blue agate occurs in veins cross-cutting dacitic lavas and/or as loose fragments at Aetochori, Evros district (Figure 10a) [17]. Agate with various colors is also found in the volcanic environments of Lefkimi/Evros (Figure 10b), and Limnos Island. Chalcedony is present at Kornofolia/Evros and at Petsofas/Lesvos, where it forms typical botryoidal and stalactitic aggregates up to 10 cm, varying in color from pale to deep blue and pink to purple (Figure 10c,d). Red jasper occurs in the Sapes area in the form of up to 40 cm thick horizons hosted in smectite-altered pyroclastics (Figure 10e,f) [60]. Jasper and agate from Lesvos Island have been described by Thewalt et al. [61].

Figure 10. Field and hand specimens photographs demonstrating gem microcrystalline and amorphous silica varieties from Greece: (**a**) blue agate from Aetochori, Evros; (**b**) agate from Lefkimi, Evros; (**c**) chalcedony from Kornofolia, Evros; (**d**) blue chalcedony from Petsofas, Lesvos Island; (**e**) red jasper layer within smectite-altered pyroclastics, Sapes, Rhodopi; (**f**) red jasper from Sapes, Rhodopi; (**g**,**h**) brown and green opalized wood from Moudros, Limnos Island; (**i**) red opal from Agioi Theodoroi, Milos Island. Photographs 10**a**,**c** courtesy Berthold Ottens.

5.9. Opal

Opal occurs in several varieties and colors (deep red, yellow, black, orange and green) at Lesvos and Limnos Islands being a constituent of fossilized wood [13,62–64], as well as at various localities at Evros, Milos, Limnos and Lesvos Islands, where it is considered to be part of either silica sinters or steam-heated alteration zones. At the Moudros area, Limnos Island several horizons of opaline silicification within the pyroclastic rocks host opalized wood [64,65]. These horizons represent either thin-bedded, lacustrine-fluviatile intercalations between the pyroclastic formations and were deposited during erosional periods that lasted between the volcanic activity phases, or true silica sinter deposition. In some places the fossiliferous horizons overly an alunitic alteration zone and hydrothermal breccias rich in natroalunite occur. In other cases the opaline horizon is intercalated between fresh to weak argillicaly altered rocks. Silicification within the horizons varies from red, green, white, to black colored, the white color resulting from total depletion in iron oxides (Figure 10g–i). Representative X-ray powder diffraction diagrams from Greek opals are given in Figure 11, demonstrating the transformation of opal-C to quartz in some samples.

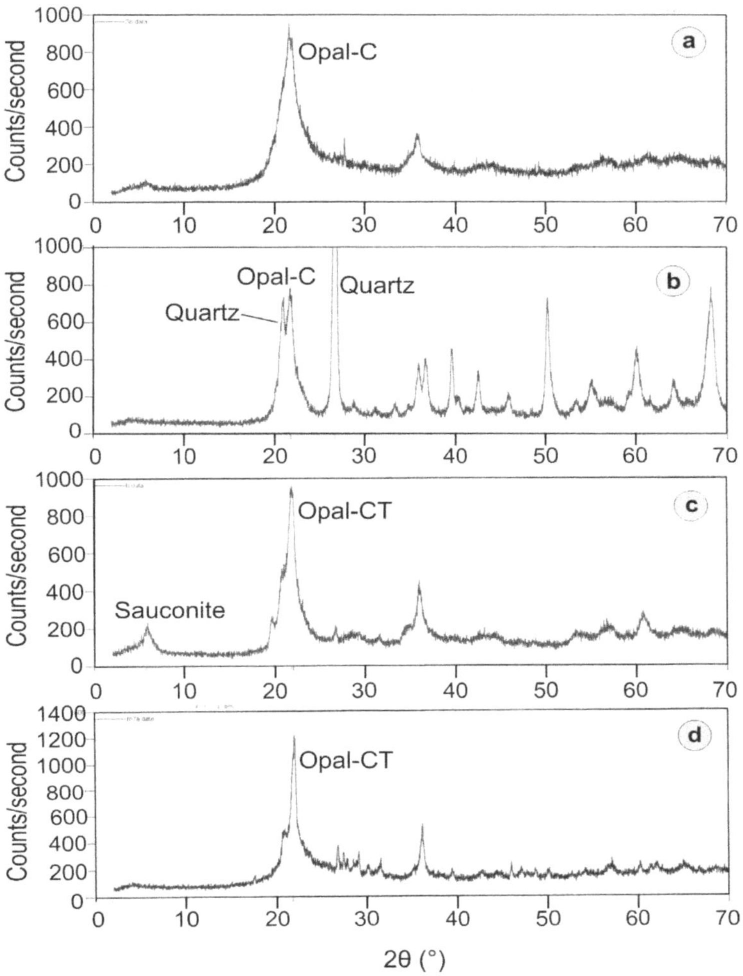

Figure 11. X-ray powder diffraction diagrams of various Greek opals (**a**,**b**) brown-orange opal-C and quartz from Lykofi Evros; (**c**) green opal-CT (color due to sauconite admixtures) from Sapes; (**d**) red-brown opal-CT from Moudros, Limnos Island.

5.10. *Feldspars*

Gem-quality albite is present as idiomorphic transparent crystals, up to 6 cm in size, grown onto clear and smoky quartz crystals from Evia and Crete Islands (Figure 12a,b). Adularia constitutes well-developed crystals up to 3 cm in size, which occupy alpine-type fissures along with quartz mainly from Evia and Thassos areas.

5.11. *Titanite*

Titanite in large yellow crystals (up to 3 cm), accompanies zoisite-bearing calc-silicate layers within amphibolites of the Rhodope massif at Therapio, Evros [66]. Green titanite crystals up to 5 cm long at Syros Island, belongs to the retrograde mineral assemblage of eclogites. Titanite in honey colored crystals up to 1 cm, accompany orthoclase and schorlomite in the Maronia endoskarn. Finally, yellow/green colored titanite is associated with adularia in amphibolite-hosted fissures at Thassos Island [51] (Figure 12c). All localities include locally transparent gemmy-material.

5.12. *Vesuvianite*

The endoscarns from Kimmeria and Maronia host large, up to 10 cm long, partly facetable pale green colored vesuvianite crystals (Figure 12d). Violet-colored vesuvianite in crystals up to 1 cm occur in rodingitized gabbros at Kymi, Evia Island [17].

Figure 12. Field and hand specimens photographs demonstrating occurrences/crystals of various gemstones from Greece: (**a**,**b**) albite from Krieza, Evia Island; (**c**) titanite and quartz from Trikorfo, Thassos Island; (**d**) vesuvianite from the Maronia skarn; (**e**) spinel in marble from Gorgona, Xanthi; (**f**) apatite and K-feldspar from potassically-altered monzonite, Fakos, Limnos Island; (**g**) purple fluorite coated with iron oxides, Ampelos, Samos Island; (**h**) azurite, Kamariza, Lavrion deposit; (**i**) blue smithsonite within marble at the interface with Fe-oxides, Kamariza, Lavrion; (**j**–**l**) colored varieties of smithsonite, Kamariza, Lavrion deposit; Photographs 12b,d,e,f,h are courtesy of Berthold Ottens.

5.13. Spinel

Spinel at Gorgona, Xanthi occurs either as isolated crystals or as rims around corundum. The color of spinel ranges from blue to green and brown (Figure 12e). Transparent blue-colored octahedral spinel crystals up to 1 cm occur at Maronia in association with phlogopite and grossular.

5.14. Tourmaline

Tourmaline is abundant in quartz veins and lenses at Trikorfo and Thymonia/Thassos, in black crystals (schorl) reaching up to 10 cm. However this material is not suitable as a gemstone. Similar crystals occur in pegmatites near Nevrokopi-Drama and at Naxos Island. Nickel-bearing brown tourmaline (dravite) accompanies corundum in the marbles at Gorgona, Xanthi and contains up to 4.4 wt. % NiO, much higher than the Ni content reported in tourmaline from Samos and nickeloan tourmaline from the Berezovskoe gold deposit by Henry and Dutrow [67] and Baksheev and Kudryavtseva [68] respectively. Crystals are generally small (up to 1 cm) but are locally transparent and of gem-quality.

5.15. Apatite, Fluorite, Azurite, Turquoise

Transparent apatite in well-shaped crystals, 1 cm long, accompany phlogopite and orthoclase in the potassic alteration zone at Fakos, Limnos Island (Figure 12f). The carbonate-replacement deposits at Lavrion and Serifos are well-known for large fluorite (up to 20 cm) crystals with colors varying from blue to purple and green. Gem-quality fluorite is also found in volcanic rocks in Samos and Lesvos Islands, where monominerallic fluorite veins cross-cut epithermally altered silicified zones and propylitically altered lavas. The veins are banded and the voids are filled with deep violet and green colored crystals up to 5 cm respectively (Figure 12g). Azurite associated with malachite is widespread in the Lavrion deposit in crystals up to 4 cm as well as in large massive aggregates suitable for cabochon cutting (Figure 12h). Turquoise in gem-quality occurs in the oxidation zone of the Vathi porphyry Cu-Au deposit, Kilkis area [69].

5.16. Smithsonite

The Lavrion smithsonite occurs in many different forms and colorations (Figure 12i–l). Samples have pale blue or green and yellow colors due to both solid solutions and impurities of other minerals [11,70]. The Lavrion smithsonites were intensively exploited during the last one hundred years with a total production of about 1.2 Mt [71]. They occur in cavities in the marbles where they form colorful botryoidal or stalactitic aggregates and/or monominerallic bodies as cavity fillings. The microanalyses suggest that the total content of minor and trace elements in smithsonite is relatively low. The determined elements are Ca, Mg, Fe, Mn, Cu and Cd (Table 4).

Table 4. Fluctuation of trace element content (wt. %) of selected smithsonite samples from Lavrion (modified after Katerinopoulos et al. [11]).

Color	Ca	Mg	Fe	Mn	Cu	Cd	Pb	Visible Inclusions
Yellowish white	0.1–1.3	bd	0.2–0.8	0–0.2	bd	bd	0.6-1.6	-
Yellowish grey	1.1–1.8	bd	0.4–0.9	0–0.2	bd	bd	bd	-
Light yellow	0.3–1.5	bd	0.8–1.9	0–0.2	bd	bd	bd	-
Yellowish brown	0.2–1.5	0.3–1.8	0.3–5.8	0.3–0.5	bd	bd	bd	Chalcophanite
Yellowish green	0.8–1.4	bd	0.4	0.3–0.5	bd	0.2	0–0.7	Greenockite
Greenish yellow	0.3	bd	bd	bd	bd	1.1	2.3	Greenockite
Light blue	0.3–1.2	bd	bd	bd	0.3–2.6	0.3–0.7	bd	-
Bluish white	0.6–0.9	bd	bd	bd	1.1–2.8	bd	bd	-
Light green	0.5–0.7	bd	bd	0.3	0.9–3.0	bd	bd	-
Green	0.2–0.5	bd	bd	0.2–0.3	1.3–2.4	bd	bd	-

As Cd^{2+} is colorless, the yellow ("turkey fat") color of some smithsonites proved to be related only to the presence of greenockite inclusions. Indeed, some samples containing Cd up to 0.7 wt. % but no greenockite inclusions have colors other than yellow. The absence of Cu-rich impurities in blue and green smithsonite and the relatively high amount of Cu indicate that there is a limited solid solution between $ZnCO_3$ and the hypothetical $CuCO_3$ molecule (up to 3.0 wt.% Cu or 6.3 mol % $CuCO_3$) (see also Boni et al. [72]; Frisch et al. [73]). This is in accordance to the findings of Samouchos et al. [74] who suggested that the blue color in Lavrion smithsonite is due to Cu substituting for Zn in the structure. Iron substitution of zinc is limited (up to 1.9 wt. % Fe, except for one sample) and turns the color of smithsonite to a pale yellow. Orange and brown colors are related to impurities of Mn-Fe-Pb oxides and hydroxides. The role of manganese is unknown as its content is very low and the influence on the color is weak.

6. Discussion

Figure 13 represents a hypothetical schematic model, where gemstone occurrences in Greece are related to the various geological environments (regional metamorphic-metasomatic, alpine-type fissures, plutonic-subvolcanic intrusions and pegmatites, zones of contact metamorphism and peripheral volcanic rocks). The hypothetical depths are speculated on the basis of geological criteria. A Late Cretaceous (~ 81.6 ± 3.5 Ma) eclogite-facies metamorphism in the Eastern Rhodopes, confirm previous data that multiple subduction events took place between ~ 200 and ~ 40 Ma along this section of the southern European plate boundary [75]. In the Cyclades, following an initial compressional phase of the Alpine Orogeny related to eclogite-facies metamorphism of the rocks at ~55–49 Ma [45,76], subsequent exhumation of high-P rocks was accompanied by a regional Barrovian-type metamorphism that locally reached partial melting conditions [77,78]. The onset of post-orogenic extensional deformation in the different tectono-metamorphic terranes occurred at ~42 Ma for the Rhodope Massif and ~35 Ma for the Cyclades [23]. These early compressional and later syn-orogenic and post-orogenic extensional events were responsible for the formation of most gemstones found in both regional metamorphic rocks, as well as in late alpine-fissures in Greece. For ruby and jadeite the term "Plate tectonic gemstones" has been proposed by Stern et al. [5], since they generally form as a result of the plate tectonic processes subduction and collision.

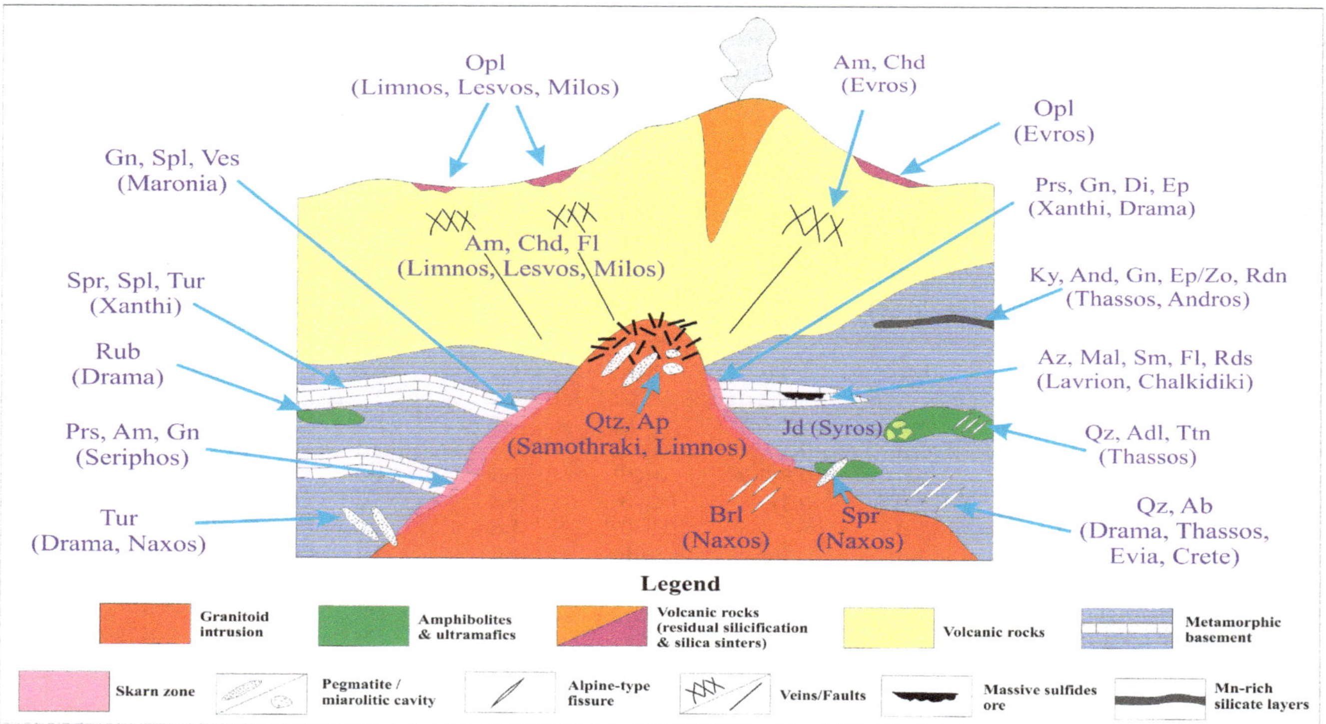

Figure 13. Hypothetical model presenting the various environments related to crystallization of gemstones in Greece. Abbreviations: Ab: Albite, Adl: Adularia; Am: Amethyst, And: Andalusite; Ap: Apatite; Az: Azurite; Brl: Beryl, Chd: Chalcedony, Di: Diopside, Epid: Epidote, Fl: Fluorite, Gn: Garnet, Jd: Jadeite; Ky: Kyanite; Prs: Prase; Rdn: Rhodonite; Rds: Rhodochrosite; Sm: Smithsonite; Spr: Sapphire, Spl: Spinel; Ttn: Titanite; Tur: Tourmaline, Ves: Vesuvianite.

Formation of the jadeitite bodies at Syros Island involved fluid interaction in and around serpentinized peridotite, derived from subduction zone devolatilization [27]. Rodingitization of doleritic and gabbroic dykes in the Rhodope massif and the Pelagonian zone is associated with metasomatic alteration processes, which resulted in the formation of garnet and vesuvianite [79]. According to Koutsovitis et al. [79], rodingitization took place during the exhumation of the mafic–ultramafic mantle wedge rocks within a serpentinitic subduction channel.

Gem corundums in Greece cover a variety of geological environments. The pargasite-schist hosted ruby deposit at Paranesti/Drama area and marble-hosted pink to blue sapphires in Xanthi area occur along the UHP-HP Nestos suture zone and are classified as metamorphic s.s hosted in mafics/ultramafics and marbles respectively [18]. At the Trikorfo area (Thassos Island, Rhodope massif, Northern Greece), which represents a unique mineralogical locality with Mn-bearing minerals such as kyanite, andalusite, garnet and epidote, localized metasomatic reactions occurred during the exhumation of the HP unit locally containing layers enriched in Mn with a strong local fO_2 buffering [19]. Metasomatic reactions first occurred under ductile conditions in an extensive context close to the kyanite-andalusite stability curve and continued until purely brittle conditions as indicated by the presence of late alpine-type veins cross-cutting the metamorphic foliation [19].

Based on the existing data, it can be proposed that the formation of the studied alpine-type fissures in Greece and their mineralogy are closely related to the extensional tectonics accompanying the exhumation of the metamorphic rocks. This was a favored environment for the crystallization of gem-quality quartz and albite. A decrease in pressure and temperature around the fissures during exhumation caused oversaturation in SiO_2 and precipitation of quartz and other mineral constituents in the fissures. Leaching of chemical components from the host rocks after their interaction with the hydrothermal fluids could also explain fluid enrichment and deposition of alpine-type minerals in the fissures studied. The blue sapphires hosted in metabauxites from southern Naxos and Ikaria, fill open-spaces in fissures cross-cutting the metamorphic foliation. They display atypical magmatic signatures indicating a hydrothermal origin and were formed during late extensional conditions.

The development of magmatic-hydrothermal and epithermal systems during the Tertiary, is closely associated with crystallization of a large variety of gemstones. Several gemstones related to the Tertiary magmatic activity are developed within and around the magmatic centres in W. Thrace (Kimmeria, Kassiteres-Sapes, Maronia-Perama, Soufli-Dadia-Levkimi, Drama, Samothraki), Limnos, Lesvos, Serifos and Milos Islands. Initial deep mineralization occurred where plutonic bodies intruded within the basement rocks. Residual fluids, enriched in volatiles, resulted in the emplacement of pegmatitic bodies and to crystallization of beryl (aquamarine) [80,81] and sapphires (in desilicated pegmatites intruding ultramafic rocks) at Naxos Island. Metasomatic processes caused by the release of magmatic-hydrothermal fluids from the intrusive rocks, led to the deposition of various gemstones in the intrusive bodies as well as in the surrounding rocks.

The mineralization in skarns started with the deposition of anhydrous minerals, some of them of gem-quality such as garnets, spinel, titanite and diopside, and ended in the deposition of gem-varieties of quartz, epidote, and vesuvianite in a retrograde stage and under temperature decrease and meteoric water incursion. Contemporaneously, magmatic-hydrothermal fluids circulating within the granitoids were responsible for the deposition of quartz within miarolitic cavities and within quartz veins associated with feldspar, and gemmy apatite.

Later, the systems shifted from a magmatic-hydrothermal dominated stage to a geothermal one, due to the increasing incursion of meteoric waters. Changes in the physico-chemical conditions of the ascending geothermal fluids were the major factors controlling the mineral deposition of the silica varieties (quartz, amethyst, chalcedony and opal) in this geothermal-epithermal environment. Opalized horizons in Thrace and Limnos could represent silica sinter deposition from alkali or neutral geothermal fluids [82]. Amethyst, indicative of oxidizing conditions [83], was probably deposited as a result of mixing of the ascending hydrothermal fluids with meteoric and/or seawater.

Finally, gem-quality smithsonite in the supergene oxidation zone of the Lavrion/Attik carbonate-replacement deposit, was formed as a result of reactivity of carbonate minerals wit acidic, zinc-rich solutions, derived from the destruction of sphalerite [11,84,85]. According to the abov authors, the neutralization of the acidic solutions deep in the marbles produces enough CO_2 to mak smithsonite the stable phase at pH values near neutral.

All Greek corundum, especially the vivid-colored varieties, are translucent to opaque, suitable fo cabochons (Figure 14). Greek corundums are characterized by a wide color variation ranging fron deep red, pink, purple, and blue to colorless, with crystal sizes of up to 5 cm, homogeneity of th color hues and transparency and should be further examined for their suitability as potential facete gemstones. Similarly, at Trikorfo/Thassos crystals of kyanite, green andalusite, garnet (grossular an spessartine) and red zoisite–clinozoisite are large (up to several cm), show vivid colors, and are suitabl for cabochons.

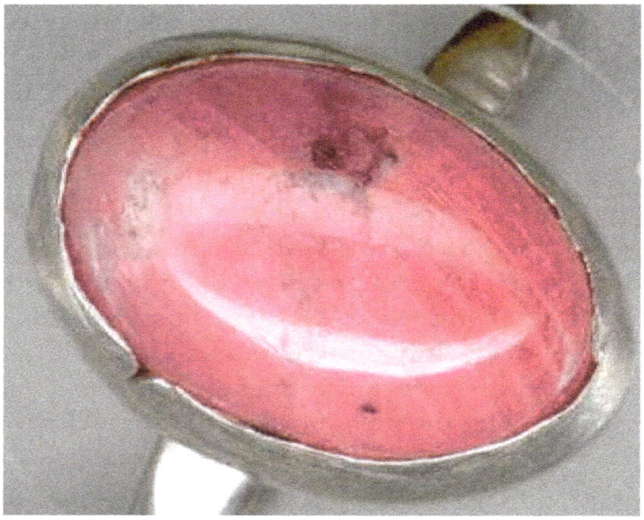

Figure 14. Cabochon quality ruby from Paranesti Drama. Photograph courtesy of V. Melfos.

Figure 15. Various faceted and cabochon gemstones from Greece. Green Mn-andalusite from Thassos (2.2 cm length) in the left, dark green epidote from Kimmeria (1.3 cm) in the center, ruby from Paranesti (4 cm) in the right of the photograph. The rest are silica varieties. Photo courtesy of Anastasios Tsinidis.

As such, the Trikorfo locality can be regarded as a promising area for the exploration of transparent, facetable gemstones. Their genesis due to metasomatic reactions also underlines the important role of metasomatism for gemstone formation in general, as previously noted in the literature [2,19]. According to Tarantola et al. [19], Thassos Island (Rhodope, Greece) can now be classified as the second locality worldwide where Mn-rich orange kyanite is reported, after that of Loliondo (Tanzania) deposit. Finally, silica varieties provide excellent facetable material or cabochons (Figure 15). Future exploration in addition to detailed studies of trace and minor element chemistry using LA-ICP-MS analysis and gemological evaluations are required in order to establish the potential for economic exploitation and to consider them as marketable gems.

7. Conclusions

In the accretionary Hellenides Orogen, gemstones occur in various rock types of mainly four tectonometamorphic units, the Rhodope- and the Attico-Cycladic massifs, the Pelagonian zone, and the Phyllites-Quartzites of Crete Island. In crystalline rocks, two groups of gemstones are distinguished, those formed during regional metamorphism and those associated with late alpine-type fissures. The first group includes Mn-bearing silicates (Mn-andalusite, spessartine, Mn-grossular, Mn-clinozoisite, Mn-zoisite and orange-colored Fe-Mn-kyanite) hosted in Carboniferous ortho- and paragneisses in both Thassos and Paros Islands (Rhodope- and Attico-Cycladic massifs respectively), as well as gem corundums in Xanthi-Drama areas/Rhodope massif and Naxos-Ikaria Islands (Attico-Cycladic massif). In the Xanthi-Drama area, corundum mineralization (sapphires and rubies) is distributed within marbles and eclogitic amphibolites oriented parallel to the (UHP-HP) Nestos suture zone. In the Attico-Cycladic massif, blue sapphires are found within marble-hosted metabauxites (Ikaria and Naxos Island). Alpine-type fissures in Greece contain gem-quality quartz (green quartz, amethyst, smoky and colorless quartz), albite and titanite. Host lithologies are ortho- and paragneisses and metabasites in the Rhodope- (Drama, Thassos Island) and the Attico-Cycladic (Pentelikon Mt, Evia, Ios Islands) Massifs, and metaquartzites (Crete Island). Metamorphic/metasomatic processes within a subduction channel, resulted in the formation of the jadeitite bodies at Syros Island and of garnet and vesuvianite in rodingite bodies at the Rhodope massif and the Pelagonian zone.

The Tertiary magmatic-hydrothermal environments in Greece (granitoids, pegmatites, skarns and carbonate-replacement deposits, and volcanic rocks) may also provide facet-quality material of several species (beryl, sapphire, garnet, vesuvianite, diopside, epidote, fluorite, rhodochrosite, quartz varieties and silica microcrystalline species). The Naxos pegmatites are prospective for aquamarine and blue, purple to pink colored sapphires. Miarolitic cavities and quartz veins cross-cutting granitoids (Samothraki Island, Maronia and, Kimmeria) contain gem-quality colorless and smoky quartz. The endo- and exoskarns of Kimmeria, Kresti and Maronia (Rhodope massif), Serifos Island (Attico-Cycladic massif) and Kos Island, host facetable grossular-andradite garnets, vesuvianite, epidote, as well as prase and amethyst crystals. Rhodochrosite and fluorite occur in the carbonate-replacement deposits of Olympias/Chalkidiki and Lavrion/Attika respectively. Gemstones associated with hydrothermally-altered volcanic rocks include amethyst, chalcedony, opal, fossilized wood and fluorite in Sapes, Soufli areas (Rhodope massif), Lesvos, Limnos and Samos Islands (northeastern and central Aegean volcanic arc) and Milos Island (Attico-Cycladic massif-south Aegean volcanic arc). Finally, the supergene oxidation of the Lavrion/Attika carbonate-replacement deposit contains gem-quality smithsonite in several colorations. Future work aims towards a gemological evaluation of the Greek gemstones and finally estimating the potential for their possible exploitation.

Author Contributions: P.V. and C.M. collected the studied samples. P.V. assisted by C.M., I.G., G.G., A.T., V.M., S.K., A.K., and A.M., evaluated the mineralogical and oxygen isotopic data. P.V. wrote the manuscript.

Acknowledgments: Authors would like to thank Stefanie Heidrich for her kind help on microanalyses in the Institute of Mineralogy and Petrography, University of Hamburg and Berthold Ottens and Anastasios Tsinidis for kindly providing photographs of Greek mineral specimens and gemstones respectively.

References

1. Fritsch, E.; Rondeau, B. Gemology: The Developing Science of Gems. *Elements* **2009**, *5*, 147–152. [CrossRef
2. Groat, L.A.; Laurs, B.M. Gem Formation, Production, and Exploration: Why Gem Deposits Are Rare and What is Being Done to Find Them. *Elements* **2009**, *5*, 153–158. [CrossRef]
3. Groat, L.A. (Ed.) *Geology of Gem Deposits*; Mineralogical Association of Canada Short Course 37: Quebec City QC, Canada, 2007; 276p.
4. Groat, L.A. Gemstones. *Am. Sci.* **2012**, *100*, 128–137.
5. Stern, R.J.; Tsujimori, T.; Harlow, G.; Groat, L.A. Plate tectonic gemstones. *Geology* **2013**, *41*, 723–726 [CrossRef]
6. Kievlenko, E.Y. *Geology of Gems*; Soregaroli, A., Ed.; Ocean Pictures Ltd.: Littleton, CO, USA, 2003; 432p.
7. Tsirambides, A.E. *The Minerals Wealth of Greece*; Giachoudi Pulb.: Thessaloniki, Greece, 2005; 391p.
8. Andronopoulos, V.; Tsoutrelis, C. *Investigation of a Red Corundum Occurrence in Xanthi, Greece*; Unpublished Report; Greek Geol. Survey: Athens, Greece, 1964; 8p.
9. Papastamatiou, I. The Emery of Naxos. *Geol. Geoph. Res.* **1951**, *1*, 37–68.
10. Paraskevopoulos, G. Die Entstehung der Manganlagerstaetten auf der Insel Paros, Griechenland. *N. Jb. Min Abh.* **1958**, *90*, 367–380.
11. Katerinopoulos, A.; Solomos, Ch.; Voudouris, P. Lavrion smithsonites: A mineralogical and mineral chemical study of their coloration. In *Mineral Deposit Research: Meeting the Global Challenge*; Mao, J., Bierlein, F.P., Eds Springer: Berlin, Germany, 2005; pp. 983–986.
12. Gauthier, G.; Albandakis, N. Minerals from the Serifos skarn, Greece. *Mineral. Rec.* **1991**, *22*, 303–308.
13. Voudouris, P.; Katerinopoulos, A. New occurences of mineral megacrysts in tertiary magmatic-hydrotherma and epithermal environments in Greece. *Doc. Nat.* **2004**, *151*, 1–21.
14. Voudouris, P.; Katerinopoulos, A.; Melfos, V. Alpine-type fissure minerals in Greece. *Doc. Nat.* **2004**, *151* 23–45.
15. Voudouris, P.; Melfos, V.; Katerinopoulos, A. Precious stones in Greece: Mineralogy and geologica environment of formation. In Proceedings of the Understanding the Genesis of Ore Deposits to Meet the Demand of the 21st Century, Moscow, Russia, 21–24 August 2006; p. 6.
16. Voudouris, P.; Graham, I.; Melfos, V.; Sutherland, L.; Zaw, K. Gemstones in Greece: Mineralogy and crystallizing environment. In Proceedings of the 34th IGC conference, Brisbane, Australia, 5–10 August 2012
17. Ottens, B.; Voudouris, P. *Griechenland: Mineralien-Fundorte-Lagerstätten*; ChristianWeise Verlag: Munchen Germany, 2018; 480p.
18. Voudouris, P.; Mavrogonatos, C.; Graham, I.; Giuliani, G.; Melfos, V.; Karampelas, S.; Karantoni, V.; Wang, K. Tarantola, A.; Zaw, K.; et al. Gem Corundum Deposits of Greece: Geology, Mineralogy and Genesis. *Minerals* **2019**, *9*, 49. [CrossRef]
19. Tarantola, A.; Voudouris, P.; Eglinger, A.; Scheffer, C.; Trebus, K.; Bitte, M.; Rondeau, B.; Mavrogonatos, C. Graham, I.; Etienne, M.; et al. Metamorphic and Metasomatic Kyanite-Bearing Mineral Assemblages of Thassos Island (Rhodope, Greece). *Minerals* **2019**, *9*, 252. [CrossRef]
20. Pouchou, J.-L.; Pichoir, F. Quantitative Analysis of Homogeneous or Stratified Microvolumes Applying the Model "PAP". In *Electron Probe Quantitation*; Heinrich, K.F.J., Newbury, D.E., Eds.; Springer: Boston, MA USA, 1991; pp. 31–75. ISBN 978-1-4899-2617-3.
21. Mattey, D.P. LaserPrep: An Automatic Laser-Fluorination System for Micromass 'Optima' or 'Prism' Mass Spectrometers. *Micromass Appl. Note* **1997**, *207*, 8.
22. Jolivet, L.; Brun, J.P. Cenozoic geodynamic evolution of the Aegean region. *Int. J. Earth. Sci.* **2010**, *99*, 109–138 [CrossRef]
23. Jolivet, L.; Faccenna, C.; Huet, B.; Labrousse, L.; Le Pourhiet, L.; Lacombe, O.; Lecomte, E.; Burov, E.; Denèle, Y. Brun, J.-P.; et al. Aegean tectonics: Strain localization, slab tearing and trench retreat. *Tectonophysics* **2013** *597*, 1–33. [CrossRef]
24. Ring, U.; Glodny, J.; Thomson, S. The Hellenic subduction system: High-pressure metamorphism, exhumation normal faulting and large-scale extension. *Earth Plan. Sci.* **2010**, *38*, 45–76. [CrossRef]
25. Brun, J.-P.; Sokoutis, D. Kinematics of the Southern Rhodope Core Complex (North Greece). *Int. J. Earth Sci* **2007**, *96*, 1079–1099. [CrossRef]

26. Pe-Piper, G.; Piper, D.J.W. The igneous rocks of Greece. The anatomy of an orogen. In *Beiträge der regionalen Geologie der Erde 30*; Stuttgart: Berlin, Germany, 2002; 573p.
27. Harlow, G.E.; Sorensen, S.S.; Sisson, V.B. Jade. In *The Geology of Gem Deposits*; Groat, L.A., Ed.; Mineralogical Association of Canada Short Course Series: Quebec City, QC, Canada, 2007; Volume 37, pp. 207–254.
28. Wang, K.K.; Graham, I.T.; Lay, A.; Harris, S.J.; Cohen, D.R.; Voudouris, P.; Belousova, E.; Giuliani, G.; Fallick, A.E.; Greig, A. The Origin of A New Pargasite-Schist Hosted Ruby Deposit From Paranesti, Northern Greece. *Can. Mineral.* **2017**, *55*, 535–560. [CrossRef]
29. Wang, K.K.; Graham, I.T.; Martin, L.; Voudouris, P.; Giuliani, G.; Lay, A.; Harris, S.J.; Fallick, A. Fingerprinting Paranesti Rubies through Oxygen Isotopes. *Minerals* **2019**, *9*, 91. [CrossRef]
30. Papanikolaou, D. Contribution to the geology of the Aegean Sea: The Island of Andros. *Ann. Geol Pays Hell.* **1978**, *29*, 477–553.
31. Stalder, H.A. Petrographische und mineralogische Untersuchungen im Grimselgebiet (Mittleres Aarmassiv). *Schweiz Miner. Petr. Mitt.* **1964**, *44*, 187–398.
32. Mullis, J. Growth conditions of quartz crystals from Val d'Illiez (Valais, Switzerland). *Schweiz Miner. Petr. Mitt.* **1975**, *55/3*, 419–430.
33. Niedermayer, G. Alpine Kluefte. *Min. Welt* **1993**, *6*, 57–59.
34. Ottens, B.; Tsinidis, T.; Voudouris, P. Amethyst and smoky quartz: Scepter crystals from Kato Nevrokopi, Greece. *Lapis* **2016**, *41*, 10–21. (In German)
35. Voudouris, P.; Melfos, V.; Mavrogonatos, C.; Tarantola, A.; Götze, J.; Alfieris, D.; Maneta, V.; Psimis, I. Amethyst occurrences in Tertiary volcanic rocks of Greece: Mineralogical, fluid inclusion and oxygen isotope constraints on their genesis. *Minerals* **2018**, *8*, 324. [CrossRef]
36. Liati, A. Corundum-and zoisite-bearing marbles in the Rhodope Zone, Xanthi area (N. Greece): Estimation of the fluid phase composition. *Mineral. Petrol.* **1988**, *38*, 53–60. [CrossRef]
37. Voudouris, P.; Graham, I.; Melfos, V.; Zaw, K.; Lin, S.; Giuliani, G.; Fallick, A.; Ionescu, M. Gem corundum deposits of Greece: Diversity, chemistry and origins. In Proceedings of the 13th Quadrennial IAGOD Symposium, Adelaide, Australia, 6–9 April 2010; Volume 69, pp. 429–430.
38. Graham, I.; Voudouris, P.; Melfos, V.; Zaw, K.; Meffre, S.; Sutherland, F.; Giuliani, G.; Fallick, A. Gem corundum deposits of Greece: A spectrum of compositions and origins. In Proceedings of the 34th IGC Conference, Brisbane, Australia, 5–10 August 2012.
39. Feenstra, A.; Wunder, B. Dehydration of diasporite to corundite in nature and experiment. *Geology* **2002**, *30*, 119–122. [CrossRef]
40. Sutherland, F.L.; Zaw, K.; Meffre, S.; Giuliani, G.; Fallick, A.E.; Graham, I.T.; Webb, G.B. Gem corundum megacrysts from East Australia basalt fields: Trace elements, O isotopes and origins. *Aust. J. Earth Sci.* **2009**, *56*, 1003–1020. [CrossRef]
41. Harris, S.J.; Graham, I.T.; Lay, A.; Powell, W.; Belousova, E.; Zappetini, E. Trace element geochemistry and metasomatic origin of alluvial sapphires from the Orosmayo region, Jujuy Province, Northwest Argentina. *Can. Mineral.* **2017**, *55*, 595–617. [CrossRef]
42. Peucat, J.J.; Ruffault, P.; Fritsch, E.; Bouhnik-Le Coz, M.; Simonet, C.; Lasnier, B. Ga/Mg ratio as a new geochemical tool to differentiate magmatic from metamorphic blue sapphires. *Lithos* **2007**, *98*, 261–274. [CrossRef]
43. Giuliani, G.; Caumon, G.; Rakotosamizanany, S.; Ohnenstetter, D.; Rakototondrazafy, M. Classification chimique des corindons par analyse factorielle discriminante: Application a la typologie des gisements de rubis et saphirs. Chapter mineralogy, physical properties and geochemistry. *Rev. Gemmol.* **2014**, *188*, 14–22.
44. Tsujimori, T.; Harlow, G.E. Petrogenic relationships between jadeitite and associated high-pressure and low-temperature metamorphic rocks in worldwide jadeitite localities: A review. *Eur. J. Min.* **2012**, *24*, 371–390. [CrossRef]
45. Bröcker, M.; Endrers, M. U-Pb zircon geochronology of unusual eclogite-facies rocks from Syros and Tinos (Cyclades, Greece). *Geol. Mag.* **1999**, *136*, 111–118.
46. Higgins, M.D. Neolithic jadeite and eclogite axes from Greece: A possible material source on the Island of Syros, Greece. In *GAC-MAC-CSPG-CSSS*; Hallifax: Nova Scotia, Canada, 2005; Volume 30.
47. Voudouris, P.; Graham, I.; Mavrogonatos, K.; Su, S.; Papavasiliou, K.; Farmaki, M.-V.; Panagiotidis, P. Mn-andalusite, spessartine, Mn-grossular, piemontite and Mn-zoisite/clinozoisite from Trikorfo, Thassos Island, Greece. *Bull. Geol. Soc. Greece* **2016**, *50*, 2068–2078. [CrossRef]

48. Fitros, M.; Tombros, S.F.; Simos, X.C.; Kokkalas, S.; Hatzipanagiotou, K. Formation of Mn-skarn Ores a Thapsana Mines, Paros Island, Attico-Cycladic Metallogenetic Massif, Greece. In *Proceedings of the the 1st In Electronic Conference on Mineral Science, 16–31 July 2018*; Sciforum, MDPI AG: Basel, Switzerland.
49. Voudouris, P.; Katerinopoulos, A.; Magganas, A. Mineralogical geotopes in Greece: Preservation an promotion of museum specimens of minerals and gemstones. Sofia Initiative "Mineral diversity preservation" In Proceedings of the IX International Symposium on Mineral Diversity Research and Reservation, Sofia Bulgaria, 16–18 October 2017; pp. 149–159.
50. Maneta, V.; Voudouris, P. Quartz megacrysts in Greece: Mineralogy and environment of formation. *Bul Geol. Soc. Greece* **2010**, *43*, 685–696. [CrossRef]
51. Voudouris, P.; Constantinidou, S.; Kati, M.; Mavrogonatos, C.; Kanellopoulos, C.; Volioti, E. Genesis o alpinotype fissure minerals from Thassos Island, Northern Greece—Mineralogy, mineral chemistry an crystallizing environment. *Bull. Geol. Soc. Greece* **2013**, *47*, 468–476. [CrossRef]
52. Klemme, S.; Berndt, J.; Mavrogonatos, C.; Flemetakis, S.; Baziotis, I.; Voudouris, P.; Xydous, S. On the colo and genesis of prase (green quartz) and Amethyst from the Island of Serifos, Cyclades, Greece. *Mineral* **2018**, *8*, 487. [CrossRef]
53. Ottens, B.; Voudouris, P. Minerals and gold ores: The mining district of Chalkidiki, Greece. *Lapis* **2007**, 4 32–58. (In German)
54. Taylor, H.P., Jr. Oxygen and hydrogen isotope relationships in hydrothermal mineral deposits. In *Geochemistr of Hydrothermal Ore Deposits*, 2nd ed.; Barnes, H.L., Ed.; Wiley: New York, NY, USA, 1979; pp. 236–318.
55. Giggenbach, W.F. The origin and evolution of fluids in magmatic-hydrothermal systems. In *Geochemistry o Hydrothermal Ore Deposits*, 3rd ed.; Barnes, H.L., Ed.; Wiley: New York, NY, USA, 1997; pp. 737–796.
56. Sharp, Z.D.; Gibbons, J.A.; Maltsev, O.; Atudorei, V.; Pack, A.; Sengupta, S.; Shock, E.L.; Knauth, L. A calibration of the triple oxygen isotope fractionation in the SiO_2–H_2O system and applications to natura samples. *Geochim. Cosmochim. Acta* **2016**, *186*, 105–119. [CrossRef]
57. Melfos, V.; Voudouris, P. Fluid inclusions and O-isotopes of an amethyst and smoky quartz occurrence i Kato Nevrokopi, Rhodope massif, Northern Greece. In Proceedings of the 11th International Conference o Experimental Mineralogy Petrology Geochemistry, Bristol, UK, 11–13 September 2006.
58. Bitte, M.; Tarantola, A.; Scheffer, C.; Voudouris, P.; Valero, M. *Fluid Inclusion Characterization of Alpine-Typ Quartz Tension Gashes of Thassos Island (Rhodope Massif, Greece)*; European Current Research on Fluid and Mel Inclusions (ECROFI): Nancy, France, 2019.
59. Tarantola, A.; Scheffer, C.; Voudouris, P.; Vanderhaeghe, O.; Rakotomanga, S.; Rigaudier, T. *Metamorphic an Basin Fluids Trapped in Evia Island Alpine Quartz Crystals during Exhumation of the Attic-Cycladic Metamorphi Complex (Greece)*; ECROFI 24: Nancy, France, 2017.
60. Thewalt, U.; Voudouris, P. Jaspis von Sykorachi (Thrakien, Griechenland) und seine Einschlüsse. *Miner. Wel* **2019**, *5*, 92–96.
61. Thewalt, U.; Dorfner, G.; Samouchos, M. Chalcedon, Achat und Jaspis vom westlichen Kustenbereich de Golfs von Kalloni, Lesbos, Griechenland. *Miner. Welt* **2018**, *5*, 90–96.
62. Kelepertzis, A.E.; Velitzelos, E. Oligocene swamp sediments of Lesvos Island, Greece (Geochemistry an Mineralogy). *Facies* **1992**, *27*, 113–118. [CrossRef]
63. Velitzelos, E.; Zouros, N. The Petrified Forest of Lesvos. In *Protected Natural Monument, Engineering Geolog and the Environment*; Marinos, P., Koukis, G., Tsiambaos, G., Stournaras, G., Eds.; Balkema: Rotterdam, The Netherlands, 1997; pp. 3037–3043.
64. Voudouris, P.; Velitzelos, D.; Velitzelos, E.; Thewald, U. Petrified wood occurences in western Thrace an Limnos Island: Mineralogy, geochemistry and depositional environment. *Bull. Geol. Soc. Greece* **2007**, 40 238–250. [CrossRef]
65. Papp, A. Erläuterungen zur Geologie der Insel Lemnos. *Ann. Geol. Pays Hell.* **1953**, *5*, 1–33.
66. Kassoli-Fournaraki, A.; Michailidis, K.; Zanas, I.; Zachos, S. Titanite-rich carbonates from the Therapio are in Thrace, northern Greece: constraints of the mineral assemblage formation. *Schw. Mineral. Petr. Mitt.* **1995** *75*, 387–398.
67. Henry, D.J.; Dutrow, B.L. Compositional zoning and element partitioning in nickeloan tourmaline fror metamorphosed karstbauxite from Samos, Greece. *Am. Mineral.* **2001**, *86*, 1130–1142. [CrossRef]

68. Baksheev, I.A.; Kudryavtseva, O.E. Nickeloan tourmaline from the Berezovskoe gold deposit, Middle Urals, Russia. *Can. Mineral.* **2004**, *42*, 1065–1078. [CrossRef]
69. Sklavounos, S.; Ericsson, T.; Filippidis, A.; Michailidis, K.; Kougoulis, C. Chemical, X-ray and Mossbauer investigation of a turquoise from the Vathi area volcanic rocks, Macedonia, Greece. *Neues Jahrb. Mineral. Mh.* **1992**, 469–480.
70. Voudouris, P.; Tsolakos, A.; Papanikitas, A.; Solomos, C. Neufunde aus Lavrion. *Lapis* **2004**, *29*, 13–15.
71. Marinos, G.; Petrascheck, W.E. Laurium. Geological and geophysical research. *Greece Inst. Geol. Subsurf. Res.* **1956**, *4*, 1–247.
72. Boni, M.; Gilg, H.A.; Aversa, G.; Balassone, G. The "Calamine" of Southwest Sardinia: Geology, Mineralogy, and Stable Isotope Geochemistry of Supergene Zn Mineralization. *Econ. Geol.* **2003**, *98*, 731–748. [CrossRef]
73. Frisch, P.L.; Lueth, V.W.; Hlava, P.F. The colors of smithsonite: A microchemical investigation. New Mexico. *Geology* **2002**, *24*, 132–133.
74. Samouhos, M.; Zavašnik, J.; Rečnik, A.; Godelitsas, A.; Chatzitheodoridis, E.; Sanakis, Y. Spectroscopic and nanoscale characterization of blue-colored smithsonite ($ZnCO_3$) from Lavrion historical mines (Greece). *Per. Mineral.* **2015**, *84*, 373–388.
75. Miladinova, I.; Froitzheim, N.; Nagel, T.J.; Janák, M.; Georgiev, N.; Fonseca, R.O.C.; Sandmann, S.; Münker, C. Late Cretaceous eclogite in the Eastern Rhodopes (Bulgaria): Evidence for subduction under the Sredna Gora magmatic arc. *Int. J. Earth Sci.* **2018**, *107*, 2083–2099. [CrossRef]
76. Laurent, V.; Huet, B.; Labrousse, L.; Jolivet, L.; Monie, P.; Augier, R. Extraneous argon in high-pressure metamorphic rocks: Distribution, origin and transport in the Cycladic Blueschist Unit (Greece). *Lithos* **2017**, *272*, 315–335. [CrossRef]
77. Altherr, R.; Kreuzer, H.; Wendt, I.; Lenz, H.; Wagner, G.A.; Keller, J.; Harre, W.; Höhndorf, A. A late Oligocene/early Miocene high temperature belt in the Attic-Cycladic crystalline complex (SE Pelagonian, Greece). *Geol. Jahrb.* **1982**, *23*, 97–164.
78. Keay, S.; Lister, G.; Buick, I. The timing of partial melting, Barrovian metamorphism and granite intrusion in the Naxos metamorphic core complex, Cyclades, Aegean Sea, Greece. *Tectonophysics* **2001**, *342*, 275–312. [CrossRef]
79. Koutsovitis, P.; Magganas, A.; Pomonis, P.; Ntaflos, T. Subduction-related rodingites from East Othris, Greece: Mineral reactions and physicochemical conditions of formation. *Lithos* **2013**, *172–173*, 139–157. [CrossRef]
80. Jansen, J.B.H.; Schuiling, R.D. Metamorphism on Naxos; petrology and geothermal gradients. *Am. J. Sci.* **1976**, *276*, 1225–1253. [CrossRef]
81. Siebenaller, L. Fluid Circulations during Collapse of an Accretionary Prism: Example of the Naxos Island Metamorphic Core Complex (Cyclades, Greece). Ph.D. Thesis, University of Lorraine, Lorraine, France, 2008.
82. Henley, R.W.; Ellis, A.J. Geothermal systems ancient and modern: A geochemical review. *Earth Sci. Rev.* **1983**, *19*, 1–50. [CrossRef]
83. Fournier, R.O. The behaviour of silica in hydrothermal solutions. Geology and geochemistry of epithermal systems. *Rev. Econ. Geol.* **1985**, *2*, 45–61.
84. Hitzman, M.W.; Reynolds, N.A.; Sangster, D.F.; Allen, C.R.; Carman, C.E. Classification, Genesis, and Exploration Guides for Nonsulfide Zinc Deposits. *Econ. Geol.* **2003**, *98*, 685–714. [CrossRef]
85. Boni, M.; Large, D. Non sulfide Zinc Mineralization in Europe: An Overview. *Econ. Geol.* **2003**, *98*, 715–729. [CrossRef]

8

Gem-Quality Tourmaline from LCT Pegmatite in Adamello Massif, Central Southern Alps, Italy: An Investigation of its Mineralogy, Crystallography and 3D Inclusions

Valeria Diella [1,*], Federico Pezzotta [2], Rosangela Bocchio [3], Nicoletta Marinoni [1,3], Fernando Cámara [3], Antonio Langone [4], Ilaria Adamo [5] and Gabriele Lanzafame [6]

1. National Research Council, Institute for Dynamics of Environmental Processes (IDPA), Section of Milan, 20133 Milan, Italy; nicoletta.marinoni@unimi.it
2. Natural History Museum, 20121 Milan, Italy; fpezzotta@yahoo.com
3. Department of Earth Sciences "Ardito Desio", University of Milan, 20133 Milan, Italy; rosangela.bocchio@unimi.it (R.B.); fernando.camara@unimi.it (F.C.)
4. National Research Council, Institute of Geosciences and Earth Resources (IGG), Section of Pavia, 27100 Pavia, Italy; langone@igg.cnr.it
5. Italian Gemmological Institute (IGI), 20123 Milan, Italy; ilaria.adamo@guest.unimi.it
6. Elettra-Sincrotrone Trieste S.C.p.A., Basovizza, 34149 Trieste, Italy; gabriele.lanzafame@elettra.eu

* Correspondence: valeria.diella@cnr.it

Abstract: In the early 2000s, an exceptional discovery of gem-quality multi-coloured tourmalines hosted in Litium-Cesium-Tantalum (LCT) pegmatites, was made in the Adamello Massif, Italy. Gem-quality tourmalines had never been found before in the Alps, and this new pegmatitic deposit is of particular interest and worthy of a detailed characterization. We studied a suite of faceted samples by classical gemmological methods, and fragments were studied with Synchrotron X-ray computed micro-tomography, which evidenced the occurrence of inclusions, cracks and voids. Electron Microprobe combined with Laser Ablation analyses were performed to determine major, minor and trace element contents. Selected samples were analysed by single crystal X-ray diffraction method. The specimens range in colour from colourless to yellow, pink, orange, light blue, green, amber, brownish-pink, purple and black. Chemically, the tourmalines range from fluor-elbaite to fluor-liddicoatite and rossmanite: these chemical changes occur in the same sample and affect the colour. Rare Earth Elements (REE) vary from 30 to 130 ppm with steep Light Rare Earth Elemts (LREE)-enriched patterns and a negative Eu-anomaly. Structural data confirmed the elbaitic composition and showed that high manganese content may induce the local static disorder at the O(1) anion site, coordinating the Y cation sites occupied, on average, by Li, Al and Mn^{2+} in equal proportions, confirming previous findings. In addition to the gemmological value, the crystal-chemical studies of tourmalines are unanimously considered to be a sensitive recorder of the geological processes leading to their formation, and therefore, this study may contribute to understanding the evolution of the pegmatites related to the intrusion of the Adamello pluton.

Keywords: granitic pegmatite; gem-quality tourmaline; Adamello Massif; Central Alps; Italy

---

1. Introduction

"Tourmaline" is considered one of the most beautiful gemstones, because it occurs in a large spectrum of colours, as well as in multi-coloured crystals [1]. In Italy, gem-quality tourmaline is known to be from the historic locality, at present rather exhausted, of Elba Island (Tyrrhenian sea)

from which it derived the root-name elbaite, given to Li-bearing and sodium-rich tourmalines of the alkali group [2]. For a very long time, Elba Island has remained the only known locality providing gem-quality tourmaline crystals in Italy. Nevertheless, since 2001, an additional significant locality has been discovered and described in the Italian Alps, which is the Adamé valley, along the western border of the late Alpine Adamello tonalite massif, in the central Southern Alps [3].

In the Adamé valley, gem-quality tourmaline occurs in multi-coloured elongated crystals up to 7 cm in length in the miarolitic cavities of LCT pegmatite (i.e., a pegmatite with prevailing minerals of lithium, caesium and tantalum) hosted in contact with metamorphic sandstones, belonging to a Permian-Mesozoic sedimentary sequence. The locality is included in the Adamello Park natural reservation. The miarolitic pegmatite has been found on the steep slope of Forcel Rosso pass, disrupted in large blocks in an ancient landslide, at an altitude of about 2600 m. Tourmaline group minerals occur together with albite, smoky quartz, K-feldspar, mica (lepidolite–muscovite) and other accessory minerals such as fluorapatite, fluorite, and several Nb–Ta–Sb oxides, all very interesting from a scientific and collector point of view. In consideration of the significant potential of the pegmatite for mineral production, the locality underwent, under strict protection, an official detailed field investigation for scientific purposes, with the supervision of one of the authors (F.P.). The first results [3] show that the tourmaline crystals have a zoned composition, mostly characterized by fluor-elbaite, with fluor-liddicoatite developing at the antilogous pole of the crystals. Colours range from colourless to yellow, pink, orange, light blue, green, amber, brownish-pink, purple and black (Figure 1). Gem-quality tourmalines are present as sectors of larger crystals, and rough fragments can reach a maximum weight of 2–3 g.

Figure 1. Crystals of pink (3.5 cm long) and green (5 cm long) tourmalines from Valle Adamé, Adamello Massif. Natural History Museum of Milan collection. Photos by R. Appiani.

The collected materials are preserved in the mineralogical collections of the Museum of Natural History of Milan and at the Museum of the Adamello Natural Park. These gem tourmalines have not entered the market, because the deposit is located within the protected areas of the Adamello Natural Park and collecting is allowed only for scientific purposes. At present, only a few gems have been faceted (Figure 2).

Figure 2. Cut tourmalines analysed in this study. (**a**) 3.97 × 8.31 × 2.61 mm; (**b**) 4.73 × 6.58 × 3.35 mm; (**c**) 4.61 × 6.36 × 3.26 mm, (**d**) 5.72 × 7.16 × 4.03 mm, corresponding to samples 13, 14, 15 and 16 described below in Table 1. Photos by M. Chinellato and F. Picciani.

As mentioned above, tourmaline represents one of the most beautiful gemstones and may b characterized through multi-methodological methods including non-invasive and non-destructiv techniques, such as Fourier Transform Infrared Spectroscopy and Raman Spectroscopy, when a conservative treatment of sample is required ([4] and references therein). Many studies recognize a significant petrological interest of this mineral, as it can contribute to the reconstruction of the evolutio of the history of a crystalline basement, e.g., [5].

In view of this, and considering that the Adamello Massif is one of the most geologically studie portions of the Alpine chain, our study aims to provide a complete mineralogical characterization o these recently discovered gem tourmalines, as well as an interesting contribution to the knowledge o local pegmatite mineralization. We investigated a suite of cut and rough samples selected from th mineralogical collections of the Natural History Museum of Milan, through gemmological analyses electron microprobe chemical analyses (EPMA) and laser ablation-inductively coupled plasma–mas spectrometry (LA-ICP-MS). The unit cell parameters and refinement of crystal structure of thre selected tourmalines were carried out by the single-crystal X-ray diffraction method. Selected rough samples have also been examined by means of synchrotron X-ray computed micro-tomography (X-μCT), which made it possible to obtain a 3D visualisation of the inner objects (pore, inclusions, etc. within a volume [6,7]. This technique does not require any sample preparation and makes it possible to overcome the problem of deriving the 3D results from the traditional 2D data obtained by traditiona imaging techniques such as optical and scanning electron microscopy. In the present study, a fully characterisation of inclusions in terms of size, shape and orientation was performed.

2. Background Information

Tourmaline is a ring-silicate crystallizing in the acentric $3m$ point group (ditrigonal pyramidal with the $R3m$ space group and a general formula of $XY_3Z_6T_6O_{18}(BO_3)_3V_3W$ [8]. The unit cell consists of a six-fold ring of tetrahedra (T sites, occupied primarily by Si) on top of a concentric arrangement of three Y-site and six Z-site octahedra. The X site is nine-coordinated and situated centred over o the six-fold ring and can be occupied by Na^+, Ca^{2+}, K^+ cations or be vacant. The Y and Z sites may contain a large variety of cations such as Mg^{2+}, Fe^{2+}, Fe^{3+}, Li^{+}, Al^{3+}, Mn^{2+}, Cu^{2+}, Ti^{4+}, Cr^{3+}, $V^{3+}$and Al^{3+}, Mg^{2+}, Fe^{2+}, Fe^{3+}, Cr^{3+}, V^{3+}, respectively. Three planar triangular boron-centred polyhedra (BO_3) are further present, which are roughly coplanar with the groups of octahedra in the unit cell, and roughly perpendicular to the *c* axis. The *V* and *W* sites are anion sites that are occupied by OH^- or O^{2-} (or both) at the *V* site, and OH^-, F^-, or O^{2-} at the *W* site. The Cl contents are generally negligible. Given the range of elements and heterovalent substitutions that can be accommodated by the tourmaline structure, the high number of different end-member compositions constitute a super group consisting of quite a large number of species, as approved by the International Mineralogical Association's Commission on New Minerals, Nomenclature and Classification [2,9–11].

3. Geological Setting

The Tertiary Adamello batholith outcrops over an area of ca. 670 km^2 in the Central Southern Alps (Northern Italy) (CNR, Italian Geological Map Adamello-Presanella, 1:50,000). It consists of four composite plutons decreasing in age from south (ca. 42 Ma) to north (ca. 31 Ma) and is composed mainly of granitoid rocks (granodiorite, tonalite, quartz diorite) with minor amounts of diorite and gabbro [12–14]. The plutons intrude into the South-Alpine crystalline basement and its Permo-Mesozoic sedimentary cover (Figure 3). The studied tourmaline crystals were collected in pegmatite dykes occurring in the thermo-metamorphic contact aureole between the igneous rocks and the surrounding sediments, at the northern slope of the Foppa mountain (2752 m) in a gully of Forcel Rosso (or "Vallone del Forcel Rosso").

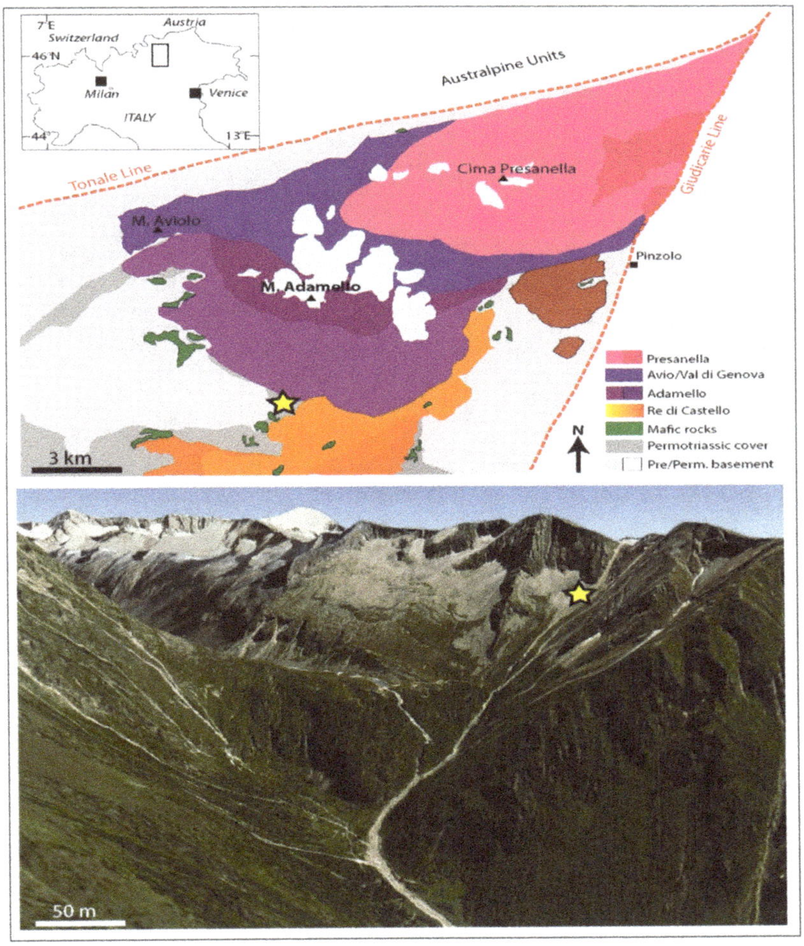

Figure 3. Geological sketch map of the Adamello pluton and a view of the area of Forcel Rosso. Photo by F. Pezzotta. The stars indicate the location of the studied pegmatite dikes.

The pegmatites are sub-horizontal and hosted in meta-sandstones enriched in accessory tourmaline. The larger veins are up to approximately 1.5 m in thickness, and some tens of meters long. Despite their relatively small size, these pegmatites display a quite strong asymmetric zoning with a layered lower unit composed by fine to medium grain size assemblage of feldspars and quartz, with accessory muscovitic mica, schorl and spessartine, and an upper coarse-grained unit composed of an assemblage of quartz and feldspars, rich in schorl and muscovitic mica, occasionally displaying cores with milky white quartz masses, lepidolite overgrowth at the rim of muscovite blades, granular lepidolite masses and multi-coloured tourmalines. Miarolitic cavities are not common, but they can be locally abundant, with a diameter achieving several decimetres in length, at the coarse-grained cores of the veins.

Accessory minerals in cavities include pink to green fluor-apatite, purple and pale green fluorite, late-stage calcite, several pyrochlore group minerals and various Ta–Nb oxides [3].

4. Materials and Methods

Sixteen samples of tourmaline from different miarolitic cavities were selected for this study, and they are described in Table 1.

Table 1. Description of the studied tourmalines.

Sample	Cavity Name/Description *	Colour	Analytical Techniques	
1	Cavity 1	Pink	Electron microprobe (WDS) (°)	Single crystal XRD (°)
2	Cavity boulder 2	Light blue	Electron microprobe (WDS) (°)	Single crystal XRD (°)
3	Pizio cavity	Colourless to brown	Electron microprobe (WDS) (°)	
4: sample cut perpendicular to the *c*-axis	Pizio cavity	Pink green	Electron microprobe (WDS) (°)	
5	Cavity 3	Light green yellow	Electron microprobe (WDS) (°)	
6	Cavity 4	Yellow to colourless	Electron microprobe (WDS) (°)	
7	Cavity 5			Synchrotron X-ray computed μ-tomography
8	Cavity 6			Synchrotron X-ray computed μ-tomography
9	"Black quartz" cavity	Pink green	Electron microprobe (WDS) (°)	
10: double terminated sample	Inv.# M36742 cavity	Browm green	Electron microprobe (WDS) (°)	
11: homologous terminated sample	Inv.# M36742 cavity	Green to colourless	Electron microprobe (WDS) (°)	
12: sample cut perpendicular to the *c*-axis	Inv.# M36742 cavity	Blue	Electron microprobe (WDS) (°)	Single crystal XRD (°)
13	Cut stone (0.731 ct) from pocket	Green brown	Specific gravity	Refractive index
14	Cut stone (0.848 ct) from pocket	Green brown	Specific gravity	Refractive index
15	Cut stone (0.838 ct) from pocket	Green brown	Specific gravity	Refractive index
16	Cut stone (1.225 ct) from pocket	Pale green	Specific gravity	Refractive index

* Cavity numbers refer to the unpublished field note-book made during the collecting campaigns performed by the Natural History Museum of Milan in 2001 and 2002. The reported inventory numbers are the ones given to specimens catalogued in the Mineralogical Collections of the Natural History Museum of Milan. ° Wavelength Dispersive Spectroscopy (WDS); X-ray Diffractometry (XRD).

Four faceted samples (13–16) were examined by standard gemmological methods at the Italian Gemmological Institute in Milan in order to determine their refractive index, hydrostatic specific gravity and microscopic features. The refractive indices were measured with a Krüss refractometer (A. Krüss Optronic, Hamburg, Germany) using ordinary light source with a sodium filter (589 nm) and methylene iodide as a contact liquid (n = 1.80). A Mettler hydrostatic balance was used to determine the specific gravity in bi-distilled water.

Two grains (7 and 8) were used for synchrotron X-ray computed micro-tomography. The samples, with a size of about 1.5 mm, were imaged at the SYRMEP beamline of ELETTRA synchrotron facility (Trieste, Italy). The computer micro-tomography experiments were performed using a polychromatic X-ray beam (white beam mode. With this configuration, the outcome beam from the storage ring was intercepted before the monochromator and pre-filtered with 1.5 mm of Si and 1.0 mm of Al. A water-cooled, 16 bit, 2028 × 2048 pixel microscope Charge-Coupled Device (CCD) camera, coupled with a 100 micron thick LuAG scintillator screen was used as detector. The mean energy was set to 28 keV. Pixel size was set at 2 μm/pixel, yielding a field of view of 5 × 5 mm.

The software suite Syrmep Tomo Project (STP) [15,16] was used to reconstruct two-dimensional axial slices from the sample projections, applying the filtered back projection algorithm [17,18]. Before image reconstruction, a single-distance phase retrieval algorithm was applied to the projection images [19] using the STP software and setting the δ/β ratio to 15 (see [15,16] for details about the phase retrieval process).

Ten of the selected fragments, 1–6 and 9–12, were embedded in epoxy resin, polished and prepared for EMPA and LA-ICP-MS analysis. Optical images of the samples were performed in advance using a stereo microscope Leica M205 C in reflected light. The backscattered electron images and quantitative chemical analyses of major and minor elements were obtained with the JEOL JXA-8200 electron microprobe in wavelength dispersion mode (EMPA-WDS) at the laboratory of the Department of Earth Sciences of the University of Milan under the following conditions: 15 kV accelerating voltage, 5 nA beam current, and a count time of 60 s on peak and 30 s on the background, with a 1 μm diameter beam. The Kα wavelengths and natural standards of pure metals were used for calibration: F (hornblende), Ti (ilmenite), Mn and Zn (rhodonite), K (K-feldspar), Na (omphacite), Fe (fayalite), Ca, Si and Al (grossular), Mg (olivine), Cr and V (pure elements). The raw data were corrected for matrix effects using a conventional ΦρZ routine in the JEOL software package. Both Fe and Mn were calculated as Fe^{2+} and Mn^{2+}. The structural formula was calculated on the basis of 31 anions (O, OH and F), and Li_2O, B_2O_3 and H_2O were calculated based on the assumed elbaite stoichiometry.

Rare earth and selected trace elements were determined by laser ablation–inductively coupled plasma–mass spectroscopy (LA-ICP-MS) at the IGG-CNR Laboratory of Pavia. The instrument consisted of a Quantel Brilliant 213 nm Nd:YAG laser (New Wave Research) coupled to a Perkin Elmer DRCe quadrupole ICPMS. The spot size was 55 μm, using NIST SRM 610 glass as an external standard and Si as an internal standard, as analysed by microprobe. Precision and accuracy estimated on the basaltic glass standard BCR2 and NIST612 were better than 10%.

Three fragments (~500 μm) from samples 1, 2, and 12 were cut for single crystal X-ray diffraction measurements. Data were collected with an Xcalibur-Oxford Diffraction diffractometer (Oxford Diffraction Ltd., Abingdon, UK) equipped with a CCD, using graphite-monochromatized MoKα radiation and operated at 50 kV and 30 mA. To maximize the reciprocal space coverage, a combination of ω and φ scans was used, with a step size of 1° and an exposure time per frame of 3–5 s. Intensity data were then integrated and corrected for Lorentz polarization effects, using the computer program CrysAlis [20]. An empirical absorption correction was applied using CrysAlis [20]. The structure was refined using SHELX-97 [21]. Scattering curves for neutral chemical species were used at all sites. No peaks larger than $\pm 0.9\ e^-/Å^3$ were present in the final difference—the Fourier maps of the electron density and residual maxima and minima were equally balanced. The final agreement index (R_1) was 0.016–0.018/0.017–0.019 (obs/all) for 97 refined parameters.

5. Results

5.1. Gemmological Properties and 3D Visualisation of the Tourmaline Inclusions

From a gemmological point of view, the four faceted tourmalines are all transparent, with a colou ranging from green to brown (Figure 2). The specific gravity results from 3.09 to 3.13 g/cm^3, which i typical of elbaitic tourmaline [22]. The optic character is uniaxial negative, and the refractive indice are $n_\omega = 1.640$ and $n_\varepsilon = 1.620$ with a birefringence of 0.020, which is in agreement with the values o elbaite [22]. The samples show a strong dichroism, in the green and brown hues, and they are all iner to ultraviolet radiation. Microscopic observations revealed that the samples contain a few inclusion typically fluid inclusions, often with a fringed aspect, also called "thrichites" in gemology [23].

More details were provided by the 3D visualisation, which made it possible to display interna features and to quantify the porosity, cracks and voids of fragments from the prismatic portion o samples 7 and 8.

The volume rendering of the tourmaline sample is reported in Figure 4a, whereas the cros sectional slices, displaying the general appearance of the microstructural features in the grains, ar highlighted in Figure 4b,c. In particular, the reconstructed 3D images show that tourmalines ar characterised by pores and cracks, the detail of whose volume rendering is displayed in Figure 4d.

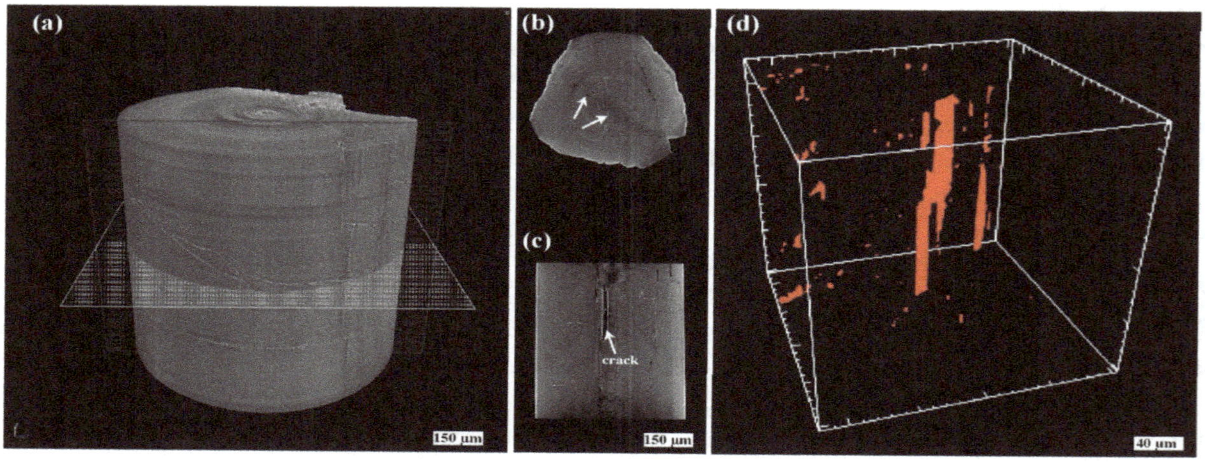

Figure 4. Synchrotron X-ray computed micro-tomography of tourmaline 7: (**a**) volume rendering of the sample; (**b**) reconstructed axial slice; (**c**) the cross section of the stack of reconstructed slices; (**d**) volume rendering of pores and cracks, a detail.

The total measured porosity accounts for about 0.9 vol.% and is given as the sum of the voids/pores (0.1 vol.%) and cracks (0.8 vol.%), respectively. Please note that the distinction between voids and cracks, which are characterised by the same grey scale values, was performed by considering some morphological features as described in [24]. In particular, the voids appear as spherical bubbles with a small surface/volume ratio (normalized to a sphere) and a low aspect ratio as well. On the contrary, cracks show a medium to medium-high surface/volume ratio, together with high values of aspect ratio.

The rendering pointed out that the voids appear spherical in shape, with an equivalent diameter ranging from 10 to 15 µm. Morphometric data showed that they are homogeneously distributed within the entire volume and are mostly isolated without any preferred orientation.

Tourmaline crystals show a predominance of crack spacing from the centre to the outer part of the crystals, with a non-homogenous spatial distribution and a heterogeneous size ranging from 20 to 200 µm. The fractures are mostly flat with a regular surface and appear oriented preferentially parallel ɔ the z-axis. Finally, the results of the skeleton analysis pointed out that the crack connectivity is null, ecause they lack a connected network.

.2. Chemical Composition

Electron microprobe analyses (EMPA) and backscattered electron (BKSE) images were performed o chemically characterize 10 of the selected samples. Analytical profiles from core to rim or from rim o rim were carried out with step width depending on the crystal size. LA-ICP-MS analyses were erformed on the same samples in the areas previously analysed by EMPA-WDS.

The values of Li_2O measured by LA-ICP-MS indicated that tourmalines contain an average of 1.8 wt % of Li_2O (ranging from 0.7 to 2.7 wt %), and the result is comparable to that calculated by stoichiometry rom the results of the electron microprobe analyses. The studied tourmaline samples exhibit extremely onvoluted chemical zoning and are very similar in composition, all classifiable as Li-bearing tourmalines: elbaite ($Na(Li_{1.5},Al_{1.5})Al_6(BO_3)_3Si_6O_{18}(OH)_4$), rossmanite ($\gamma(LiAl_2)Al_6(BO_3)_3Si_6O_{18}(OH)_4$), fluor-elbaite $Na(Li_{1.5},Al_{1.5})Al_6(BO_3)_3Si_6O_{18}(OH)_3F$) and fluor-liddicoatite ($Ca(LiAl_2)Al_6(BO_3)_3Si_6O_{18}(OH)_3F$). Minor foitite ($\gamma Fe^{2+}{}_2Al)Al_6(BO_3)_3Si_6O_{18}(OH)_4$), a rare tourmaline iron rich in Y-site and with more han 50% vacant X-sites [25], was also found.

The complete set of chemical analyses is graphically plotted in the ternary diagram of X-site composition (Figure 5).

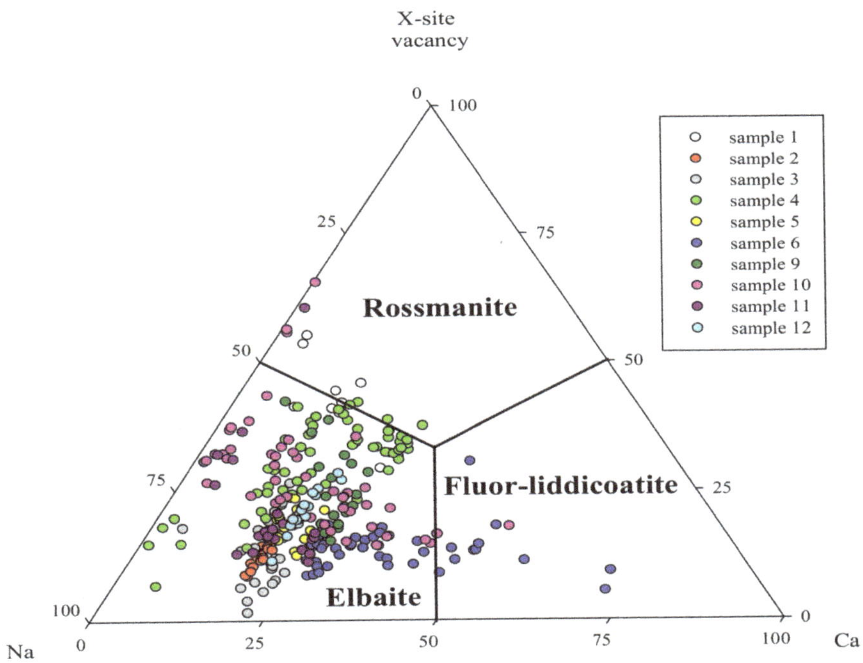

Figure 5. Ternary diagram of X-site composition of all analysed tourmalines.

Most analyses fall in the elbaite/fluor-elbaite field, although a few correspond to rossmanite (samples 1, 4, 10, 11) and fluor-liddicoatite (samples 6, 10). Foitite was found at the analogous pole of the crystals of tourmalines 10, 11 and as fibrous overgrowth on sample 9b. The foitite analyses are not reported in the diagram where they would fall in the rossmanite field due to their >50% vacant in X-site [25].

Selected electron microprobe analyses are reported in Table 2.

Table 2. Representative electron microprobe analyses of the studied tourmalines.

Sample			1		2	3a	3b				4				5
	Pink Elbaite	Pink Rossmanite	Pink Elbaite	Pink Rossmanite	Light Blue Elbaite	Colourless Elbaite	Brown Elbaite	Rim-Green Elbaite	Core-Pink Elbaite	Core-Pink Elbaite	Core-Pink Elbaite	Inter-Pink Elbaite	Rim-Green Elbaite	Elbaite	Light Green Elbaite
					Average (20 pts)	Average (10 pts)	Average (15 pts)								Average (20 pts)
Point Number °	4	6	8	9				16	10	24	23	56	54	33	
SiO_2 (wt %)	36.34	36.63	36.28	37.06	36.2	36.17	36.54	36.99	36.79	36.67	36.59	37.04	37.28	36.71	36.29
TiO_2	bdl	bdl	0.06	bdl	0.03	0.03	0.07	0.04	bdl	bdl	0.04	bdl	0.02	0.24	0.03
B_2O_3 *	11.06	11.18	11.02	11.27	10.78	10.87	10.77	10.98	10.94	10.93	10.91	11.04	11.05	10.64	10.9
Al_2O_3	43.3	44.73	42.95	44.91	38.83	40.28	37.86	40.32	41.41	41.36	41.34	42.14	40.72	36.01	40.16
V_2O_3	0.02	bdl	bdl	bdl	0.01	bdl	0.02	0.02	0.02	bdl	bdl	0.01	bdl	0.04	0.01
Cr_2O_3	bdl	bdl	bdl	bdl	0.02	0.01	bdl	bdl	bdl	0.03	0.01	0.03	bdl	0.01	bdl
FeO	0.02	bdl	bdl	0.03	1.00	0.73	1.43	0.83	bdl	bdl	0.01	0.01	0.52	4.75	0.77
MgO	0.03	bdl	bdl	0.01	0.02	0.02	0.01	0.01	bdl	0.01	0.01	0.01	bdl	bdl	0.01
MnO	0.52	0.21	0.55	0.27	3.79	2.79	3.99	2.82	0.16	0.16	0.13	0.12	2.02	2.45	2.93
ZnO	0.09	bdl	bdl	bdl	0.05	0.03	0.08	bdl	0.14	0.08	0.05	0.03	0.10	0.50	0.05
CaO	0.95	0.28	0.98	0.27	1.08	0.98	1.23	0.78	1.52	1.65	1.57	1.23	1.27	0.36	1.18
Li_2O *	1.87	1.74	1.91	1.75	1.73	1.77	1.76	1.76	2.12	2.16	2.15	2.09	1.95	1.5	1.80
Na_2O	1.38	1.39	1.42	1.35	2.19	2.05	2.25	1.98	1.26	1.29	1.31	1.39	1.72	2.73	1.98
K_2O	0.01	bdl	0.01	0.01	0.02	0.02	0.01	0.01	0.01	0.01	0.02	bdl	bdl	0.03	0.01
H_2O*	3.53	3.83	3.53	3.78	3.12	3.27	3.13	3.24	3.22	3.24	3.17	3.34	3.26	3.00	3.24
F	0.61	0.07	0.57	0.24	1.27	1.00	1.25	1.16	1.17	1.11	1.25	0.98	1.16	1.41	1.10
Total	99.73	100.07	99.27	100.94	100.13	100.02	100.48	100.95	98.77	98.7	98.57	99.45	101.08	100.39	100.47
O=F	0.26	0.03	0.24	0.1	0.53	0.42	0.53	0.49	0.49	0.47	0.53	0.41	0.49	0.59	0.46
Total *	99.47	100.04	99.03	100.84	99.6	99.6	99.96	100.46	98.27	98.23	98.04	99.04	100.59	99.79	100
						Structural formula based on 31 anions (O, OH, F)									
Si	5.71	5.692	5.723	5.714	5.835	5.783	5.905	5.856	5.846	5.831	5.829	5.831	5.864	5.996	5.788
Al	0.29	0.308	0.277	0.286	0.165	0.217	0.097	0.144	0.154	0.169	0.171	0.169	0.136	0.004	0.212
T sum	6.000	6.000	6.000	6.000	6.000	6.000	6.000	6.000	6.000	6.000	6.000	6.000	6.000	6.000	6.000
B	3.000	3.000	3.000	3.000	3.000	3.000	3.000	3.000	3.000	3.000	3.000	3.000	3.000	3.000	3.000
Al (Z)	6.000	6.000	6.000	6.000	6.000	6.000	6.000	6.000	6.000	6.000	6.000	6.000	6.000	6.000	6.000
Cr	-	-	-	-	-	-	-	-	-	-	-	-	-	-	-
Al	1.727	1.884	1.709	1.875	1.211	1.373	1.081	1.379	1.601	1.583	1.59	1.65	1.413	0.928	1.336
Ti	-	-	0.007	0.001	0.004	0.004	0.012	0.005	-	-	0.004	-	0.002	0.03	0.004
V	0.003	-	-	-	0.001	0.001	0.002	0.002	0.003	-	-	0.001	-	0.006	0.001
Cr	-	-	-	-	0.002	0.001	0.001	-	bdl	0.004	0.002	0.003	-	0.001	-
Mg	0.007	0.001	-	0.002	0.004	0.004	0.003	0.003	bdl	0.003	0.003	0.002	-	-	0.003
Mn	0.069	0.028	0.073	0.035	0.518	0.377	0.527	0.378	0.022	0.021	0.017	0.016	0.269	0.339	0.396
Fe^{2+}	0.003	-	-	0.003	0.134	0.097	0.231	0.11	0.001	-	0.001	0.001	0.069	0.649	0.102
Zn	0.01	-	-	-	0.006	0.004	0.013	-	0.016	0.009	0.006	0.003	0.012	0.061	0.006
Li *	1.181	1.087	1.212	1.084	1.12	1.139	1.129	1.122	1.358	1.38	1.375	1.325	1.235	0.987	1.151
Y sum	3.000	3.000	3.000	3.000	3.000	3.000	3.000	3.000	3.000	3.000	3.000	3.000	3.000	3.000	3.000
Ca	0.159	0.046	0.165	0.044	0.187	0.168	0.201	0.132	0.259	0.281	0.268	0.207	0.214	0.064	0.202
Na	0.42	0.419	0.434	0.404	0.684	0.635	0.707	0.608	0.388	0.396	0.406	0.424	0.525	0.865	0.613
K	0.002	-	0.002	0.003	0.004	0.003	0.003	0.003	0.002	0.002	0.004	-	-	0.005	0.002
Vacancy	0.418	0.535	0.399	0.55	0.125	0.194	0.089	0.257	0.351	0.32	0.322	0.368	0.261	0.066	0.183
X sum	1.000	1.000	1.000	1.000	1.000	1.000	1.000	1.000	1.000	1.000	1.000	1.000	1.000	1.000	1.000
OH	3.695	3.967	3.715	3.884	3.353	3.492	3.363	3.419	3.412	3.442	3.37	3.512	3.423	3.272	3.445
F	0.305	0.033	0.285	0.116	0.647	0.508	0.637	0.581	0.588	0.558	0.63	0.488	0.577	0.728	0.555

Table 2. *Cont.*

Sample			6		9a	9b					10				11	12
								Analogous Green			Antilogous Dark Green					
	Light Yellow Elbaite	Light Yellow Elbaite	Light Yellow Fluor-Liddicoatite	Colourless Flour-Liddicoatite	Pink Elbaite	Green Elbaite	Foitite	Elbaite	Elbaite	Elbaite	Fluor-Liddicoatite	Elbaite	Elbaite	Rossmanite	Green Elbaite	Blue Elbaite
					Average (9 pts)	Average (10 pts)									Average (21 pts)	Average (20 pts)
Point Number °								4	10	12	36	42	43	45		
SiO_2 (wt %)	35.70	36.00	36.71	37.50	37.12	37.32	35.82	37.38	38.13	37.51	37.60	37.82	37.13	37.06	36.95	37.10
TiO_2	0.14	0.17	0.09	0.09	0.01	0.17	bdl	bdl	bdl	0.18	0.03	0.05	0.20	0.07	0.18	0.17
B_2O_3 *	10.84	10.83	10.94	10.99	11.07	11.00	10.39	10.92	11.09	11.02	10.99	11.04	10.98	10.76	10.96	10.93
Al_2O_3	40.53	39.68	39.90	39.43	41.59	39.78	33.55	39.33	39.60	39.74	39.03	40.09	39.60	37.57	39.86	39.30
V_2O_3	0.01	0.04	0.01	bdl	0.01	0.02	bdl	0.03	bdl	0.10	0.02	bdl	0.04	bdl	0.01	bdl
Cr_2O_3	bdl	bdl	bdl	0.01	0.01	0.01	0.01	0.01	bdl	bdl	0.02	bdl	0.05	bdl	0.01	0.01
FeO	0.34	0.54	0.22	0.16	0.08	0.71	11.52	1.68	0.44	0.85	0.34	0.88	1.25	3.40	1.18	0.66
MgO	0.01	0.01	0.01	bdl	0.01	0.01	2.57	0.03	0.01	0.01	0.02	0.05	0.04	0.02	0.03	0.03
MnO	2.38	2.51	1.29	0.99	1.60	2.05	0.17	3.68	1.82	1.82	1.45	2.89	1.87	5.67	2.31	3.10
ZnO	bdl	bdl	0.08	bdl	0.03	0.06	0.10	0.12	0.12	0.02	0.08	0.09	0.19	0.18	0.02	0.03
CaO	1.49	1.58	2.67	2.94	1.16	1.53	0.02	0.13	1.96	1.62	3.05	0.27	1.47	0.07	1.00	1.21
Li_2O *	1.90	1.97	2.31	2.44	1.96	2.05	0.14	1.45	2.20	2.05	2.38	1.71	1.98	0.91	1.87	1.88
Na_2O	1.81	1.97	1.32	1.05	1.59	1.77	1.48	1.97	1.50	1.54	0.99	2.10	1.84	1.37	2.08	1.88
K_2O	bdl	0.02	0.01	bdl	0.01	0.01	0.01	0.01	bdl	0.01	0.01	bdl	0.01	0.02	0.02	0.02
H_2O *	3.29	3.17	3.23	3.21	3.39	3.25	3.58	3.77	3.45	3.22	3.18	3.81	3.28	3.71	3.33	3.22
F	0.95	1.19	1.16	1.23	0.91	1.14	bdl	bdl	0.79	1.23	1.29	bdl	1.08	bdl	0.95	1.17
Total	99.39	99.67	99.94	100.03	100.54	100.88	99.37	100.52	101.11	100.90	100.49	100.81	101.00	100.82	100.77	100.70
O=F	0.40	0.50	0.49	0.52	0.38	0.48	bdl	bdl	0.33	0.52	0.54	bdl	0.45	bdl	0.40	0.49
Total *	98.99	99.17	99.46	99.52	100.16	100.40	99.37	100.52	100.78	100.38	99.95	100.81	100.55	100.82	100.37	100.21
Structural formula based on 31 anions (O, OH, F)																
Si	5.726	5.779	5.831	5.931	5.830	5.894	5.994	5.948	5.978	5.918	5.946	5.952	5.875	5.985	5.858	5.898
Al	0.274	0.221	0.169	0.069	0.170	0.106	0.006	0.052	0.022	0.082	0.054	0.048	0.125	0.015	0.142	0.102
T sum	6.000	6.000	6.000	6.000	6.000	6.000	6.000	6.000	6.000	6.000	6.000	6.000	6.000	6.000	6.000	6.000
B	3.000	3.000	3.000	3.000	3.000	3.000	3.000	3.000	3.000	3.000	3.000	3.000	3.000	3.000	3.000	3.000
Al (*Z*)	6.000	6.000	6.000	6.000	6.000	6.000	6.000	6.000	6.000	6.000	6.000	6.000	6.000	6.000	6.000	6.000
Cr	-	-	-	-	-	-	-	-	-	-	-	-	-	-	-	-
Al	1.387	1.287	1.301	1.282	1.528	1.299	0.612	1.323	1.296	1.307	1.220	1.387	1.260	1.136	1.305	1.263
Ti	0.017	0.020	0.011	0.010	0.001	0.020	-	-	-	0.021	0.004	0.006	0.024	0.008	0.022	0.020
V	0.002	0.005	0.001	-	0.001	0.002	-	0.004	-	0.012	0.003	-	0.005	-	0.001	0.001
Cr	-	-	-	0.001	0.001	0.001	0.002	0.001	-	-	0.003	-	0.006	-	0.001	0.001
Mg	0.004	0.003	0.003	0.001	0.001	0.003	0.641	0.008	0.002	0.002	0.005	0.012	0.009	0.006	0.008	0.007
Mn	0.323	0.341	0.173	0.133	0.213	0.275	0.024	0.496	0.242	0.243	0.194	0.385	0.251	0.776	0.310	0.417
Fe^{2+}	0.046	0.072	0.029	0.021	0.010	0.093	1.612	0.224	0.058	0.112	0.045	0.116	0.165	0.459	0.157	0.087
Zn	-	-	0.010	-	0.004	0.007	0.013	0.014	0.014	0.002	0.010	0.010	0.022	0.022	0.003	0.003
Li *	1.223	1.271	1.473	1.552	1.240	1.299	0.096	0.930	1.389	1.300	1.517	1.083	1.259	0.594	1.193	1.200
Y sum	3.000	3.000	3.000	3.000	3.000	3.000	3.000	3.000	3.000	3.000	3.000	3.000	3.000	3.000	3.000	3.000
Ca	0.256	0.272	0.454	0.498	0.195	0.259	0.003	0.022	0.329	0.274	0.517	0.046	0.249	0.012	0.171	0.207
Na	0.563	0.613	0.407	0.321	0.485	0.542	0.480	0.608	0.456	0.471	0.303	0.641	0.564	0.429	0.640	0.581
K	0.001	0.003	0.002	0.000	0.002	0.002	0.002	0.002	-	0.001	0.002	-	0.003	0.003	0.003	0.004
Vacancy	0.181	0.112	0.137	0.180	0.317	0.197	0.515	0.368	0.214	0.254	0.179	0.313	0.184	0.555	0.187	0.209
X sum	1.000	1.000	1.000	1.000	1.000	1.000	1.000	1.000	1.000	1.000	1.000	1.000	1.000	1.000	1.000	1.000
OH	3.519	3.396	3.417	3.385	3.547	3.429	4.000	4.000	3.608	3.386	3.355	4.000	3.460	4.000	3.522	3.412
F	0.481	0.604	0.583	0.615	0.453	0.571	-	-	0.392	0.614	0.645	-	0.540	-	0.478	0.588

* calculated on the assumed elbaite stoichiometry. bdl = below the detection limit (0.01 wt % for all oxides). ° point number corresponding to the number in the sample images, pts = points.

Colour changes are generally controlled by a combination of concentration and oxidation states c chromophore elements such as Ti, V, Fe and Mn [26]: in samples 3, 10 and 11, a higher content of iro is detected in the portion typically brown-black. Sample 10 (Figure 6a,b) is the only one including th analogous and antilogous pole and shows compositional variations with changes in Mn, Fe (Figure 6 and Ca. This last displays the highest values in the central part of the crystal (Figure 6d). Titaniur ranges from 0 to 0.04 a.p.f.u., whereas V and Cr are very low or below the detection limit. Rossmanit is present both at the analogous and antilogous poles (Figure 6b).

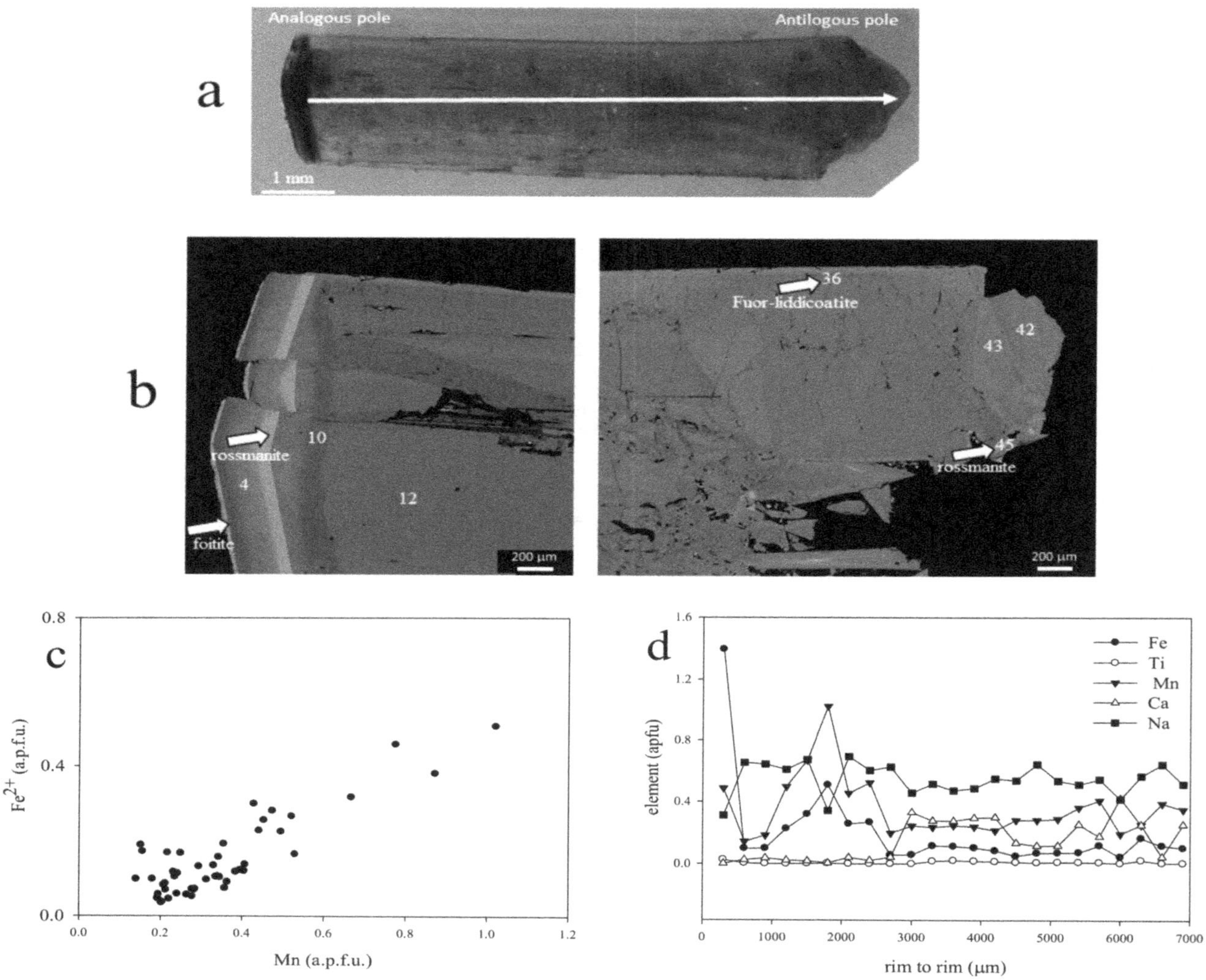

Figure 6. Optical microscope (**a**), BKSE (**b**) images, (**c**) variation of Fe vs. Mn and (**d**) an example of zoning patterns of different elements of sample 10. The points in BKSE images correspond to the points reported in Table 2.

Sample 3 (Figure 7a), along the *c*-axis, shows a colourless (a) and a brown (b) zone with an iron content of up to 0.106 and up to 0.762 a.p.f.u., respectively. Sample 6 shows an elbaite composition with a decrease of Mn and Na together with an increase of Ca from light yellow to near colourless zones where a fluor-liddicoatite composition is identified (Figure 7b,c). Ti and Fe are very low, and therefore manganese content seems to be mainly responsible of the colour changes.

Samples 2 and 12 have a homogeneous composition with the higher content of Mn that may induce the deep blue colour (up to 0.548 and 0.443 a.p.f.u., respectively).

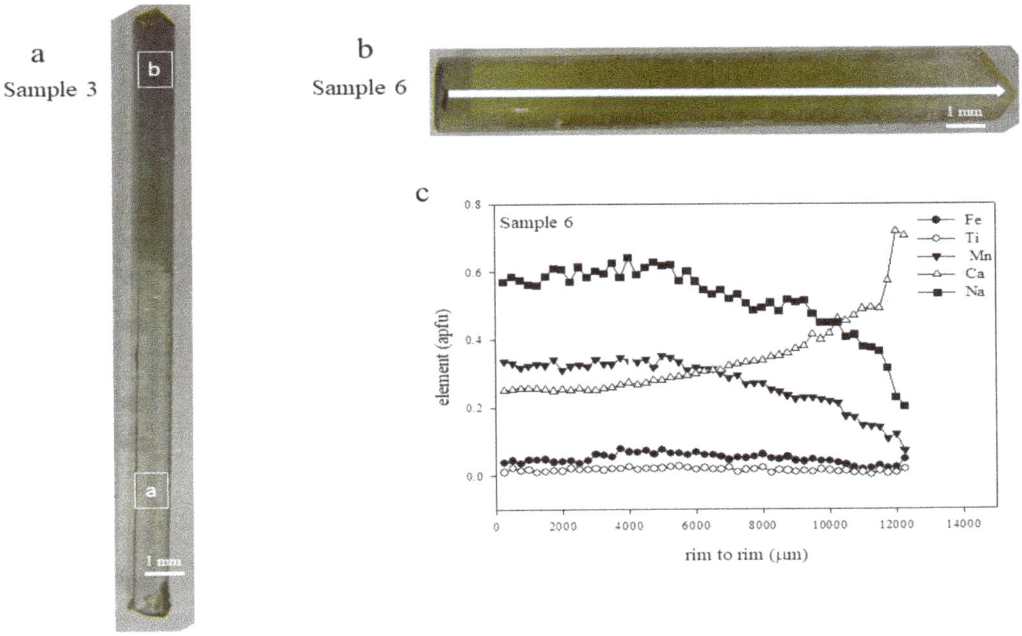

Figure 7. Optical microscope image of sample 3 (**a**) and sample 6 (**b**); zoning patterns of different elements of sample 6 (**c**).

The pink tourmaline (sample 1, Figure 8a) is nearly devoid of iron and presents the lowest Mn content with two different compositional zones: one with higher Ca + Mn (on average, 0.27 a.p.f.u., i.e., pts 4 and 8), the second one with lower Ca + Mn (on average, 0.12 a.p.f.u.), Mn < 0.04 a.p.f.u. and higher mole fraction of rossmanite (i.e., pts 6 and 9). Lighter-coloured areas included in rossmanite resulted in being muscovite (Figure 8b).

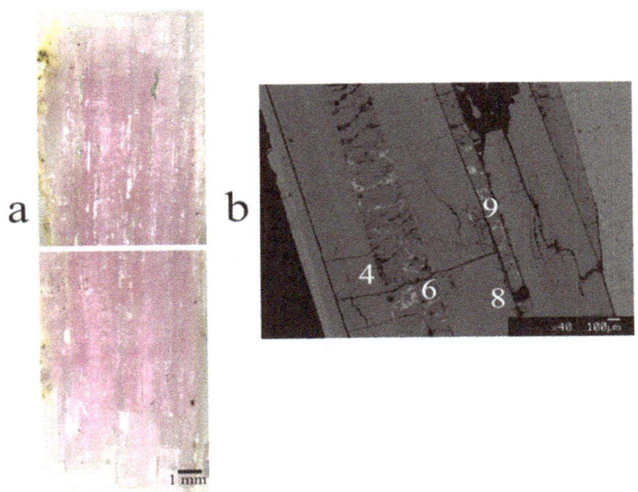

Figure 8. Optical microscope (**a**) and BKSE (**b**) images of sample 1. The points in BKSE image correspond to the points reported in Table 2. Points 6 and 9 correspond to rossmannite.

Sample 4 (Figure 9a) occurs in the same geode of sample 3 and has been cut perpendicular to the *c*-axis. The slice reveals a complex growth history characterized by a central part, pink in colour, depleted in Mn and Fe, but enriched in Ca with respect to the green rim, with a composition ranging from fluor-liddicoatite-rich elbaite to elbaite. The lighter-coloured zone at the bottom of the sample (Figure 9b, e.g., point 33, analysis 33 in Table 2) presents an enrichment in Fe^{2+} with respect to Mn^{2+} (Figure 9c). In Figure 9d, an example of the variation from rim to rim of these elements, including Ti, Ca and Na, is shown.

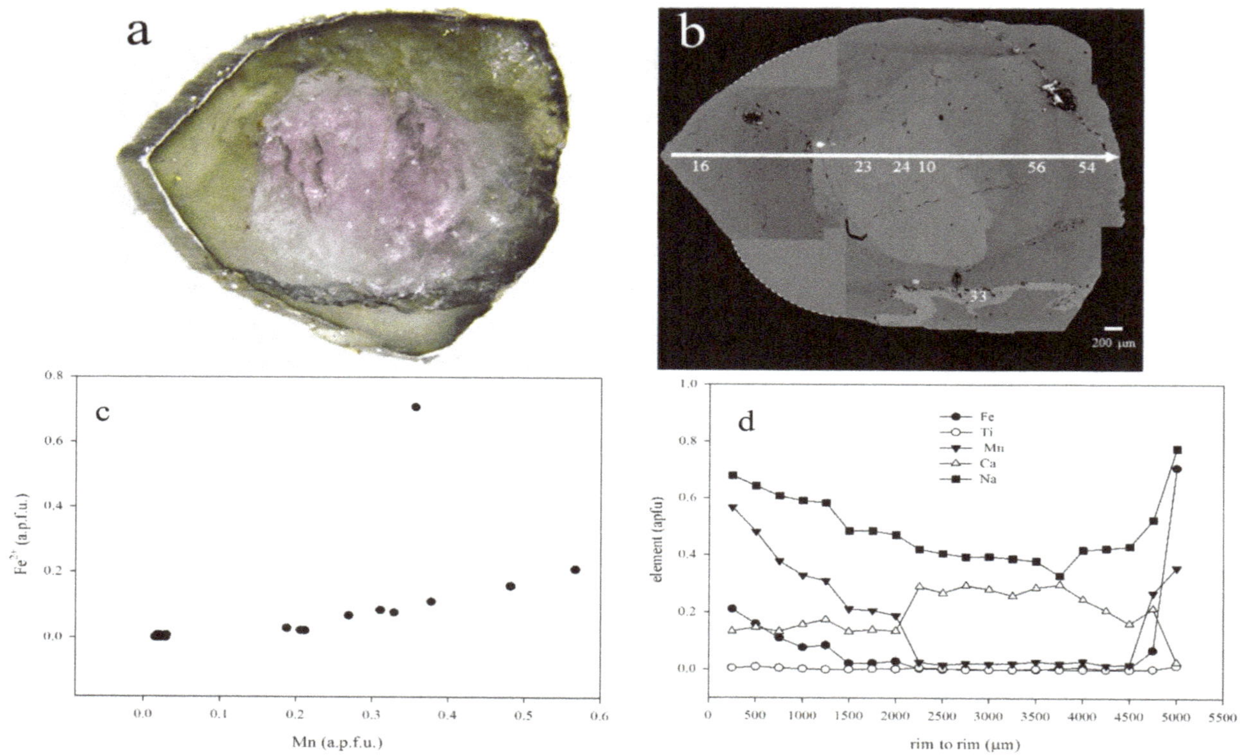

Figure 9. Optical microscope (**a**) and BKSE (**b**) images, (**c**) variation of Fe vs. Mn and (**d**) an example of zoning patterns of different elements of sample 4. The points in BKSE image correspond to the points reported in Table 2.

Sample 9 (Figure 10a,b) comes from a geode containing black quartz, zircon and uraninite and displays a change from pink (sample 9a) to green colour (sample 9b) that seems to be due to manganese, iron and titanium enrichment. In sample 9b, a thin overgrowth with foitite composition (see Figure 10c and Table 2) is present.

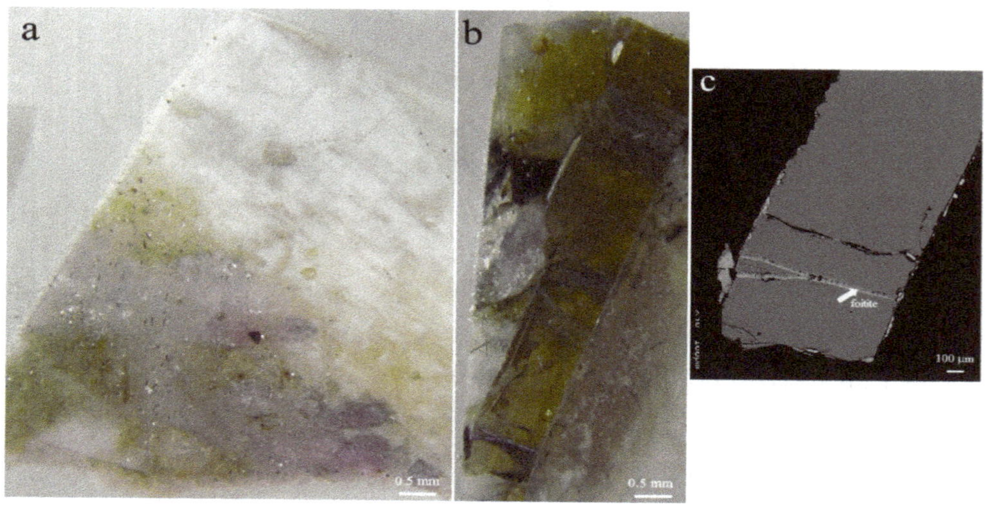

Figure 10. Optical microscope images of sample 9a (**a**), 9b (**b**) and BSKE image of sample 9b (**c**) where the arrow indicates the foitite.

The averaged LA-ICP-MS analyses of selected trace elements: Be, Sc, V, Cu, Ga, Ge, Sr, Y, Ta, Pb, Th, U and REE of the tourmalines are reported in Table 3.

Table 3. LA–ICP–MS analyses of trace elements in studied tourmalines (ppm).

Sample	1		2		3a		3b		4		5	
	Average (4 pts)	st dev	Average (4 pts)	st dev	Average (4 pts)	st dev	Average (5 pts)	st dev	Average (8 pts)	st dev	Average (5 pts)	st dev
Li	8546.20	463.86	7985.55	182.95	7859.06	75.06	7919.98	324.03	8899.59	1020.41	8073.58	322.46
Be	52.19	14.18	37.62	2.83	28.92	2.96	21.29	8.82	99.38	122.07	31.05	9.50
Sc	3.03	0.55	1.83	0.28	1.82	0.35	2.79	0.45	1.63	0.52	2.12	0.16
V	0.38	0.32	1.08	0.20	0.82	0.12	13.95	7.92	8.56	5.26	3.17	4.77
Cu	196.20	27.01	19.23	1.05	16.10	1.41	16.04	12.69	71.89	62.62	22.16	3.19
Ga	181.73	28.06	40.83	2.51	45.27	1.04	59.90	13.21	78.28	26.48	44.19	6.70
Ge	9.11	2.22	20.47	1.68	19.85	1.52	15.65	2.90	15.91	2.78	21.57	2.77
Sr	0.73	0.23	31.66	1.23	19.88	1.72	231.22	390.52	93.98	234.91	29.42	3.66
Y	0.25	0.17	0.46	0.05	0.35	0.03	0.24	0.15	0.22	0.08	0.39	0.20
Ta	4.63	3.57	1.07	0.14	0.28	0.01	0.56	0.31	3.18	5.43	0.44	0.40
Pb	11.94	4.76	43.78	2.22	35.05	1.19	2882.66	3775.66	184.51	453.38	163.22	71.23
Th	3.58	2.95	0.79	0.06	0.55	0.06	0.16	0.14	6.98	15.66	0.52	0.32
U	0.36	0.29	0.02	0.00	bdl		0.01	0.01	0.79	1.15	0.03	
La	44.13	18.50	24.29	1.46	13.58	0.35	22.86	14.36	9.50	4.54	17.07	4.19
Ce	46.68	19.98	49.72	3.99	30.16	0.30	48.87	26.13	15.89	8.20	37.41	11.14
Pr	2.35	1.08	5.13	0.38	2.87	0.10	4.94	2.44	1.44	0.85	3.60	1.08
Nd	3.05	1.40	14.54	1.49	7.46	0.24	13.43	6.78	3.42	2.55	9.15	2.88
Sm	0.53	0.36	3.14	0.69	1.96	0.17	2.74	1.23	1.17	0.41	2.29	0.95
Eu	bdl		0.12	0.03	0.12	0.04	0.13	0.05	0.05	0.02	0.10	0.04
Gd	0.27	0.02	0.93	0.19	0.64	0.13	0.68	0.40	0.42	0.16	0.72	0.38
Tb	0.05	0.04	0.11	0.02	0.09	0.02	0.08	0.05	0.05	0.02	0.09	0.04
Dy	0.15	0.17	0.36	0.11	0.24	0.05	0.24	0.17	0.14	0.04	0.26	0.17
Ho	bdl		bdl		0.03	0.02	0.02	0.01	0.02	0.01	0.02	0.01
Er	0.20	0.26	bdl		0.05	0.001	bdl		0.06	0.03	bdl	
Tm	bdl		0.02	0.005	0.01	0.005	0.01	0.00	bdl		bdl	
Yb	0.10	0.10	bdl				bdl		0.17	0.15	bdl	
Lu	bdl		bdl		bdl		bdl		bdl		0.01	
Σ REE	97.50		98.36		57.21		94.00		32.34		70.72	

Table 3. *Cont.*

Sample	6		9a		9b		10		11		12	
	Average (7 pts)	st dev	Average (3 pts)	st dev	Average (3 pts)	st dev	Average (8 pts)	st dev	Average (5 pts)	st dev	Average (4 pts)	st dev
Li	9993.23	1095.08	8494.71	249.68	8767.86	145.96	8444.96	1786.74	7175.63	2373.44	8271.11	399.03
Be	20.20	5.78	23.17	2.81	27.37	7.68	22.63	13.36	14.62	11.00	19.95	2.56
Sc	1.96	0.40	2.96	0.25	3.09	0.24	2.26	0.46	3.47	0.84	1.65	0.47
V	0.37	0.09	0.48	0.18	8.88	3.03	0.84	0.63	1.21	0.54	0.54	0.09
Cu	21.30	9.49	148.85	58.16	31.74	3.36	24.96	16.28	17.92	12.74	30.46	6.96
Ga	68.20	25.78	121.62	49.09	35.29	1.08	104.43	26.62	91.33	12.88	45.58	5.77
Ge	12.01	2.63	7.05	0.84	16.59	1.50	6.97	2.13	7.10	1.54	11.74	3.80
Sr	104.28	95.41	4.37	1.46	168.96	25.82	30.01	25.73	175.48	356.03	16.88	4.55
Y	0.36	0.05	0.46	0.08	0.48	0.20	0.65	0.62	0.31	0.11	0.21	0.12
Ta	1.11	0.36	1.06	0.39	1.02	0.43	1.04	1.47	0.16	0.10	0.56	0.25
Pb	135.44	129.63	13.36	1.50	64.72	10.31	40.34	60.43	881.33	710.51	27.57	6.20
Th	0.23	0.09	1.07	0.40	0.49	0.16	1.55	1.67	0.52	0.25	0.32	0.12
U	0.03	0.02	0.06	0.01	bdl		0.06	0.02	0.06	0.06	0.02	0.02
La	22.46	4.14	23.09	7.25	33.99	10.92	18.37	19.28	9.75	8.73	14.74	5.05
Ce	46.61	14.32	49.63	12.21	66.11	21.77	35.44	44.29	20.35	18.36	31.63	11.11
Pr	4.65	1.67	4.84	0.58	6.41	2.04	3.98	5.18	2.03	1.80	3.15	1.06
Nd	12.45	5.70	11.78	0.79	18.64	5.80	13.82	16.80	7.01	4.65	8.26	2.74
Sm	2.83	1.40	3.86	0.57	3.76	1.18	4.74	5.45	1.73	1.19	1.91	0.70
Eu	0.13	0.05	0.07	0.03	0.32	0.05	0.13	0.16	0.11	0.03	0.09	0.02
Gd	0.84	0.44	1.31	0.35	1.49	0.59	1.66	1.44	0.50	0.38	0.49	0.16
Tb	0.08	0.02	0.15	0.02	0.17	0.02	0.21	0.18	0.07	0.05	0.07	0.03
Dy	0.24	0.12	0.48	0.17	0.41	0.14	0.61	0.61	0.21	0.10	0.16	0.05
Ho	bdl		0.02	0.01	0.04	0.02	0.06	0.04	0.03	0.03	bdl	
Er	0.05	0.03	0.05	0.04	0.07	0.03	bdl		0.05	0.02	0.08	0.04
Tm	0.01	0.01	0.02	0.01	bdl		0.03	0.01	0.02	bdl	bdl	
Yb	bdl		bdl		bdl		bdl		0.15	0.20	0.10	0.01
Lu	0.02	0.01	bdl		bdl		0.02	0.01	0.04	0.05	bdl	
Σ REE	90.37		95.29		131.42		79.06		42.06		60.67	

bdl = below detection limit, pts = points, st dev = standard deviation.

In general, the concentration of trace elements is variable and does not show any significant trend or correlation with the colour changes. The exceptions are represented by the higher values of Pb (up to 880 and 2880 ppm in sample 11 and 3b, respectively) corresponding to Mn (Fe)-enriched points and of Cu (up to 196 ppm) in samples 1, 4, 9a, determined in the pink part of the crystals. Variable concentration of Ga (41–182 ppm) and Sr (1–231 ppm) was observed. The Y content, generally correlated with HREE, is very low, less than 1 ppm in all samples.

The REE content in tourmalines from granitic pegmatites is generally low (<30 ppm) while in the examined samples, the total REE content ranges from 30 to 130 ppm with light REE-enrichment and Ce being the most abundant element. Chondrite normalized REE patterns (plotted as mean values for each sample in Figure 11) display a general depletion in the medium and heavy, with respect to the light, rare earths. The negative Eu anomaly is probably related to the local depletion of Eu^{2+} content in the melt due to its consumption during the growth of K-feldspar, an important carrier of Eu^{2+} in magmatic rocks.

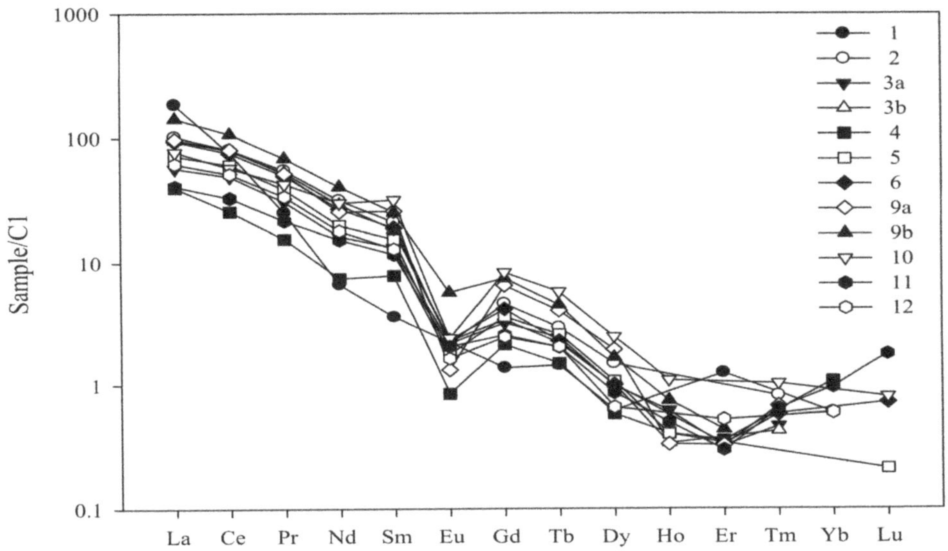

Figure 11. REE patterns normalized to the C1 values reported in [27] for all the analysed samples.

5.3. X-Ray Crystal Structure Refinement

The crystallographic data obtained by single crystal refinement confirm that the examined tourmalines belong to the *R3m* space group with the cell parameters in the range of elbaite species. Absolute structure parameter ranged from 0.01(11) to −0.10(10) and secondary extinction coefficient ranged from 0.0024(2) to 0.0049(3) (Table 4).

The selected interatomic distances, geometrical parameters and refined and observed (from chemical analyses) site-scattering values are given in Tables 5 and 6, respectively. The final atom coordinates and equivalent displacement parameters, as well as the complete set of crystallographic data (crystallographic information files and lists of observed and calculated structure factors), have been deposited in supplementary electronic material (Tables S1 and S2a–c).

The structural data confirm elbaitic compositions: <Y–O> ranges from 1.989 to 2.030 Å compared to a calculated value of 2.015 Å for an ideal elbaite (using ionic radii of [11]; <Z–O> ranges from 1.907 to 1.908 Å compared to a <^{Z}Al–O> grand mean value of 1.906 Å (Figure 3 of [28]); site scattering at X sites (8.73–12.43 electrons per formula unit or a.p.f.u., Table 6) is compatible with a dominant Na occupancy. The high quality of the reported structure refinements makes it possible to discuss the site assignment of cations among the different sites of the studied crystals. Following [11] and using his Equation (4) [$^{Z}Al = -0.1155 + 1.1713\cdot{}^{[6]}Al - 0.0522\cdot{}^{[6]}Al^2$; Al = Al − ^{T}Al (a.p.f.u.)], it is possible to estimate the ^{Z}Al occupancy just from chemical data. Applying this equation to the data reported in Table 2 produces a slight ^{Z}Al deficiency of 0.126–0.439 a.p.f.u.. However, this is in contrast with single crystal XRD data that provide refined scattering values of 12.85(5)–12.96(3) a.p.f.u. for the Z site, implying a maximum of 0.12 ^{Z}Li atoms per formula unit or just pure Al Z sites, considering 3σ. It is, therefore, highly improbable that any Mn^{2+} (or Fe^{2+}) could have been disordered into the Z-sites. The agreement between observed (SC-XRD) and calculated (EMPA-WDS) site scattering at X-sites is poorer (9–10%, Table 6), probably due to the high chemical variability of crystals, because the occupancy of X site is the main chemical vector observed in the studied tourmalines.

It is worthwhile to note the high values of *U* equivalent for the O(1) (from 0.0291(7) to 0.057(2) Å^2 compared to the mean value of 0.007–0.010 Å^2 for the other anion sites; see Tables S1 and S2a–c). This is very probably due to static disorder at the O(1) anion site. All three studied crystals show large and flat [parallel to (0001)] thermal ellipsoids, making the estimation of the electron density at the O(1) site inaccurate (see Figure 12). We tried a split model, but it was unsuccessful. A similar delocalization of electron density was reported by [29] (compare their Figure 1a with our Figure 12b) for manganese-bearing elbaitic compositions. Burns and co-workers [29] interpreted the large anisotropic displacements as positional disorder, rather than thermal vibration, due to the 6 possible local arrangements at the three *Y* sites around the O(1) site; the three principal *Y* cations (Al, Li and Mn^{2+}) have very different ionic radii (0.547(3), 0.751(9), 0.809(1) Å, respectively; values from [11]). Considering the composition of the Y sites (close to AlLiMn) of our tourmalines, the ideal average trimer would promote a distorted environment for the O(1), confirming in a new set of samples the behaviour already reported by [29].

Table 4. Crystal data and structure refinement for tourmalines 1, 2, 12.

Sample	1	2	12
Temperature	293(2) K	293(2) K	293(2) K
Wavelength	0.71073 Å	0.71073 Å	0.71073 Å
Crystal system	Trigonal	Trigonal	Trigonal
Space group	*R*3*m*	*R*3*m*	*R*3*m*
Unit cell dimensions	a = 15.8283(3) Å	a = 15.8951(8) Å	a = 15.8909(2) Å
	c = 7.09392(18) Å	c = 7.1216(4) Å	c = 7.1163(3) Å
Volume	1539.16(5) Å^3	1558.22(15) Å^3	1556.26(7) Å^3
Z	3	3	3
Absorption coefficient	1.004 mm^{-1}	1.056 mm^{-1}	1.020 mm^{-1}
F(000)	1395	1430	1422
Crystal size (mm^3)	0.29 × 0.61 × 0.47	0.35 × 0.51 × 0.65	0.40 × 0.53 × 0.70
θ range for data collection	3.23 to 36.11°.	3.22 to 29.04°.	3.22 to 35.95°.
Index ranges	$-26 \leq h \leq 26$, $-24 \leq k \leq 24$, $-9 \leq l \leq 9$	$-16 \leq h \leq 13$, $-20 \leq k \leq 20$, $-9 \leq l \leq 9$	$-25 \leq h \leq 26$, $-25 \leq k \leq 25$, $-9 \leq l \leq 9$
Reflections collected	14,418	3806	14,500
Independent reflections	1571	914	1261
R(int)	0.0277	0.0216	0.0334
Completeness to θ = 35.95°	89.30%	95.20%	72.70%
Refinement method	Full-matrix least-squares on F2	Full-matrix least-squares on F2	Full-matrix least-squares on F2
Data/restraints/parameters	1571/1/97	914/1/97	1261/1/97
Goodness-of-fit on F^2	1.161	1.076	1.09
Final *R* indices [*I* > 2σ (*I*)]	*R*1 = 0.0159 w*R*2 = 0.0400	*R*1 = 0.0175 w*R*2 = 0.0452	*R*1 = 0.0179 w*R*2 = 0.0448
R indices (all data)	*R*1 = 0.0168 w*R*2 = 0.0403	*R*1 = 0.0177 w*R*2 = 0.0453	*R*1 = 0.0186 w*R*2 = 0.0451
Absolute structure parameter	0.04(7)	0.01(11)	−0.10(10)
Extinction coefficient	0.0049(3)	0.0041(3)	0.0024(2)
Largest diff. peak and hole	0.461 and −0.439 e.Å^{-3}	0.530 and −0.497 e.Å^{-3}	0.889 and −0.709 e.Å^{-3}

Table 5. Bond lengths [Å] and angles [°] for tourmalines 1, 2, 12.

Sample	1	2	12
T–O(6)	1.6092(9)	1.6049(15)	1.6021(14)
T–O(7)	1.6103(7)	1.6134(13)	1.6110(10)
T–O(4)	1.6208(5)	1.6263(8)	1.6260(7)
T–O(5)	1.6363(5)	1.6412(9)	1.6389(8)
<*T*–O>	1.619	1.621	1.62
V (Å^3)	2.173	2.181	2.173
TQE	1.0017	1.0023	1.0022
TAV	6.812	9.502	9.0511
B–O(2)	1.3630(18)	1.352(4)	1.361(3)
B–O(8) (×2)	1.3798(10)	1.386(2)	1.3850(15)
<*B*–O>	1.374	1.375	1.377
X–O(2) (×3)	2.4570(18)	2.426(3)	2.425(2)
X–O(5) (×3)	2.7422(13)	2.748(2)	2.7492(17)
X–O(4) (×3)	2.8095(13)	2.810(2)	2.8078(18)
<*X*–O>	2.67	2.661	2.661
V (Å^3)	31.14	31.419	31.086
Y–O(2) (×2)	1.9620(9)	1.9824(15)	1.9812(14)
Y–O(6) (×2)	1.9591(9)	2.0150(15)	2.0129(13)
Y–O(1)	1.9577(14)	2.028(2)	2.024(2)
Y–O(3)	2.1346(14)	2.171(2)	2.1678(19)
<*Y*–O>	1.989	2.032	2.03
V (Å^3)	10.125	10.796	10.759
OQE	1.0252	1.0253	1.0253
OAV	79.39	79.54	79.7
Z–O(6)	1.8643(8)	1.8506(15)	1.8514(12)
Z–O(7)	1.8816(8)	1.8823(14)	1.8834(12)
Z–O(8)	1.8877(8)	1.8865(14)	1.8847(11)
Z–O(8)	1.9008(8)	1.9114(14)	1.9098(11)
Z–O(7)	1.9415(7)	1.9551(14)	1.9551(11)
Z–O(3)	1.9647(6)	1.9617(11)	1.9612(10)
<*Z*–O>	1.907	1.908	1.908
V (Å^3)	9.039	9.079	9.075
OQE	1.0154	1.0137	1.0137
OAV	52.08	45.82	45.75
O(3) –H(3)	0.80(3)	0.71(4)	0.72(4)

Table 6. Observed (Single Crystal XRD) and calculated (EMPA) site scattering.

Site	(Electrons Per Site, eps) for Tourmalines 1, 2, 12		
	1	2	12
X (obs)	8.73	12.43	11.81
X (calc)	7.95	11.34	10.61
Y (obs)	9.49	11.98	11.26
Y (calc)	9.25	11.97	11.12
Z (obs)	12.96	12.83	12.85
Z (calc)	13	13	13

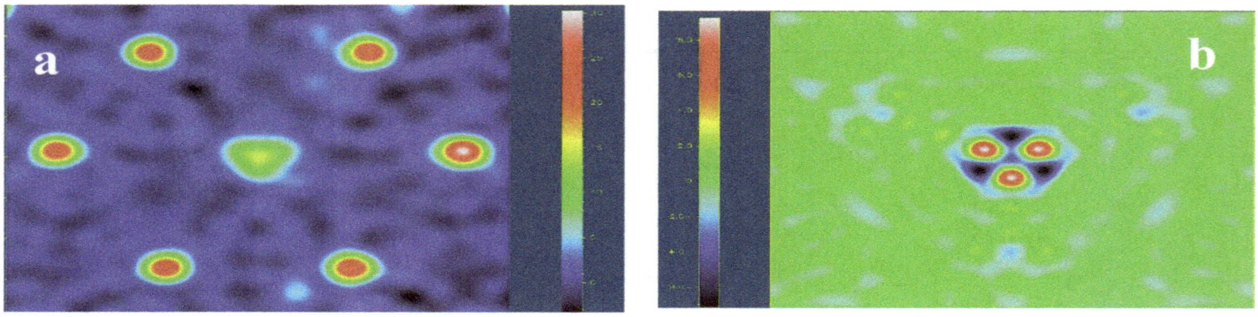

Figure 12. Sample 2: Electron density at O(1). The high displacement parameters and oblate behaviour of the O(1) anion site in the three samples are related to the trilobated shape of the maxima at that position. This is ascribable to local static disorder due to the presence of three different size cations as Al < Li < Mn^{2+} in the three Y coordinating sites for every O(1) anion site. (**a**) is a F_o Fourier synthesis; (**b**) is a F_o–F_c Fourier synthesis.

6. Discussion and Conclusions

The tourmalines discovered for the first time in the miarolitic LCT pegmatites at the western border of Adamello Massif may be considered gem materials due to their attractive pink to green-brown hue and transparency. Crystals providing the gem-quality requirements are available, despite the presence of cracks and voids well evidenced by synchrotron X-ray computed micro-tomography.

The studied tourmalines resulted principally in fluor-elbaite, along with minor fluor-liddicoatite foitite and rossmanite that likely represent the final stages of tourmaline compositional evolution.

The chemical variations, reflecting the environment physicochemical changes during their growth resulted both in a zoning from nearly black to green to blue to pink elbaite to rossmanite (sample 10) and in a reverse geochemical trend at the latest stages by a pink core, colourless to white midsection and green rim in sections cut perpendicularly to the *c*-axis (sample 4). In the non-homogeneous samples two different zones, possibly indicating the occurrence of a different and separated generations of tourmaline, can be distinguished: the first can be described as having high Ca + Mn or with a continuous fractionation trend of Mn vs. Fe, and the second one with poor Ca + Mn (sample 1) or with a marked enrichment in Fe (sample 4 and 10).

Mn and Fe are the main factors controlling the colour of the Li-rich tourmaline. Iron is always lower than manganese, increases in the darker zone and is virtually absent in the near colourless and pink tourmalines. One of the hypotheses on the intensifying pink hues in tourmaline is the presence of manganese in the 3+ oxidation state. Despite the uranium and thorium contents being very low (6.98 and 0.79 ppm), the presence, as in sample 9, of uraninite and its natural γ-radiation could be responsible for the oxidization change from Mn^{2+} to Mn^{3+}. From our results in tourmalines 1 and 4, it seems that a very low amount (<0.08 a.p.f.u.) of Mn may also be enough to give the bright pink colour. Cu^{2+} in combination with other cations may modify the resulting colour in tourmalines [1] and, interestingly, the pink parts of samples 1, 4, 9a, corresponding to Ti, Fe and Mn depleted points, contain a relatively high content of Cu.

The REE content is slightly higher than the literature data for tourmalines from relatively primitive NYF (Niobium, Yttrium, Fluorine) and mixed NYF-LCT pegmatites [30]. The chondrite normalized REE patterns are very similar and present a negative Eu^{2+} anomaly, probably due to the local depletion of Eu^{2+} in the melt.

The structural data confirm elbaitic compositions of the examined samples. The positional disorder found at the O(1) anion positions may be due to the high content of manganese entering into the Y sites together with cations of very different charge and radius (Li and Al).

The presence of gem-quality tourmalines in miarolitic cavities suggests the enrichment in volatiles and other exotic elements in pegmatite melt, as well as a shallow level formation in the thermo-metamorphic aureole of the Adamello pluton. The pegmatitic liquids could be exsolved from the granitoid magmas during the latest stages of crystallization of the pluton or, alternatively, generated by partial melting of the including metasedimentary sequence.

Further field exploration could turn up additional reserves of gem-quality tourmalines, and their study could contribute to a better understanding of the formation environment of the pegmatitic swarms at the border of Adamello Massif.

Supplementary Materials
Table S1: Atom coordinates of the three tourmaline crystals of this study, Table S2a: Anisotropic displacement parameters (Å^2 × 10^3) for tourmaline 1. The anisotropic displacement factor exponent takes the form: $-2\pi^2[h^2a^{*2}U^{11} + \ldots + 2\,h\,k\,a^*\,b^*\,U^{12}]$, Table S2b: Anisotropic displacement parameters (Å^2 × 10^3) for tourmaline 2. The anisotropic displacement factor exponent takes the form: $-2\pi^2[h^2a^{*2}U^{11} + \ldots + 2\,h\,k\,a^*\,b^*\,U^{12}]$, Table S2c: Anisotropic displacement parameters (Å^2 × 10^3) for tourmaline 12. The anisotropic displacement factor exponent takes the form: $-2\pi^2[h^2a^{*2}U^{11} + \ldots + 2\,h\,k\,a^*\,b^*\,U^{12}]$.

Author Contributions: V.D. performed EPMA analyses; F.P. collected the research material; N.M. and G.L. performed the Synchrotron X-ray computed micro-tomography; F.C. performed the SC-XRD analysis; A.L. collected LA-ICP-MS data; I.A. characterized the gemmological materials; V.D., F.P., R.B., N.M., F.C. analysed the data and wrote the paper. All authors contributed to the final version and gave their approval for submission.

Acknowledgments: We are thankful to Jan Cempirek for fruitful suggestions on electron microprobe data.

References

1. Pezzotta, F.; Laurs, B.M. Tourmaline: The kaleidoscopic gemstone. *Elements* **2011**, *7*, 333–338. [CrossRef]
2. Henry, D.J.; Novák, M.; Hawthorne, F.C.; Ertl, A.; Dutrow, B.L.; Uher, O.; Pezzotta, F. Nomenclature of the tourmaline-supergroup minerals. *Am. Mineral.* **2011**, *96*, 895–913. [CrossRef]
3. Pezzotta, F.; Guastoni, A. Adamello: La pegmatite LCT della valle Adamè. *Riv. Mineral. Ital.* **2002**, *3*, 127–142.
4. Mercurio, M.; Rossi, M.; Izzo, F.; Cappelletti, P.; Germinario, C.; Grifa, C.; Petrelli, M.; Vergara, A.; Langella, A. The characterization of natural gemstones using non-invasive FT-IR spectroscopy: New data on tourmalines. *Talanta* **2018**, *178*, 147–159. [CrossRef]
5. Van Hinsberg, V.J.; Henry, D.J.; Marschall, H.R. Tourmaline: An ideal indicator of its host environment. *Can. Mineral.* **2011**. [CrossRef]
6. Marinoni, N.; Voltolini, N.; Mancini, L.; Vignola, P.; Pagani, A.; Pavese, A. An investigation of mortars affected by alkali-silica reaction by X-ray synchrotron microtomography: A preliminary study. *J. Mater. Sci.* **2009**, *44*, 5815–5823. [CrossRef]
7. Marinoni, N.; Voltolini, N.; Broekmans, M.A.T.M.; Mancini, L.; Monteiro, P.J.M.; Rotiroti, N.; Ferrari, E.; Bernasconi, A. A combined synchrotron radiation micro computed tomography and micro X-ray diffraction study on deleterious alkali-silica reaction. *J. Mater. Sci.* **2015**, *50*, 7985–7997. [CrossRef]
8. Hawthorne, F.C.; Henry, D.J. Classification of the minerals of the tourmaline group. *Eur. J. Mineral.* **1999**, *11*, 201–215. [CrossRef]
9. Bosi, F.; Andreozzi, G.B.; Skogby, H.; Lussier, A.J.; Abdu, Y.; Hawthorne, F.C. Fluor-elbaite, $Na(Li_{1.5}Al_{1.5})Al_6(Si_6O_{18})(BO_3)_3(OH)_3F$, a new mineral species of the tourmaline supergroup. *Am. Mineral.* **2013**, *98*, 297–303. [CrossRef]
10. Bosi, F.; Skogby, H.; Ciriotti, M.E.; Gadas, P.; Novák, M.; Cempírek, J.; Všianský, D.; Filip, J. Lucchesiite, $CaFe_3^{2+}Al_6(Si_6O_{18})(BO_3)_3(OH)_3O$, a new mineral species of the tourmaline supergroup. *Miner. Mag.* **2017**. [CrossRef]
11. Bosi, F. Tourmaline crystal chemistry. *Am. Mineral.* **2018**, *103*, 298–306. [CrossRef]
12. Dupuy, C.; Dostal, J.; Fratta, M. Geochemistry of the Adamello Massif (Northern Italy). *Contrib. Mineral. Pet.* **1982**, *80*, 41–48. [CrossRef]
13. Zhang, C.; Gieré, R.; Stünitz, H.; Brack, P.; Ulmer, P. Garnet-quartz intergrowths in granitic pegmatites from Bergell and Adamello, Italy. *Schweiz. Mineral. Petrogr. Mitt.* **2001**, *81*, 89–113. [CrossRef]
14. Fiedrich, A.M.; Bachmann, O.; Ulmer, P.; Deering, C.D.; Kunze, K.; Leuthold, J. Mineralogical, geochemical, and textural indicators of crystal accumulation in the Adamello Batholith (Northern Italy). *Am. Mineral.* **2017**, *102*, 2467–2483. [CrossRef]

15. Brun, F.; Mancini, L.; Kasae, P.; Favretto, S.; Dreossi, D.; Tromba, G. Pore 3D: A software library fo quantitative analysis of porous media. *Nucl. Inst. Methods Phys. Res. A* **2010**, *615*, 326–332. [CrossRef]
16. Brun, F.; Massimi, L.; Fratini, M.; Dreossi, D.; Billé, F.; Accardo, A.; Pugliese, R.; Cedola, A. SYRMEP Tomo Project: A graphical user interface for customizing CT reconstruction workflows. *Adv. Struct. Chem. Imag* **2017**, *3*, 1–9. [CrossRef]
17. Herman, G.T. *Image Reconstruction from Projections*; Academic Press: New York, NY, USA, 1980.
18. Kak, A.C.; Slaney, M. *Principles of Computerized Tomographic Imaging*; IEEE, Institute for Electrical and Electronic Engineers Press: Piscataway, NY, USA, 1988.
19. Paganin, D.; Mayo, S.C.; Gureyev, T.E.; Miller, P.R.; Wilkins, S.W. Simultaneous phase and amplitude extraction from a single defocused image of a homogeneous object. *J. Microsc.* **2002**, *206*, 33–40. [CrossRef]
20. Agilent CrysAlis Computer Program, Agilent Technologies, XRD Products 2012. Available online: http //www.agilent.com/chem (accessed on 11 December 2018).
21. Sheldrick, G.M. Crystal structure refinement with SHELXL. *Acta Crystallogr.* **2015**, *C71*, 3–8.
22. O'Donoghue, M. *Gems*, 6th ed.; Butterworth-Heinemann: Oxford, UK, 2006.
23. Gübelin, E.J.; Koivula, J.I. *Photo Atlas of Inclusions in Gemstones*; Opinio Publishers: Basel, Switzerland, 2005 Volume 2.
24. Voltolini, M.; Marinoni, N.; Mancini, L. Synchrotron X-ray computed microtomography investigation o a mortar affected by alkali silica reaction: A quantitative characterization of its microstructural features *J. Mater. Sci.* **2011**, *46*, 6633–6641. [CrossRef]
25. MacDonald, D.J.; Hawthorne, F.C.; Grice, J.D. Foitite, a new alkali deficient tourmaline: Description and crystal structure. *Am. Mineral.* **1993**, *78*, 1299–1303.
26. Brown, C.D.; Wise, M.A. Internal zonation and chemical evolution of the Black Mountain granitic pegmatite Maine. *Can. Mineral.* **2001**, *39*, 45–55. [CrossRef]
27. Anders, E.; Grevesse, N. Abundances of the elements: Meteoritic and solar. *Geochim. Cosmochim. Acta* **1989** *53*, 197–214. [CrossRef]
28. Bosi, F.; Andreozzi, G.B. A critical comment on Ertl et al. (2012): "Limitations of Fe^{2+} and Mn^{2+} site occupancy in tourmaline: Evidence from Fe^{2+}- and Mn^{2+}-rich tourmaline". *Am. Mineral.* **2013**, *98*, 2183–2192 [CrossRef]
29. Burns, P.C.; MacDonald, D.J.; Hawthorne, F.C. The crystal-chemistry of manganese-bearing elbaite. *Can Mineral.* **1994**, *32*, 31–41.
30. Čopjaková, R.; Škoda, R.; Vašinová Galiová, M.; Novák, M. Distributions of Y + REE and Sc in tourmaline and their implications for the melt evolution; examples from NYF pegmatites of the Třebíč Pluton, Moldanubian Zone, Czech Republic. *J. Geosci.* **2013**, *58*, 113–131. [CrossRef]

9

Enigmatic Alluvial Sapphires from the Orosmayo Region, Jujuy Province, Northwest Argentina: Insights into their Origin from in situ Oxygen Isotopes

an T. Graham [1,*], Stephen J. Harris [1], Laure Martin [2], Angela Lay [1] and Eduardo Zappettini [3]

1 PANGEA Research Centre, School of Biological, Earth and Environmental Sciences, University of NSW, Sydney 2052, Australia
2 Centre for Microscopy Characterisation and Analysis, The University of Western Australia, Perth 6009, Australia
3 Argentine Geological-Mining Survey (SEGEMAR), Buenos Aires C1067ABB, Argentina
* Correspondence: i.graham@unsw.edu.au

Abstract: This study sought to investigate in situ oxygen isotopes ($\delta^{18}O$) within alluvial colorless-white to blue sapphires from the Orosmayo region, Jujuy Province, NW Argentina, in order to provide additional constraints on their origin and most likely primary geological environment. Analyses were conducted using the in situ SIMS oxygen isotope technique on the same grains that were analyzed for their mineral inclusions and major and trace element geochemistry using EMPA and LA–ICP–MS methods in our previous study. Results show a significant range in $\delta^{18}O$ across the suite, from +4.1‰ to +11.2‰. Additionally, akin to their trace element chemistry, there is significant variation in $\delta^{18}O$ within individual grains, reaching a maximum of 1.6‰. Both the previous analyses and $\delta^{18}O$ results from this study suggest that these sapphires crystallized within the lower crust regime, involving a complex interplay of mantle-derived lamprophyres and carbonatites with crustal felsic rocks and both mantle- and crustal-derived metasomatic fluids. This study reinforces the importance of the in situ analysis of gem corundums, due to potential significant variation in major and trace element chemistry and ratios and even oxygen isotope ratios within discrete zones in individual grains.

Keywords: sapphires; corundum; in situ oxygen isotopes; Orosmayo Argentina; lamprophyre; secondary ion mass spectrometry (SIMS); carbonatite

1. Introduction

Since the seminal paper of Giuliani et al. (2005) [1], oxygen isotopes have been widely used in order to help constrain the likely geological environment of formation for gem sapphires and rubies, especially those from alluvial deposits whose primary formation is unknown [1–5], such as in the case of the sapphires from this study.

Although worldwide corundum oxygen isotope values cover a wide range from −27‰ (Khitostrov, Russia) to +23‰ (Mong Hsu, Myanmar), most are in the range of +3‰ to +21‰ [1,3,6–9]. This criterion has often been used to determine the geological origin of colored corundum, and especially the gem corundums, rubies and sapphires. However, there are only a few studies on in situ oxygen isotope analysis of gem corundums [10,11]. The next research frontier is the use of in situ oxygen isotopic compositions combined with detailed LA–ICP–MS and multivariate statistical methods to help "fingerprint" the geographic location (i.e., geographic typing) of gem corundums.

As there are a number of possible geological environments for gem corundum formation [3], any information which can help provide constraints on their origin is of great use. Trace element geochemistry of gem corundums has been widely used for this in the past, especially when distinc and/or unusual chemistry can provide even site-specific signatures useful for geographic typing [10–13

In this study, we investigated the primary origin of the little-known alluvial colorless-white to blue sapphires from the Orosmayo region, Jujuy Province, NW Argentina using in situ oxygen isotope analysis. The discovery of alluvial sapphires in this region of NW Argentina dates back to 1991 during the exploration for alluvial gold in the Orosmayo region. These sapphires were first described by Zappettini and Mutti (1997) [14], and then in a more detailed study on the sapphires themselves by Harris et al. (2017) [15] using SEM, EMPA, and LA–ICP–MS techniques. A summary of the main findings of Harris et al. (2017) [15] is given in Table 1.

Table 1. Summary of main findings of Harris et al. (2017) [15] for the Orosmayo sapphire suite (bdl= below detection limit).

Color range	white, pale to deep blue, pale lilac, pale pink
Color distribution	oscillatory, sector (most common), star sapphires
Surface features	percussion marks, corrosion features
Mineral Inclusions	apatite, rutile, magnetite, phlogopite, trikalsilite, pyrite, baryte, monazite- (La), xenotime-(Y), zircon
Ga/Mg	0.3–10.4
Fe/Mg	4–117
Ti	47–8670
V	15–399
Cr	bdl-506
Ga	38–239
Mg	11–253
Elevated trace elements	Si, P, Ca, Y, La, Ce, Nd, Zn, Pb, Be, Nb, Ta, Sn, Th, U

Orosmayo (located at approximately 22°15′ S–66°25′ W) is located on the western slope of the Sierra de Rinconada, Jujuy Province, within the San Juan de Oro Basin of northwest Argentina (Figure 1; [14,15]). The area forms part of the *Puna* of northwestern Argentina which connects with the Altiplano of Bolivia to form a plateau [14,16]. The sapphires were sampled from stream sediments located at approximately 22°30′ S and 66°30′ W (see Harris et al. 2017; [15]).

2. Materials and Methods

The same exact grain mounts used in the previous study of Harris et al. (2017) [15] were also used in this study with the addition of a standard (PAR-1; see Wang et al. 2019 [11]), which was embedded into each polished mount before they were then repolished. This study analyzed 16 sapphire grains (AC-1 through to AC-16) obtained from the 37 original grains analyzed by Zappettini and Mutti (1997) [14]. The sapphires from this region range in size from a few mm up to 10 mm and all represent a single group in terms of their overall morphology and colors. However, for this study, the smaller grains were used in order to analyze as many grains as possible. The selected grains were mounted in two epoxy discs, with eight grains on each disc, and polished prior to analysis.

Secondary Ionization Mass Spectrometry (SIMS)

The oxygen isotope analyses were made using secondary ion mass spectrometry (SIMS) for the analysis of different areas of selected sapphire grains in situ to measure oxygen isotope ratios with per mil (‰) level precision. For this technique, oxygen ($^{18}O/^{16}O$) isotope ratios were analyzed using a Cameca IMS 1280 multi-collector ion microprobe within the Centre for Microscopy, Characterisation and Analysis (CMCA), University of Western Australia (UWA). The materials examined included the 16 variously colored and zoned sapphire grains (AC-1 to AC-16), which were embedded in the middle of one-inch-diameter resin mounts and polished flat. Results were calibrated using the standard PAR-1 of Wang et al. 2019 [11] embedded in a separate mount and measured before and after unknown mounts. Each analysis point is clearly marked on the sample images (Figure 2a,b).

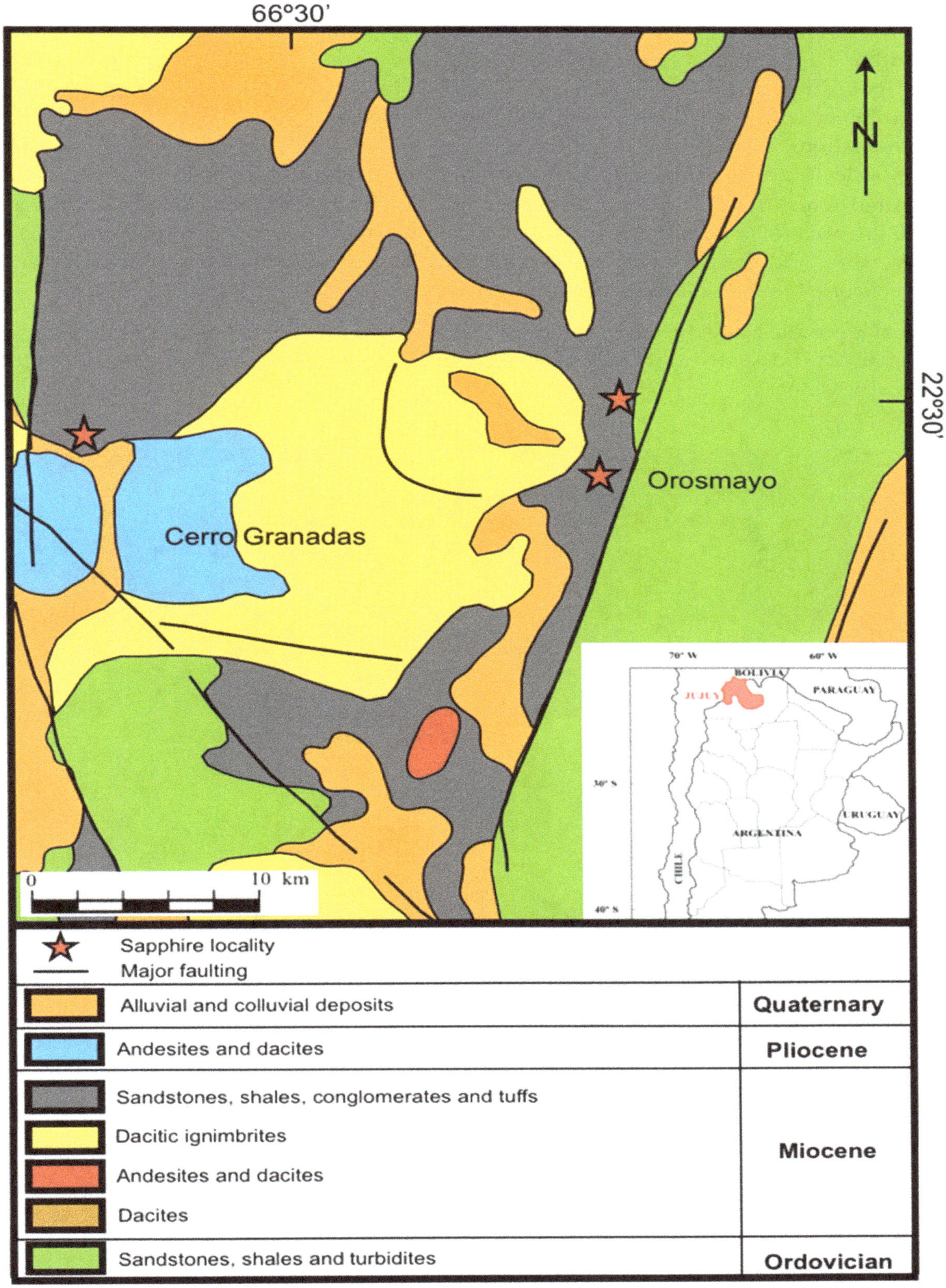

Figure 1. Map showing the location of Orosmayo in northwest Argentina (adapted from Harris et al. 2017) [15].

The sample mounts were carefully cleaned with detergent, distilled water, and ethanol in an ultrasonic bath and then coated with gold (30 nm in thickness) prior to SIMS O isotope analysis. For the oxygen isotopic analyses, secondary ions were sputtered from the sample by bombarding its surface with a Gaussian Cs^+ beam and a total impact energy of 20 keV. The surface of the sample was rastered with a 2.5 nA primary beam over a 15 × 15 μm area. An electron gun was used to ensure charge compensation during the analyses. Secondary ions were admitted in the double focusing mass spectrometer within a 100 μm entrance slit and focused in the center of a 4000 μm field aperture (100× magnification). They were energy-filtered using a 30 eV band pass with a 5 eV gap towards the high-energy side. ^{16}O and ^{18}O were collected simultaneously in multicollection mode in Faraday Cup detectors fitted with 10^{10} Ω and 10^{11} Ω, respectively. Each analysis included pre-sputtering over a 20 μm × 20 μm area for 30 s, and the automatic centering of the secondary ions in the field aperture, contrast aperture, and entrance slit, and consisted of 20 four-second cycles which give an average internal precision of ~0.16‰ (2 standard error (SE)).

External reproducibility during the analytical sessions was evaluated by repeating analyses in one single fragment of PAR-1. External reproducibility in this fragment was 0.3‰ and 0.4‰ (2σ) during the analytical sessions.

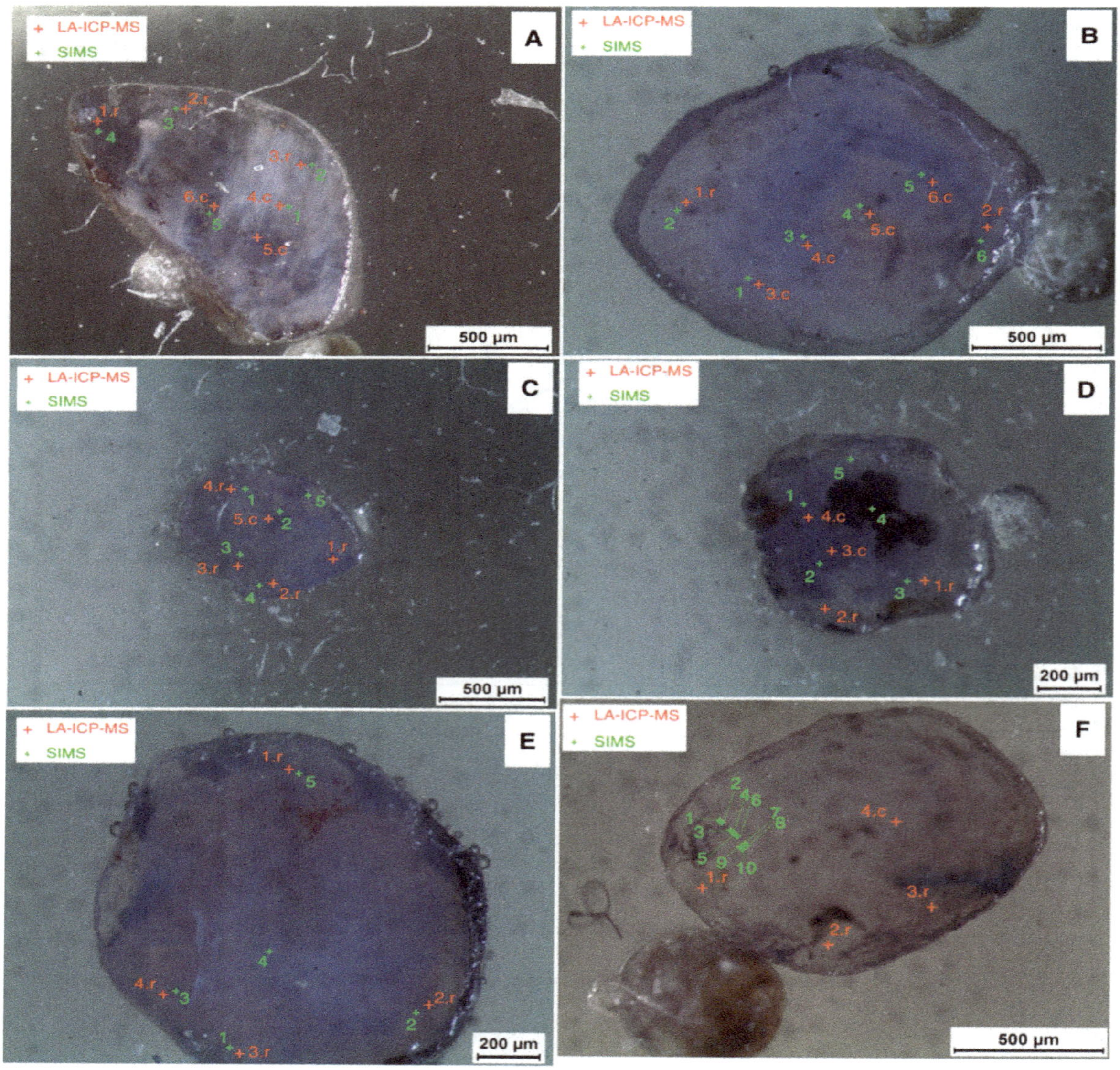

(a)

Figure 2. *Cont.*

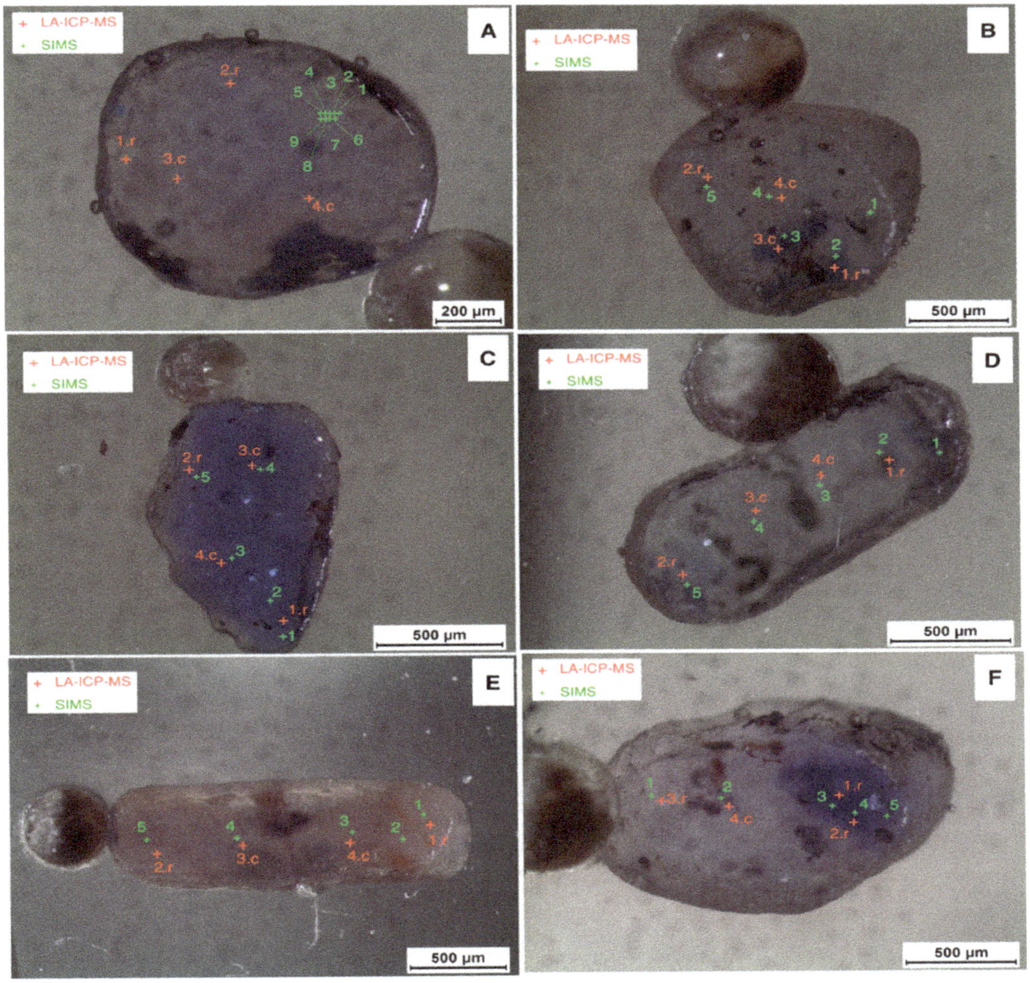

(b)

Figure 2. Images of the analyzed sapphires with both LA–ICP–MS and SIMS analysis spots indicated (**a**). A = AC-1, B = AC-2, C = AC-3, D = AC-4, E = AC-5 and F = AC-7. (**b**) A = AC-10, B = AC-11, C = AC-13, D = AC-14, E = AC-15 and F = AC-16).

3. Results

The high precision of oxygen isotope measurements using the SIMS method resulted in spot-to-spot reproducibility of 0.3‰ (2σ) during the analytical run, and individual spot analyses typically had errors of 0.5‰ (2σ). The oxygen isotope ratios $\delta^{18}O$ (reported relative to Vienna standard mean ocean water or VSMOW) are presented in Table 2. A summary of the results for each grain, along with characteristics of that grain are shown in Table 3. The complete SIMS oxygen isotope analytical results for the Orosmayo sapphires and PAR-1 ruby are given in Supplementary Materials Table S1. There was large variation between the different grains analyzed, overall from +4.1 to +11.2 $\delta^{18}O$‰, and a measurable variation within sample AC-3 up to 1.6‰ (Table 3). Although this variation is caused by the one outlier point (spot 3, also see Figure 3), sample AC-1 also showed a variation of 1.5‰ without any clear outliers (Table 3). There was no consistent zonation in terms of core to rim variation amongst the sapphires analyzed. This would suggest much more complex growth mechanisms than simply from core to rim, which is also suggested by the occurrence of complex sector-zoned sapphires and these have the highest oxygen isotope values. Other methods, such as a detailed cathodoluminescence study, would help in understanding the growth zonation and possible growth mechanisms. The lowest values correspond to apparently unzoned blue sapphires and the highest values mostly occur within sector zoned sapphires.

Table 2. $\delta^{18}O$ oxygen isotope results for grains AC-1 to AC-5, and AC-7, AC-10 to AC-11, AC-13 to AC-16.

Sapphire	Analysis Point	$\delta^{18}O$	2σ abs	Part of Crystal
AC-1	1	8.66	0.49	outer core
	2	7.52	0.49	outer rim
	3	7.66	0.47	outer rim
	4	8.08	0.48	outer rim
	5	8.97	0.46	inner rim
AC-2	1	5.24	0.48	outer rim
	2	5.68	0.50	outer rim
	3	5.76	0.46	outer core
	4	5.41	0.48	inner core
	5	5.43	0.46	outer core
	6	5.5	0.48	outer rim
AC-3	1	5.77	0.48	inner rim
	2	5.28	0.58	outer core
	3	4.12	0.50	inner rim
	4	5.61	0.48	outer rim
	5	5.4	0.46	outer rim
AC-4	1	5.69	0.51	inner rim
	2	5.33	0.49	inner rim
	3	4.97	0.48	outer rim
	4	5.89	0.49	outer core
	5	5.92	0.48	outer rim
AC-5	1	7.8	0.45	outer rim
	2	8.42	0.45	outer rim
	3	8.05	0.46	outer rim
	4	7.52	0.47	outer core
	5	8.71	0.47	outer rim
AC-7	1	6.09	0.45	outer rim
	2	5.42	0.46	outer rim
	3	6.14	0.50	inner rim
	4	6.02	0.47	inner rim
	5	6.81	0.47	inner rim
	6	5.42	0.49	inner rim
	7	6.59	0.46	inner rim
	8	6.66	0.47	inner rim
	9	5.56	0.51	inner rim
	10	5.47	0.48	inner rim
AC-10	1	9.28	0.46	inner rim
	2	9.42	0.46	inner rim
	3	9.44	0.45	inner rim
	4	9.02	0.46	inner rim
	5	9.28	0.45	inner rim
	6	8.95	0.47	inner rim
	7	9.27	0.45	inner rim
	8	9.21	0.46	inner rim
	9	9.02	0.45	inner rim
AC-11	1	11.22	0.46	inner rim
	2	10.94	0.45	outer rim
	3	10.88	0.46	outer core
	4	11.21	0.45	inner core
	5	10.65	0.46	inner rim
AC-13	1	5.68	0.44	outer rim
	2	5.46	0.48	inner rim
	3	5.97	0.46	outer core
	4	5.91	0.45	outer core
	5	5.96	0.45	outer rim
AC-14	1	6.18	0.47	outer rim
	2	6.23	0.47	outer core
	3	6.51	0.44	core
	4	6.68	0.48	core
	5	6.29	0.44	outer rim
AC-15	1	9.38	0.45	outer rim
	2	9.57	0.47	inner rim
	3	9.36	0.46	outer core
	4	8.91	0.45	outer core
	5	9.36	0.43	outer rim
AC-16	1	9.18	0.48	outer rim
	2	8.84	0.45	outer core
	3	8.65	0.45	outer core
	4	8.94	0.46	inner rim
	5	9.00	0.44	outer rim

Table 3. Summary of $\delta^{18}O$ results for the 12 samples analyzed.

Sapphire	Range	Range Variation	Average	Characteristics
AC-1	7.52–8.97	1.45	8.18	star sapphire
AC-2	5.24–5.76	0.52	5.50	magmatic zoning
AC-3	4.12–5.77	1.65	5.24	blue sapphire
AC-4	4.97–5.92	0.95	5.56	sector zoning
AC-5	7.52–8.71	1.19	8.10	blue-lilac
AC-7	5.42–6.81	1.39	6.02	pale lilac
AC-10	8.95–9.44	0.49	9.21	blue-lilac
AC-11	10.65–11.22	0.57	10.98	sector zoning
AC-13	5.46–5.97	0.51	5.80	deep blue
AC-14	6.18–6.68	0.50	6.38	sector zoning
AC-15	8.91–9.57	0.66	9.32	star sapphire
AC-16	8.65–9.18	0.53	8.92	sector zoning

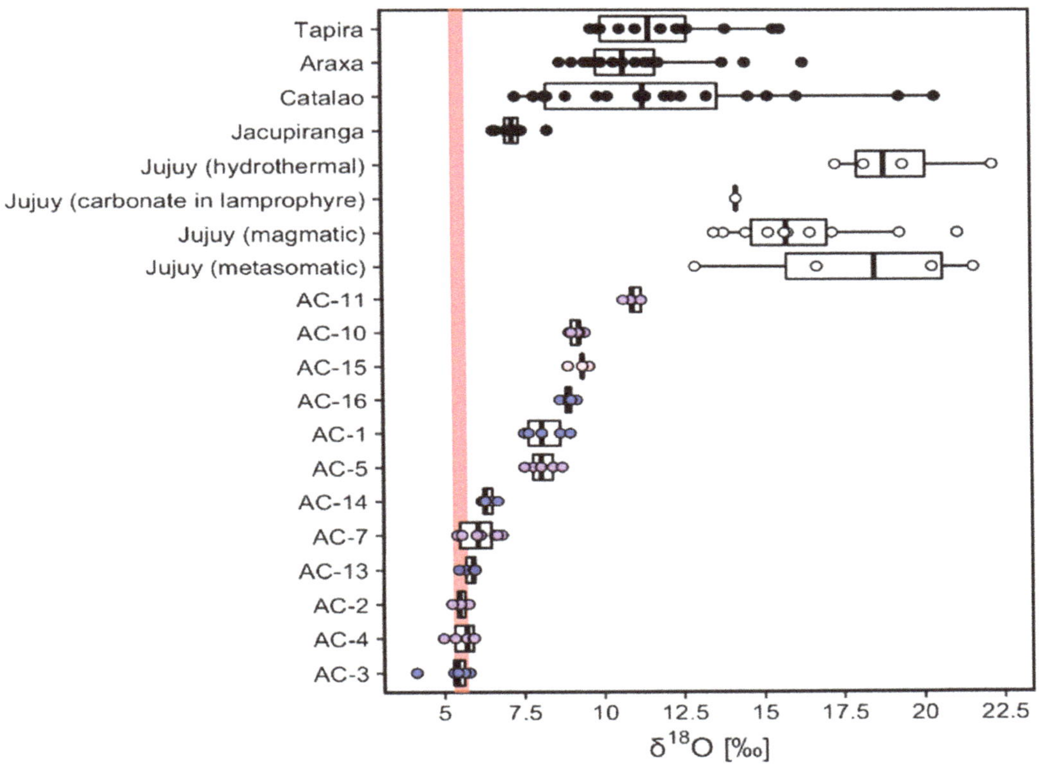

Figure 3. Diagram showing the range in oxygen isotope values for Jujuy sapphires (this study) in comparison to carbonatites and lamprophyres from the Jujuy region [17] and those of carbonatites from Brazil [18] and mantle $\delta^{18}O$ value (red band).

4. Discussion

4.1. Oxygen Isotope Groupings

Although there is a continuous spectrum in terms of trace element chemistry and $\delta^{18}O$ values, five main groupings are apparent: Group 1 ($n = 4$) comprises the blue sapphires and has $\delta^{18}O$‰ 4.1–5.9; Group 2 ($n = 2$) comprises pale lilac-blue sapphires with $\delta^{18}O$‰ 5.4–6.7; Group 3 ($n = 2$) consists of one star sapphire (AC-1) and a blue-lilac sapphire (AC-5) with $\delta^{18}O$‰ 7.5–8.9; Group 4 ($n = 3$) comprises a pale blue sapphire (AC-10), a star sapphire (AC-15) and a sector zone colorless-blue sapphire (AC-16) with $\delta^{18}O$‰ 8.6–9.6 and Group 5 ($n = 1$) is a sector zoned colorless-blue sapphire (AC-11) with $\delta^{18}O$‰ 10.6–11.2. These groupings, along with a summary of their trace element chemistry based on

Harris et al. (2017) [15] are shown in Table 4. Collectively, there is a range in $\delta^{18}O$ from 4.1‰ to 11.2‰ and a range in Ga/Mg of 0.3–10.4 (average range of 0.73–7.7) but no correlation or apparent relationshi between Ga/Mg and the oxygen isotope values. Importantly, Baldwin et al. (2017) [19] found tha sapphire megacrysts from the Siebenbirge Volcanic Field of Germany had a wide Ga/Mg of 1.0–23. and ascribed this to the influence of a carbonatitic melt in the crystallization of the sapphires. The range in Ga/Mg values for Jujuy Province sapphires at 0.30–10.40 fits well within this range and, base on this alone, would tentatively suggest a similar mechanism of formation. However, interpretatio of Ga/Mg ratios in sapphires must be treated with caution as Sutherland et al. (2009) [20] found a even larger range in Ga/Mg for Australian basalt-hosted blue-green-yellow (BGY suite) sapphires wit 0.24–31.60 for the Yarrowitch field and 0.17–193 for the Barrington field. This collectively lends suppor to the views of Baldwin et al. (2017) [19] and Palke et al. (2017) [21] who suggested that Ga/Mg ratio are not good discriminators for sapphire origin.

Table 4. Groupings based on $\delta^{18}O$‰ compositions with their main geochemical characteristics from Harris et al. (2017) [15].

Group 1 (Blue Sapphires) with $\delta^{18}O$‰ from 4.12 to 5.92									
Sapphire	Mg	Ti	V	Cr	Fe	Ga	Fe/Mg	Ga/Mg	Number of analyses
AC-2	38	73	64	456	789	140	22.3	4	6
AC-3	119	1987	154	359	1024	118	9.8	1.2	5
AC-4	103	1771	230	160	843	126	9.5	1.4	4
AC-13	162	428	155	91	1481	119	9.3	0.73	4
Group 2 (Pale Blue-Lilac Sapphires) with $\delta^{18}O$‰ from 5.42 to 6.68									
Sapphire	Mg	Ti	V	Cr	Fe	Ga	Fe/Mg	Ga/Mg	Number of analyses
AC-7	87	1044	113	224	621	116	7.8	1.5	4
AC-14	20	190	16	27	1348	121	86	7.7	4
Group 3 (Star Sapphire AC-1; Blue-Lilac Sapphire AC-5) with $\delta^{18}O$‰ from 7.52 to 8.97									
Sapphire	Mg	Ti	V	Cr	Fe	Ga	Fe/Mg	Ga/Mg	Number of analyses
AC-1	259	6619	187	45	927	146	4.3	0.8	6
AC-5	42	1077	67	209	677	99	17.3	2.5	4
Group 4 (Pale Blue Sapphire AC-10; Star Sapphire AC-15; Sector Zoned Colorless to Blue Sapphire AC-16) with $\delta^{18}O$‰ from 8.65 to 9.57									
Sapphire	Mg	Ti	V	Cr	Fe	Ga	Fe/Mg	Ga/Mg	Number of analyses
AC-10	71	1037	76	89	777	105	11	1.5	4
AC-15	200	6831	163	309	1278	160	6.5	0.8	4
AC-16	38	348	63	202	581	99	17	2.9	4
Group 5 (Sector Zoned Colorless to Blue Sapphire) with $\delta^{18}O$‰ from 10.65 to 11.22									
Sapphire	Mg	Ti	V	Cr	Fe	Ga	Fe/Mg	Ga/Mg	Number of analyses
AC-11	106	1741	230	16	517	128	5	1.3	4

4.2. *What Do the Oxygen Isotopes Tell Us about the Primary Source for Jujuy Sapphires?*

A framework on the interpretation of the geological origin of gem corundums using the $^{18}O/^{16}O$ ratio first proposed by Giuliani et al. (2005) [1] is now widely adopted. Based on this framework, rubies and pink sapphires can be classified into 5 categories based on their $\delta^{18}O$ value range. 1. Mafic gneiss hosted 2.9–3.8‰; 2. Mafic-ultramafic rocks (amphibolite, serpentinite) 3.2–6.8‰; 3. Desilicated pegmatites 4.2–7.5‰; 4. Shear zones cross-cutting ultramafic lenses and pegmatites within sillimanite gneisses 11.9–13.1‰; 5. Marble-hosted rubies 16.3–23‰. This framework has been further validated by numerous subsequent corundum oxygen isotope studies [3–5,11,13,20,22–32]. A summary of worldwide oxygen isotope values for gem sapphires, along with their host rock and Ga/Mg ratios, is presented in Table 5. In a study of sapphires from the French Massif Central basaltic field, Giuliani et al. (2009) [25] found two distinctive groups. The first group with $\delta^{18}O$ from 4.4‰ to 6.8‰ were believed to have crystallized from felsic magmas in the upper mantle while the second group with higher $\delta^{18}O$ from 7.6‰ to 13.9‰ were believed to have been derived from lower crustal granulites. However, the oxygen isotope range for Jujuy sapphires spans across both of these groupings. Critically, the study of Harris et al. (2017) [15] showed that for Jujuy sapphires, there is a continuous spectrum n composition with significant trace element variation within single grains. This strongly suggests crystallization from the same source but under variable metasomatic conditions.

Table 5. Summary of worldwide oxygen isotope ranges and Ga/Mg for some sapphires (n.d = not determined).

Color	Host Rock	$d^{18}O‰$	Ga/Mg	Reference
Blue, yellow, brown	alkali basalts	4.6–5.2	5.7–11.3	[31]
Pink	kimberlites	4.3–5.4	1.9–3.9	[31]
Blue	lamprophyre	5.4–6.8	0.3–0.5	[21]
Blue, yellow, green, colorless, brown	alkali basalts	2.7–6.9	n.d	[29]
Blue, green, yellow	alkali basalts	4.5–5.6	3–18	[13]
Blue, grey-pink, pink	alkali basalts	3.8–5.9	0.9–4.8	[26]
Blue, violet	lamprophyre	4.5–7.7	0.2–0.6	[33]
Blue, green, pink, colorless	alkali basalts	4.4–6.8	n.d	[25]
Blue, green, grey, white	alkali basalts	7.6–13.9	n.d	[25]
Blue, grey, brown	alkali basalts	n.d	1.0–23.4	[19]
Blue, violet	alluvial	2.6–6.8	0.6–9.0	[34]
Blue, green	alkali basalts	4.3–6.4	n.d	[5]
Blue, grey, violet, red	syenite pegmatites	2.0–10.6	n.d	[5]
Blue, green, yellow	alkali basalts	1.34–39.29	5.3–5.4	[20]
Blue, green, yellow	alkali basalts	6.6–193.0	5.2–5.8	[20]
Blue, green, yellow, pink	alkali basalts	4.3–8.4	n.d	[1]
Blue, grey	amphibolite	4.2–4.9	n.d	[1]
Blue	desilicated pegmatite in mafic rocks	10.9–11.2	n.d	[1]
Yellow, green, colorless, brown	syenite/anorthoclasite	5.2–7.8	n.d	[1]
Blue, grey, colorless	skarn	7.7–10.7	n.d	[1]

In a study of carbonatites from Brazilian alkaline complexes, Santos and Clayton (1995) [18] found that the non-fenitized carbonatites from Jacupiranga had significantly lower oxygen isotope ratios (6.6–7.3) compared to the fenitized (i.e., K and Na metasomatism; Elliott et al. 2018, [35]) carbonatites of Araxa (9.7–16.3), Catalao (7.3–19.3), and Tapira (9.7–15.4). Zappettini et al. (1998) [17] analyzed a range of carbonatites from Jujuy Province and found that oxygen isotope values for magmatic carbonatites (13.5–21.1), metasomatic carbonatites (12.9–21.6), and hydrothermally altered carbonatites (17.3–22.2) overlapped each other but that magmatic carbonatites had lower values in general. They also analyzed one carbonatized lamprophyre sample which gave a value of 14.2 within the field of the magmatic carbonatites. The range in oxygen isotope compositions for Jujuy sapphires is much lower than all of these analyzed carbonatites and the lamprophyre from this region [17]. However, Harmon et al. (1981) [36] analyzed volcanic rocks from Cerro Galan in Catamarca Province, 400 km to the south of Jujuy, and found that these calc-alkaline volcanics had a range in oxygen isotope values from 8.3 to 10.7, and had formed directly from mantle-derived magmas but with likely partial crustal assimilation during ascent. This range is well within that of Groups 3–5 from the Jujuy suite. Additionally, the Brazilian fenitized carbonatites have a very similar range and would tentatively suggest that there was significant metasomatic influence in the Jujuy sapphire-forming environment, especially for the sapphires with the highest oxygen isotope ratios (Figure 3). It also suggests influence from an additional underlying deep source that had lower values than those for the carbonatites previously analyzed from Jujuy Province. This metasomatic influence would have been highly variable, as suggested by the wide range in oxygen isotope ratios and trace element contents and ratios for the Jujuy sapphires as shown by Harris et al. (2017) [15].

As suggested by Baldwin et al. (2017) [19], a possible origin of sapphires from carbonatitic melts could be identified by undertaking oxygen isotope measurements on the sapphires as, in general, carbonate-associated sapphires generally have higher isotopic ratios [1], as seen for the higher range amongst the Jujuy sapphires. However, interpreting such data would be problematic as there are no data available on oxygen isotope fractionation between corundum and co-existing carbonatitic melt [19].

4.3. Origin of Jujuy Sapphires

Any model proposed for the genesis of the alluvial Jujuy sapphires needs to account for (1) a predominant metamorphic source signature, and a less prominent mafic-ultramafic association and contrasting felsic/acidic association; (2) elevated Si contents, and unusually high Ga content for metamorphic sapphires (i.e., typically >100); (3) enrichments in Be, Nb, Ta, Sn, Th, and U within syngenetic rutile growths; (4) elevated light rare earth element (LREE) concentrations; (5) a bimodal mineral inclusion suite, including sulfides, feldspar, phosphates, and likely LREE-bearing minerals; (6) the range in Ga/Mg from 0.88 to 7.33; (7) significant variation in trace element contents, ratios, and oxygen isotope ratios within single grains; and (8) a wide but continuous range in oxygen isotope values from 4.1‰ to 11.2‰.

Given the presence of LREE-bearing inclusions and elevated LREE concentrations, it is possible that these elevated contents were the result of a carbonatitic parental melt. This is possible given reports of REE-bearing carbonatites within Jujuy Province. Based on $\delta^{13}C$ and $\delta^{18}O$ isotopic studies, Zappettini et al. (1998) [17] suggested that these carbonatites formed by the crystallization of carbonate magma with subsequent formation of metasomatic and hydrothermal carbonatitic veins. This is akin to the variably fenitized carbonatites from Brazil as shown above. Similarly, nearby Rio Grande monchiquites (163 ± 9 Ma) were shown to have been derived from heterogeneous enriched lithospheric mantle metasomatized by carbonatite fluids involving significant magma mixing (Hauser et al. 2010) [37]. Similar processes, involving exchanges between a carbonatite (Si-poor) and mantle-derived acidic magma (Si- and Ga-rich) could, in turn, be responsible for sapphire-bearing lamprophyre formation (e.g., the plumasite model; Peucat et al. 2007) [38] within Jujuy Province (e.g., Zappettini 1989) [39]. Factor analysis on Jujuy sapphires by Harris et al. (2017) [15] showed a largely continuous linear grouping from the metasomatic field of Giuliani et al. (2014) [4] into their plumasite field. The one exception to this trend was the star sapphires (i.e., the sapphires with the highest oxygen isotope ratios), which formed a distinctly separate grouping within the plumasite field of Giuliani et al. (2014) [4]. This once again provides supporting evidence for a sapphire-crystallization environment involving highly variable metasomatic interaction.

Collectively, the in situ study of Harris et al. (2017) [15] and this in situ study provide crucial evidence for understanding the origin of Jujuy sapphires. Based on their study of surface features, mineral inclusions, and major and trace element chemistry, Harris et al. (2017) [15] concluded that the Jujuy sapphires crystallized via metasomatic fluid exchange between an evolved crustal felsic system and intruding coeval mantle-derived carbonatized lamprophyres as part of a regional-wide alkaline magmatic event within NW Argentina during the Lower Cretaceous. Oxygen isotope ratios for the Jujuy sapphires largely support this model but allow us to go one step further and suggest that the influence of carbonatites on sapphire crystallization was greater than previously thought, and that the localized sapphire-former crustal environment was more complex, involving episodic interactions between mantle-derived lamprophyres, mantle-derived carbonatites, lower crustal felsic magmas and, most likely, both mantle and lower crustal metasomatic fluids. As stated by Harris et al. (2017) [15], such a complex interplay of widely diverse magma and fluid compositions would induce localized desilicification reactions that would lead to localized sapphire crystallization. Based on all of the available evidence, we suggest that most of these interactions would have occurred at the crust/mantle boundary, with initial crystallization within the upper mantle (i.e., sapphires with lower oxygen isotope signatures of 4.12–6.50) then continuing crystallization within the lower crust (i.e., sapphires with oxygen isotope signatures >6.50).

The wide range in chemistry and oxygen isotope compositions but within a single comagmatic group is explained by the degree and type of interaction (i.e., carbonatite–lamprophyre vs. carbonatite–granite) between these magmas/fluids, with some crystallizing at lower crustal levels as suggested by Palke et al. [21,34] for sapphires from Montana, USA. The large within-grain variation seen in trace element contents, trace element ratios, and oxygen isotope ratios is highly suggestive of a geochemically dynamic environment with rapidly changing degrees of interaction between the various diverse fluid and magmatic components, as described above.

This model accounts for the dominant metasomatic signature, while also accounting for the mafic–ultramafic (i.e., carbonatitic and lamprophyric magmas) and felsic (i.e., mantle-derived enriched acidic magmas and/or granitic anatectic melts) associations. It also accounts for the bimodal mineral inclusion suite, enrichment in LREE, Ga/Mg ratios, and range in trace element and oxygen isotope compositions within a single sapphire group that crystallized at the same time but across the crust–mantle boundary. The oxygen isotope values and range provide additional support to the initial hypothesis of Zappettini et al. (1997) [40], who proposed that the Jujuy sapphires formed via metasomatic exchanges between REE-bearing carbonatites and an evolved felsic magma during the Cretaceous (150–110 Ma), and were then subsequently incorporated as xenocrysts within lamprophyre dykes.

5. Conclusions

In situ SIMS oxygen isotope analyses on the Orosmayo sapphire suite revealed that they have values spanning a continuous range from +4.1 to +11.2 $\delta^{18}O$‰, with measurable variation within individual sapphires. Based on their trace element geochemistry, geochemical ratios, and oxygen isotopes, although there is a continuous spectrum of compositions, the sapphires can be subdivided into five suites reflecting different degrees of interaction between the various fluid/magmatic components involved in sapphire formation. Both the previous geochemical work and this study suggest that the Jujuy sapphires crystallized through the crust–mantle boundary in a localized environment involving episodic interactions between mantle-derived lamprophyres, mantle-derived carbonatites, lower crustal felsic magmas and, most likely, both mantle and crustal metasomatic fluids. This complex interplay of widely diverse magma and fluid compositions would induce localized desilicification reactions, leading to localized sapphire crystallization. This study reinforces the importance of carbonatites on sapphire crystallization. The significant range in trace element chemistry and oxygen isotope ratios within individual grains stresses the critical importance of undertaking in situ analysis on gem corundums in order to fully understand their environment of formation and to aid in their geographic typing.

Author Contributions: I.T.G. wrote the manuscript, was responsible for the conception of this project and interpreted the results of the analyses. S.J.H. provided input into the discussion and constructed some of the figures. L.M. Provided technical input and ran the SIMS analyses. A.L. provided technical input. E.Z. collected the samples and provided geological expertise on the Jujuy region of NW Argentina.

Acknowledgments: The authors would like to thank Joanne Wilde, formerly of the School of BEES, for making the polished mounts required for SIMS analysis. The authors acknowledge the facilities and the scientific and technical assistance of the Australian Microscopy & Microanalysis Research Facility at the Centre for Microscopy, Characterisation & Analysis, The University of Western Australia, a facility funded by the University, State and Commonwealth Governments.

References

1. Giuliani, G.; Fallick, A.E.; Garnier, V.; France-Lanord, C.; Ohnenstetter, D.; Schwarz, D. Oxygen isotope composition as a tracer for the origins of rubies and sapphires. *Geology* **2005**, *33*, 249–252. [CrossRef]
2. Yui, T.-F.; Wu, C.-M.; Limtrakun, P.; Sricharin, P.; Boonsoong, A. Oxygen isotope studies on placer sapphires and ruby in the Chanthaburi-Trat alkali basalt gem field, Thailand. *Lithos* **2006**, *86*, 197–211. [CrossRef]
3. Giuliani, G.; Ohnenstetter, D.; Garnier, V.; Fallick, A.E.; Rakotondrazafy, M.; Schwarz, D. The geology and genesis of gem corundum deposits. In *Geology of Gem Deposits*; Groat, L.A., Ed.; Mineralogical Association of Canada: Quebec City, QC, Canada, 2007; Volume 37, pp. 23–78. ISBN 978092194375.

4. Giuliani, G.; Ohnenstetter, D.; Fallick, A.E.; Groat, L.; Fagan, J. The geology and genesis of gem corundum deposits. In *Geology of Gem Deposits (2nd edition)*; Groat, L.A., Ed.; Mineralogical Association of Canada: Quebec City, QC, Canada, 2014; Volume 44, pp. 29–112. ISBN 9780921294542.
5. Vysotskiy, S.V.; Nechaev, V.P.; Kissin, A.Y.; Yakovenko, V.V.; Ignat'ev, A.V.; Velivetskaya, T.A.; Sutherland, F.L.; Agoshkov, A.I. Oxygen isotopic composition as an indicator of ruby and sapphire origin: A review of Russian occurrences. *Ore Geol. Rev.* **2015**, *68*, 164–170. [CrossRef]
6. Yui, T.F.; Zaw, K.; Limtrakun, P. Oxygen isotope composition of the Denchai sapphire, Thailand: A clue to its enigmatic origin. *Lithos* **2003**, *67*, 153–161. [CrossRef]
7. Bindeman, I.N.; Schmitt, A.K.; Evans, D.A.D. Limits of hydrosphere-lithosphere interaction: Origin of the lowest-known o18O silicate rock on Earth in the Paleoproterozoic Karelian rift. *Geology* **2010**, *38*, 631–634. [CrossRef]
8. Bindeman, I.N.; Serebryakov, N.S. Geology, Petrology and O and H isotope geochemistry of remarkably ^{18}O depleted Paleoproterozoic rocks of the Belomorian Belt, Karelia, Russia, attributed to global glaciation 2.4 Ga. *Earth Planet. Sci. Lett.* **2011**, *306*, 163–174. [CrossRef]
9. Krylov, D.P. $^{18}O/^{16}O$ ratios in the corundum-bearing rocks of Khitostrov, northern Karelia. *Dokl. Earth Sci.* **2008**, *419*, 453–456. [CrossRef]
10. Sutherland, F.L.; Graham, I.T.; Harris, S.J.; Coldham, T.; Powell, W.; Belousova, E.A.; Martin, L. Unusual ruby-sapphire transition in alluvial megacrysts, Cenozoic basaltic gem field, New England, New South Wales, Australia. *Lithos* **2017**, *278*, 347–360. [CrossRef]
11. Wang, K.K.; Graham, I.T.; Martin, L.; Voudouris, P.; Giuliani, G.; Lay, A.; Harris, S.J.; Fallick, A.E. Fingerprinting Paranesti rubies through oxygen isotopes. *Minerals* **2019**, *9*, 91. [CrossRef]
12. Sutherland, F.L.; Abduriyim, A. Geographic typing of gem corundum: A test case from Australia. *J. Gemmol.* **2009**, *31*, 203–210. [CrossRef]
13. Sutherland, F.L.; Coenraads, R.R.; Abduriyim, A.; Meffre, S.; Hoskin, P.W.O.; Giuliani, G.; Beattie, R.; Wuhrer, R.; Sutherland, G.B. Corundum (sapphire) and zircon relationships, Lava Plains gem fields, NE Australia: Integrated mineralogy, geochemistry, age determination, genesis and geographical typing. *Mineral. Mag.* **2015**, *79*, 545–581. [CrossRef]
14. Zappettini, E.; Mutti, D. Zafiros de la Puna Argentina: Un Potencial recurso minero. *Recur. Miner.* **1997**, *2*, 33–62.
15. Harris, S.J.; Graham, I.T.; Lay, A.; Powell, W.; Belousova, E.; Zappettini, E. Trace element geochemistry and metasomatic origin of alluvial sapphires from the Orosmayo region, Jujuy Province, northwest Argentina. *Can. Mineral.* **2017**, *55*, 595–617. [CrossRef]
16. Chernicoff, C.J.; Richards, J.P.; Zappettini, E.O. Crustal lineament control on magmatism and mineralization in northwestern Argentina: Geological, geophysical, and remote sensing evidence. *Ore Geol. Rev.* **2002**, *21*, 127–155. [CrossRef]
17. Zappettini, E.O.; Rubiolo, D.; Hubberten, H.W. *Carbon and Oxygen Isotopes in Carbonatites from Puna, Jujuy and Salta, Argentina*; Congreso Nacional de Geologia Economica: Buenos Aires, Argentina, 1998; Volume 2. (In Spanish)
18. Santos, R.V.; Clayton, R.N. Variations of oxygen and carbon isotopes in carbonatites: A study of Brazilian alkaline complexes. *Geochim. Cosmochim. Acta* **1995**, *59*, 1339–1352. [CrossRef]
19. Baldwin, L.C.; Tomaschek, F.; Ballhaus, C.; Gerdes, A.; Fonseca, R.O.C.; Wirth, R.; Geisler, T.; Nagel, T. Petrogenesis of alkaline basalt-hosted sapphire megacrysts: Petrological and geochemical investigations of in situ sapphire occurrences from the Siebengebirge Volcanic Field, Germany. *Contrib. Mineral. Petrol.* **2017**, *172*, 1–27. [CrossRef]
20. Sutherland, F.L.; Zaw, K.; Meffre, S.; Giuliani, G.; Fallick, A.E.; Graham, I.T.; Webb, G.B. Gem-corundum megacrysts from east Australian basalt fields: Trace elements, oxygen isotopes and origins. *Aust. J. Earth Sci.* **2009**, *56*, 1003–1022. [CrossRef]
21. Palke, A.; Renfro, N.D.; Berg, R.B. Melt inclusions in alluvial sapphires from Montana, USA: Formation of sapphires as a restitic component of lower crustal melting? *Lithos* **2017**, *278*, 43–53. [CrossRef]
22. Giuliani, G.; Fallick, A.E.; Rakotondrazafy, M.; Ohnenstetter, D.; Andriamamonjy, A.; Rakotosamizanany, S.; Ralantoarison, T.; Razanatseheno, M.; Dunaigre, C.; Schwarz, D. Oxygen isotope systematics of gem corundum deposits in Madagascar: Relevance for their geological origin. *Miner. Depos.* **2007**, *42*, 251–270. [CrossRef]

23. Sutherland, F.L.; Duroc-Danner, J.M.; Meffre, S. Age and origin of gem corundum and zircon megacrysts from the Mercaderes–Rio Mayo area, South-west Colombia, South America. *Ore Geol. Rev.* **2008**, *34*, 155–168. [CrossRef]
24. Garnier, V.; Giuliani, G.; Ohnenstetter, D.; Fallick, A.E.; Dubessy, J.; Banks, D.; Vinh, H.Q.; Lhomme, T.; Maluski, H.; Pecher, A.; et al. Marble-hosted ruby deposits from central and southeast Asia: Towards a new genetic model. *Ore Geol. Rev.* **2008**, *34*, 169–191. [CrossRef]
25. Giuliani, G.; Fallick, A.; Ohnenstetter, D.; Pegere, G. Oxygen isotopes compositions of sapphires from the French Massif Central: Implications for the origin of gem corundum in basaltic fields. *Miner. Depos.* **2009**, *44*, 221–231. [CrossRef]
26. Uher, P.; Giuliani, G.; Szakall, S.; Fallick, A.; Strunga, V.; Vaculovic, T.; Ozdin, D.; Greganova, M. Sapphires related to alkali basalts from the Cerova Highlands, Western Carpathians (southern Slovakia): Composition and origin. *Geol. Carpathica* **2012**, *63*, 71–82. [CrossRef]
27. Vysotskiy, S.V.; Ignat'ev, A.V.; Levitskii, V.I.; Nechaev, V.P.; Velvetskaya, T.A.; Yakovenko, V.V. Geochemistry of stable oxygen and hydrogen isotopes in minerals and corundum-bearing rocks in Northern Karelia as an indicator of their unusual genesis. *Geochem. Int.* **2014**, *52*, 773–782. [CrossRef]
28. Sutherland, F.L.; Zaw, K.; Meffre, S.; Yui, T.F.; Thu, K. Advances in trace element "fingerprinting" of gem corundum, ruby and sapphire, Mogok Area, Myanmar. *Minerals* **2014**, *5*, 61–79. [CrossRef]
29. Rakotosamizanany, S.; Giuliani, G.; Ohnenstetter, D.; Rakotondrazafy, A.F.M.; Fallick, A.E.; Paquette, J.-L.; Tiepolo, M. Chemical and oxygen isotopic compositions, age and origin of gem corundums in Madagascar alkali basalts. *J. Afr. Earth Sci.* **2014**, *94*, 156–170. [CrossRef]
30. Zaw, K.; Sutherland, L.; Yui, T.F.; Meffre, S.; Thu, K. Vanadium-rich ruby and sapphire within Mogok Gemfield, Myanmar: Implications for gem color and genesis. *Miner. Depos.* **2015**, *50*, 25–39. [CrossRef]
31. Giuliani, G.; Pivin, M.; Fallick, A.E.; Ohnenstetter, D.; Song, Y.; Demaiffe, D. Geochemical and oxygen isotope signatures of mantle corundum megacrysts from the Mbuji-Mayi kimberlite, democratic Republic of Congo, and the Changle alkali basalt, China. *Comptes Rendus Geosci.* **2015**, *347*, 24–34. [CrossRef]
32. Wong, J.; Verdel, C. Tectonic environments of sapphire and ruby revealed by a global oxygen isotope compilation. *Int. Geol. Rev.* **2018**, *60*, 188–195. [CrossRef]
33. Palke, A.C.; Wong, J.; Verdel, C.; Avila, J.N. A common origin for Thai/Cambodian rubies and blue and violet sapphires from Yogo Gulch, Montana, U.S.A.? *Am. Mineral.* **2018**, *103*, 469–479. [CrossRef]
34. Palke, A.C.; Renfro, N.D.; Berg, R.B. Origin of sapphires from a lamprophyre dike at Yogo Gulch, Montana, USA: Clues from their melt inclusions. *Lithos* **2016**, *260*, 339–344. [CrossRef]
35. Elliott, H.A.L.; Wall, F.; Chakhmouradian, A.R.; Siegfried, P.R.; Dahlgren, S.; Weatherley, S.; Finch, A.A.; Marks, M.A.W.; Dowman, E.; Deady, E. Fenites associated with carbonatite complexes: A review. *Ore Geol. Rev.* **2018**, *93*, 38–59. [CrossRef]
36. Harmon, R.S.; Thorpe, R.S.; Francis, P.W. Petrogenesis of Andean andesites from combined O-Si isotope relationships. *Nature* **1981**, *290*, 396–399. [CrossRef]
37. Hauser, N.; Matteini, M.; Omarini, R.H.; Pimentel, M.M. Constraints on metasomatised mantle under Central South America: Evidence from Jurassic alkaline lamprophyre dykes from the Eastern Cordillera, NM Argentina. *Mineral. Petrol.* **2010**, *100*, 153–184. [CrossRef]
38. Peucat, J.J.; Ruffault, P.; Fritsch, E.; Bouhnik-La Coz, M.; Simonet, C.; Lasnier, B. Ga/Mg ratios as a new geochemical tool to differentiate magmatic from metamorphic blue sapphire. *Lithos* **2007**, *98*, 261–274. [CrossRef]
39. Zappettini, E. Geologia y Metalogenesis de la Region Comprendida Entre las Localidades de Santa Ana y Cobres, Provincias de Jujuy y Salta, Republica Argentina. Ph.D. Thesis, Facultad de Ciencias Exactas y Naturales, Universidad de Buenos Aires, Buenos Aires, Argentina, 1989.
40. Zappettini, E.; Mutti, D.; Bernhardt, H.J. Genesis de Zafiro y hercinita de la puna argentina. *Actas* **1997**, *2*, 1598–1602.

Permissions

All chapters in this book were first published in MDPI; hereby published with permission under the Creative Commons Attribution License or equivalent. Every chapter published in this book has been scrutinized by our experts. Their significance has been extensively debated. The topics covered herein carry significant findings which will fuel the growth of the discipline. They may even be implemented as practical applications or may be referred to as a beginning point for another development.

The contributors of this book come from diverse backgrounds, making this book a truly international effort. This book will bring forth new frontiers with its revolutionizing research information and detailed analysis of the nascent developments around the world.

We would like to thank all the contributing authors for lending their expertise to make the book truly unique. They have played a crucial role in the development of this book. Without their invaluable contributions this book wouldn't have been possible. They have made vital efforts to compile up to date information on the varied aspects of this subject to make this book a valuable addition to the collection of many professionals and students.

This book was conceptualized with the vision of imparting up-to-date information and advanced data in this field. To ensure the same, a matchless editorial board was set up. Every individual on the board went through rigorous rounds of assessment to prove their worth. After which they invested a large part of their time researching and compiling the most relevant data for our readers.

The editorial board has been involved in producing this book since its inception. They have spent rigorous hours researching and exploring the diverse topics which have resulted in the successful publishing of this book. They have passed on their knowledge of decades through this book. To expedite this challenging task, the publisher supported the team at every step. A small team of assistant editors was also appointed to further simplify the editing procedure and attain best results for the readers.

Apart from the editorial board, the designing team has also invested a significant amount of their time in understanding the subject and creating the most relevant covers. They scrutinized every image to scout for the most suitable representation of the subject and create an appropriate cover for the book.

The publishing team has been an ardent support to the editorial, designing and production team. Their endless efforts to recruit the best for this project, has resulted in the accomplishment of this book. They are a veteran in the field of academics and their pool of knowledge is as vast as their experience in printing. Their expertise and guidance has proved useful at every step. Their uncompromising quality standards have made this book an exceptional effort. Their encouragement from time to time has been an inspiration for everyone.

The publisher and the editorial board hope that this book will prove to be a valuable piece of knowledge for researchers, students, practitioners and scholars across the globe.

List of Contributors

Jens Götze
Institute of Mineralogy, TU Bergakademie Freiberg, Brennhausgasse 14, 09599 Freiberg, Germany

Robert Möckel
Helmholtz-Zentrum Dresden-Rossendorf, Hemholtz Institute Freiberg for Resource Technology, Chemnitzer Str. 40, 09599 Freiberg, Germany

Yuanming Pan
Department of Geological Sciences, University of Saskatchewan, Saskatoon, SK S7N 5E2, Canada

Stefan G. Bindereif, Felix Rüll and Stephan Schwarzinger
North Bavarian NMR-Centre, University of Bayreuth, 95447 Bayreuth, Germany

Clemens Schwarzinger
Institute for Chemical Technology of Organic Materials, Johannes Kepler Univeristy Linz, 4040 Linz, Austria

Panagiotis Voudouris, Constantinos Mavrogonatos, Maria Tsortanidis and Andreas Lampridis
Faculty of Geology and Geoenvironment, National and Kapodistrian University of Athens, 15784 Athens, Greece

Ian Graham and Kandy Wang
PANGEA Research Centre, School of Biological, Earth and Environmental Sciences, University of NSW, 2052 Sydney, Australia

Gaston Giuliani
Université Paul Sabatier, GET/IRD et Université de Lorraine, C.R.P.G./C.N.R.S., 54501 Vandoeuvre, CEDEX, France

Vasilios Melfos and Vilelmini Karantoni
Faculty of Geology, Aristotle University of Thessaloniki, 54124 Thessaloniki, Greece

Stefanos Karampelas
Bahrain Institute for Pearls & Gemstones (DANAT), WTC East Tower, Bahrain

Alexandre Tarantola
UMR GeoResources, Faculté des Sciences et Technologies, Université de Lorraine, 54506 Vandoeuvre-lès-Nancy, France

Khin Zaw and Sebastien Meffre
CODES Centre of Ore Deposit and Earth Sciences, University of Tasmania, Tas 7001, Australia

Stephan Klemme and Jasper Berndt
Institut für Mineralogie, Westfälische-Wilhelms Universität Münster, 48149 Münster, Germany

Stefanie Heidrich
Mineralogisch-Petrographisches Institut, Universität Hamburg, 20146 Hamburg, Germany

Federica Zaccarini
Department of Applied Geosciences and Geophysics, University of Leoben, 8700 Leoben, Austria

Anthony Fallick
Isotope Geosciences Unit, S.U.E.R.C., Glasgow G75 0QF, UK

Paula C. Piilonen, Glenn Poirier and Ralph Rowe
Mineralogy Section, Beaty Centre for Species Discovery, Research & Collections, Canadian Museum of Natured, Station D, Ottawa, Ontario K1P 6P4, Canada

F. Lin Sutherland
Geoscience, Australian Museum, 1 William Street, Sydney NSW 2010, Australia

Martin Danišík
John de Laeter Centre, The Institute for Geoscience Research (TIGeR), Curtin University, Perth WA 6845, Australia

John W. Valley
Wisconsin Secondary Ion Mass Spectrometer Laboratory (WiscSIMS), Department of Geoscience, University of Wisconsin, Madison, Wisconsin 53706, USA

Sabina Strmić Palinkaš
Department of Geosciences, Faculty of Sciences and Technology, UiT-The Arctic University of Norway in Tromsø, N-9037 Tromsø, Norway

Ladislav Palinkaš and Vladimir Bermanec
Department of Geology, Faculty of Science, University of Zagreb, Horvatovac 95, HR-10000 Zagreb, Croatia

Franz Neubauer
Department of Geography and Geology, Paris-Lodron-University of Salzburg, Hellbrunner Str. 34, A-5020 Salzburg, Austria

Ricardo Scholz
Departamento de Geologia, Escola de Minas, Universidade Federal de Ouro Preto, Ouro Preto MG-31400-000, Brazil

Sibila Borojević Šoštarić
Faculty of Mining, Geology and Petroleum Engineering, University of Zagreb, Pierottijeva 6, HR-10000 Zagreb, Croatia

Ján Štubňa and L'udmila Illášová
Gemmological Institute, Faculty of Natural Sciences, Constantine the Philosopher University in Nitra, Nábrežie mládeže 91, 949 74 Nitra, Slovakia

Peter Bačík and Jana Fridrichová
Department of Mineralogy and Petrology, Faculty of Natural Sciences, Comenius University in Bratislava, Ilkovičova 6, Mlynská dolina, 842 15 Bratislava, Slovakia

Radek Hanus
Gemological Laboratory of e-gems.cz, 110 00 Prague, Czech Republic

Stanislava Milovská
Earth Science Institute of the Slovak Academy of Sciences, Ďumbierska 1, 974 01 Banská Bystrica, Slovakia

Radek Škoda
Department of Geological Sciences, Masaryk University, Kotlářská 2, 611 37 Brno, Czech Republic

Tomáš Vaculovič
Central European Institute of Technology, Masaryk University, Kamenice 5, 625 00 Brno, Czech Republic

Slavomír Čerčanský
Department of Environmental Ecology, Faculty of Natural Sciences, Comenius University in Bratislava, Ilkovičova 6, Mlynská dolina, 842 15 Bratislava, Slovakia

Athanasios Katerinopoulos
Faculty of Geology and Geoenvironment, National and Kapodistrian University of Athens, GR-15784 Athens, Greece

Valeria Diella
National Research Council, Institute for Dynamics of Environmental Processes (IDPA), Section of Milan, 20133 Milan, Italy

Federico Pezzotta
Natural History Museum, 20121 Milan, Italy

Rosangela Bocchio and Fernando Cámara
Department of Earth Sciences "Ardito Desio", University of Milan, 20133 Milan, Italy

Nicoletta Marinoni
National Research Council, Institute for Dynamics of Environmental Processes (IDPA), Section of Milan, 20133 Milan, Italy
Department of Earth Sciences "Ardito Desio", University of Milan, 20133 Milan, Italy

Antonio Langone
National Research Council, Institute of Geosciences and Earth Resources (IGG), Section of Pavia, 27100 Pavia, Italy

Ilaria Adamo
Italian Gemmological Institute (IGI), 20123 Milan, Italy

Gabriele Lanzafame
Elettra-Sincrotrone Trieste S.C.p.A., Basovizza, 34149 Trieste, Italy

Ian T. Graham, Stephen J. Harris and Angela Lay
PANGEA Research Centre, School of Biological, Earth and Environmental Sciences, University of NSW, Sydney 2052, Australia

Laure Martin
Centre for Microscopy Characterisation and Analysis, The University of Western Australia, Perth 6009, Australia

Eduardo Zappettini
Argentine Geological-Mining Survey (SEGEMAR), Buenos Aires C1067ABB, Argentina

Index

www.ingramcontent.com/pod-product-compliance
Lightning Source LLC
LaVergne TN
LVHW060531210926
848485LV00005B/1181
9781641168052